Handbook for Small Science Centers

Do you have what it takes to ride the Hi Wire Cycle across a one-inch steel cable that stretches sixty feet across the atrium and learn about how gravity and counterbalances will keep you safely on two wheels during this unforgettable experience? COSI-Toledo, Ohio.

Handbook for Small Science Centers

Edited by

Cynthia C. Yao

WITH

Lynn D. Dierking, Peter A. Anderson,
Dennis Schatz, and Sarah Wolf

ALTAMIRA
PRESS

A Division of
ROWMAN & LITTLEFIELD PUBLISHERS, INC.
Lanham • New York • Toronto • Plymouth, UK

AltaMira Press
A Division of Rowman & Littlefield Publishers, Inc.
A wholly owned subsidiary of The Rowman & Littlefield Publishing Group, Inc.
4501 Forbes Boulevard, Suite 200
Lanham, MD 20706
www.altamirapress.com

Estover Road
Plymouth PL6 7PY
United Kingdom

British Library Cataloguing in Publication Information Available

Library of Congress Cataloguing-in-Publication Data

Handbook for small science centers / edited by Cynthia C. Yao, with Lynn D. Dierking . . . [et al.].
    p.   cm.
    Includes bibliographical references and index.
    ISBN-13: 978-0-7591-0652-9 (alk. paper)
    ISBN-10: 0-7591-0652-5 (alk. paper)
    ISBN-13: 978-0-7591-0653-6 (pbk : alk. paper)
    ISBN-10: 0-7591-0653-3 (pbk : alk. paper)
    1. Science museums—Administration.   2. Science museums—Management.
  I. Yao, Cynthia.

  Q105.H36 2006
  507′.4—dc22

                        2005035445

Printed in the United States of America

♾™ The paper used in this publication meets the minimum requirements of American National Standard for Information Sciences—Permanence of Paper for Printed Library Materials, ANSI/NISO Z39.48-1992.

Handbook for Small Science Centers is dedicated to all the unsung heroines and heroes who have devoted their lives to working and playing in science centers and museums to bring the joy of discovery and a better understanding of science and the world around us to everyone.

Museum Haiku

See understanding
Bloom in hands-on museums
As we smile and learn.

Lorraine Nadelman

I hear and I forget
I see and I remember
I do and I understand

Chinese proverb

Hands-on turns minds on

Contents

Foreword

About ten years ago, I was fortunate to be the project leader for the evaluation of Pacific Science Center's Science Carnival Consortium (SCC) project, funded by the National Science Foundation (NSF) in an effort to support the development of new and emerging science centers (the project is described by many of the people writing case study chapters in this book and was led by one of the book's coeditors, Dennis Schatz at Pacific Science Center [PSC]). The seven participating "institutions" received a core set of exhibits, attended a management seminar and received ongoing advice from PSC during the course of the program and beyond. A second miniconsortium of eight developing museums was formed to provide its participants with a six-week showing of SCC exhibits and a weeklong training experience at PSC.

The Science Carnival Consortium project was rewarding in many ways, but two reasons in particular stand out. The first reason is that I got to travel around the country and visit such places as Bemidji, Minn., (on a very cold day in January as I recall), Waterloo, Iowa, Rockford, Ill., and Ocala, Fla. While visiting these communities, and through other opportunities the project afforded, I got to meet with many folks who had the vision, tenacity, and wild optimism needed to undertake the creation, day-to-day operations, and often the expansion of a science center. I also got to meet community leaders throughout the United States who believed passionately enough in the vision to assist with fundraising and sweat equity (or often both). These experiences made a tremendous impression on me, demonstrating the important role that science centers play in their communities, no matter their size, mission, or community makeup. This is strongly evidenced by the stories in this book about institutions in the United States and Europe involved in start-up and expansion.

The second reason that this experience stands out for me is that it was through working on this project

that for the first time, in person, I met Cynthia Yao, the founder and, at that time, executive director of the Ann Arbor Hands-On Museum (HOM) in Ann Arbor, Mich. I knew of Cynthia's reputation because of her excellent track record of creating and expanding HOM, receiving National Science Foundation monies, and being an important leader in the science center and children's museum fields but had not met her personally. Cynthia graciously agreed to have her institution be one of the comparison sites for the evaluation. Since she had not had the benefit of the resources of the Science Carnival Consortium project in starting her institution, her experience was useful as a benchmark. The interview with her was enlightening and helped tremendously as we tried to tease out the benefits and effectiveness of the project. At the time, I realized what a font of knowledge Cynthia was, and so when she called me about three years ago to say she was planning to write a handbook about how to start and expand a science center, sharing not only her own experience but that of others around the country and the world, I was excited. As these chapters suggest, the individuals who start, run, and expand science centers are eager for resources, and I knew that a handbook, written specifically for them by their peers who had started their own institutions, would fill an important niche.

Fortunately, as Cynthia has proven on many occasions, her vision was actualized, and small science centers now have a useful resource written specifically for them. Participating in this collaborative effort has been another rewarding opportunity, allowing me to reconnect with many of my Science Carnival Consortium colleagues around the country, including Cassandra Henry at Science Spectrum, Karen Johnson at Discovery Science Center, Laddie Elwell at Headwaters Science Center, and Sarah Wolf at the Discovery Museum in Rockford, which was another comparison site. It is gratifying to see how these institutions have grown and expanded and continue to flourish. It is wonderful to know that they will now be able to inspire other people and communities to do the same.

The Science Carnival Consortium evaluation, completed nine years ago, focused primarily on assessing the program, but a multitude of factors involved in establishing a successful science center emerged also. Among those factors were the following:

- Establishing expectations and a clear work plan early on
- Ensuring open, two-way communication with the community and key stakeholders
- Communicating a shared vision for how the science center will enhance and serve the community

These factors are clearly evident within the many examples in this book, as is one other important factor: the benefit of establishing collaborative relationships with other organizations in the local community and with other museums in the local community or region. Although such collaborations can be time-consuming and costly for small institutions, when these relationships work, they allow institutions to better integrate themselves into their local and regional communities and allow them to do more with less. Using the shared resources of these collaborations, small- and medium-size institutions are able to access a larger audience and to accomplish tasks that would have been too challenging for a single institution.

When all is said and done, the primary purpose of this book is to gather the collaborative wisdom, experience, and insights of those who have undertaken the creation, running, and expansion of such institutions, so that new start-ups and institutions contemplating expansion do not have to reinvent the wheel but can learn from these experiences as they pursue their dream. In addition to the useful pointers, though, the book should be an invaluable resource as individuals try to convince their communities to undertake such an effort. The inspiring stories collected here document that despite the hard work, sleepless nights, and days of worry and indecision, all of it is worth it in the end.

Lynn D. Dierking
Institute for Learning Innovation
February, 2005

Preface and Acknowledgments

The idea of this book has been with me for a long time, and I am very happy to see it evolve into a wonderful collaboration. I am grateful to the fifty-five colleagues, the best in the field, who agreed to share their knowledge and best practices and to volunteer their time to write chapters on all aspects of small science center operations. *Handbook for Small Science Centers* contains an estimated 1,200 years of accumulated knowledge and experience gained by contributing authors while working in science centers. All contributors have spent their lives working behind the scenes as unsung heroines and heroes in small and large science centers. I am very pleased to note that there are more than twenty-three PhDs among the authors who have written chapters. David Taylor was in the middle of his PhD studies, before his untimely death, and his degree would have been the twenty-fourth PhD. *Handbook for Small Science Centers* is dedicated to David Taylor, Roy Shafer, and Dan Goldwater, three science center heroes, now gone, who have left their marks and made the science center world a much better place.

The *Handbook for Small Science Centers* has many fascinating stories of how some small and large science centers became realities, but more important, its focus is on what happens after the science centers open: stories of how these science centers struggled to become successful; how they became viable institutions and maintained sustainable operations; how they created innovative exhibits and programs to provide significant informal science education for their communities; and how staff conducted their work as professionals with dedication and passion. In relating these stories, the authors share their best practices as secrets of their success. It is vitally important to find good models, and indeed, I learned the most valuable lessons by visiting museums, observing practices, and emulating the best models at the beginning. Museums became my learning playground.

Best of all has been the continuous help, support, and friendship of colleagues over the years, which has been invaluable and has culminated in the creation of *Handbook for Small Science Centers*. This book is a way of giving back to science centers and museums what has been an honor and privilege to receive.

My quest to start a science center began in the late 1970s, when I was completing graduate studies in the museum practice program at the University of Michigan. I had taken time out to be home with our four young children. Becoming a student again opened a new passion for me, that of pursuing a new kind of museum filled with hands-on exhibits. I believe that interactive exhibits provide the best ways to learn by playing at a science center.

My adventure began by encounters with many wonderful museum people whom I met along the way, who were generously willing to share information and advice. In the summer of 1979, I went to the Exploratorium as an intern to be exposed to this exciting new world. I met Shab Levy, who was also visiting. He invited me to OMSI in Portland, Ore., where he was the head of exhibits, and introduced me to staff who helped with good advice. Many years later, Shab kindly donated his "refurbished like new" Giant Zipper to the Hands-On Museum.

I met Mike Templeton, who was director of the Association of Science-Technology Centers (ASTC) at the time, at the International Year of the Child (1980) conference at United Nations in New York City. A few years later, when he was a program officer at the National Science Foundation (NSF), he guided the Geometry in Our World grant, the first of six grants awarded by NSF. Geometry led to the meeting of the legendary Dan Goldwater, director of exhibits at the Franklin Institute in Philadelphia, who became our national advisor for this and other subsequent NSF grants.

In the summer of 1982, while walking on the Smithsonian Mall, I stopped to pull out a map, when a lady asked me if she could help. I told her that I was going to the Natural History Museum. She was going in that direction, and we chatted along the way. I told her that I wanted to see the new Discovery Room and was considering adding such a gallery in the newly opened Hands-On Museum (HOM). She said she knew something about this gallery and offered to send me some information. When I went to the museum, I was disappointed to see that the Discovery Room was closed. However, the Insect Zoo was open, and I arrived just in time to see the feeding time of the tarantula. I saw an observation beehive for the first time, which fascinated me. A few weeks later, I did receive materials about setting up Discovery Rooms from the lady. And, we did create a Discovery Room, which included an observation beehive.

Many years later, at an ASTC conference in Baltimore, I went to see the new exhibition on memory, a project of the Exploratorium. The person behind this exhibition was Dr. Caryl Marsh; she was the same kind lady who stopped to help me! She was the creator of the first Discovery Room at the Smithsonian! Caryl was also the visionary who created the outstanding exhibition Psychology, which we were fortunate to bring to Ann Arbor!

I am so privileged and fortunate to have met so many wonderful people in the field over the years. I met Peter Anderson on a bus at an ASTC meeting. A few years later, he sent two people from Dundee, Scotland, to Ann Arbor when Sensation was being set up. I met Michael Spock, former director of the Children's Museum in Boston, and Larry Bell, head of exhibits at the Science Museum, when we were on a committee together at the MIT Museum regarding the fate of Doc Edgerton's Strobe Alley. I met most of the talented contributing authors of *Handbook for Small Science Centers* who became my colleagues and friends by attending and participating in ASTC and American Association of Museums meetings.

Coincidence or fate... whatever it is, it has been wonderful. These fortuitous meetings and connections are an important part of my destiny with science centers and museums. I feel truly honored to be present at the beginning of the movement for hands-on museums and to be a participant in the middle of this wonderful ferment. To put it simply, I found my calling!

When the HOM was getting established and becoming known, we were getting weekly calls and visits from many people who were starting museums. There was a great need for information. There was very little written about hands-on museums and how to start these centers. One of the few important printed documents available was an essay "From Pumpkins to Coaches" written by Bonnie Pitman-Gelles, which later became *Museums, Magic & Children, Youth Education in Museums,* which ASTC published in 1981. I tried to answer those calls and was always willing to help, just as many people had helped me. Today, there is still a need, as science centers are growing all over the world. Most of these museums are small and have very modest beginnings. I know that *Handbook for Small Science Centers* will help to inspire and fill this need.

I wish to thank everyone who has participated in this book. Thanks to Mitch Allen, former publisher of AltaMira Press, who gave me the wonderful opportunity to make this book a reality. Special thanks to Susan Walters, former editor at AltaMira, who encouraged and guided me through the whole process, even after it was not her job to do so anymore. Susan is responsible for naming *Handbook for Small Science Centers.* My thanks also go to AltaMira's staff: Chris Anzalone, senior editor; Jehanne Schweitzer, senior production editor; Marcia Ryan, copy editor; Claire Rojstaczer, editorial assistant; Piper Furbush, cover designer; and Melanie Allred, marketing coordinator, for shepherding the production of *Handbook for Small Science Centers.*

Special thanks to all the contributors, who took the time to write their stories and share their experiences and best practices. They are the finest people working in science centers whom I have been privileged to know. They have generously shared their time, talents, creativity, and exciting programs, which have contributed to dynamic and vibrant science centers. They are the reasons science centers have become so popular and have become magnets for informal education and teaching science. Programs such as YouthALIVE! and the Explainer Program have provided innovative ways to get young people working in science centers and interested in science careers.

A very special thanks to my coeditors Peter Anderson, Lynn Dierking, Dennis Schatz, and Sarah Wolf, who agreed to be advisors and editors. They were wise keepers of the high quality of the content and provided their valuable editing skills, especially after we lost our editor. They are simply the most knowledgeable experts on science centers today. Everyone made the daunting task of putting together this volume a fulfilling experience for me, and together we created and achieved something extraordinary. It has been a labor of love for me as well as for all the contributing authors.

I would like to thank my family, especially my husband, Ed, who has given me such wonderful support and encouragement through thick and thin moments. He is responsible for making me take on this wonderful life-changing leap into the museum world. Thanks also to our children—Michelle, Mark, Steven, and Lisa—who endured late pickups from lessons, too many meetings, and not enough time given to their needs. I am proud that they survived and pursued science careers and became medical doctors—a radiation oncologist, a cardiologist, an emergency medicine doctor, and an electrical engineer—despite the shortcomings of their very busy mother.

Also, I am grateful to my sister, Lily Yamamoto, and Joe Cabrera at LMY Studio for the cover concept design, guidance, and solving problems of all the graphics and computer challenges I faced along the way.

Special thanks to Jean Magnano Bollinger for her contributions as a cofounder of the Ann Arbor Hands-On Museum, and a special appreciation for her faith, friendship, and support through the most difficult times for me. Thanks to the many Hands-On Museum volunteers and supporters who were willing to write memories of their time at HOM. Their support brought great meaning to my life and my work at the Hands-On Museum.

Another note of deep gratitude goes to ASTC and their wonderful supportive staff: Bonnie VanDorn, Wendy Pollock, DeAnna Beane, and everyone else there. The annual meetings at ASTC where we all look forward to participating in sessions, "power lunches" with colleagues, or just giving a hug or greeting provided inspiration for *Handbook for Small Science Centers*. This has been a wonderful family for me, and I am proud and privileged to be part of it.

I hope that the many stories in this volume will inspire others to work in science centers and to make science centers the best they can be. It has been the most rewarding life experience for me to be an active participant serving our visitors and affecting their visits, as well as sharing the knowledge, fun, and joy of learning that each person takes away with them at every moment that they are at the museum. It is the dedication of the staff working behind the scenes in science centers who make the visitors' experience memorable; the scientists who are able to bring understanding and fun to science; the educators who will find simple ways to teach complicated concepts; and the exhibit fabricators who will grit their teeth when the exhibit breaks down for the umpteenth time but will keep trying to make it better. Science center staff will spend many hours brainstorming to try again until they reach the right Aha! moment when they know it works.

Walking through the quiet science center when it is closed and missing the happy hum of visitors who make the place come alive—it is a special place that only museum staff knows. The flame of hands-on learning, long may it blaze!

Cynthia C. Yao

Know you what it is to be a child?

... it is to turn pumpkins into coaches,
and mice into horses, lowness into loftiness,
and nothing into everything.

—Francis Thompson, "Shelley"
(from *Museums, Magic & Children*)

Acronyms

AAAS	American Association for the Advancement of Science
AAHOM	Ann Arbor Hands-On Museum
AAM	American Association of Museums
AAPS	Ann Arbor Public Schools
ACM	Association of Children's Museums
ADA	Americans with Disabilities Act
ASTC	Association of Science-Technology Centers
The BA	British Association for the Advancement of Science
CILS	Center for Informal Learning and Schools
CKM	Curious Kids' Museum
COSI	Center of Science & Industry
CSTM	Canada Science and Technology Museum
ECSITE	European Collaborative for Science, Industry and Technology Exhibitions
FAST	Family Adventures in Science and Technology
FIPSE	Fund for the Improvement of Postsecondary Education
FWMSH	Fort Worth Museum of Science and History
HHMI	Howard Hughes Medical Institute
HO	Hands On!
HOM	Hands-On Museum
HSC	Headwaters Science Center
ICOM	International Council of Museums
ILE	Informal Learning Experiences
ILI	Institute for Learning Innovation
IMAX	Image Maximum (large screen theaters)
IMLS	Institute for Museum and Library Services
IMS-GOS	Institute of Museum Services-General Operating Support
LASER	Leadership and Assistance for Science Education Reform
LHS	Lawrence Hall of Science
MAP	museum assessment programs
MDO	medium-density overlay
MIT	Massachusetts Institute of Technology
NEA	National Endowment for the Arts
NEH	National Endowment for the Humanities
NESTA	National Endowment for Science, Technology and the Arts
NSF	National Science Foundation
NSTA	National Science Teachers Association
NYHS	New York Hall of Science
OIS	Opportunities in Science

OMSI	Oregon Museum of Science and Industry
OSC	Ontario Science Centre
OST	Office of Science and Technology
PERG	Program Evaluation and Research Group
PSC	Pacific Science Center
RGO	Royal Greenwich Observatory
SCC	Science Carnival Consortium
SESAME	Search for Excellence in Science and Mathematics Education
SITES	Smithsonian Institution Traveling Exhibition Service
SMEC	Science Museum Exhibit Collaborative
SMM	Science Museum of Minnesota
STEM	Science, Technology, Engineering, and Mathematics
TEAMS	Traveling Exhibitions at Museums of Science
W5	whowhatwherewhenwhy

Part I

CENTRAL CASE STUDY

CHAPTER 1

The Ann Arbor Hands-On Museum: From Dream to Reality

Cynthia C. Yao

It is not often that one gets to start a museum. I did, and the Ann Arbor Hands-On Museum became my master's thesis in reality. I experienced a wonderful adventure and a rewarding journey. This chapter tells the Hands-On Museum story through the personal perspectives of some of the individuals who played significant roles in the museum's creation and growth. Some of these people worked at the museum, and some participated in its programs—all were affected by the Hands-On Museum. This chapter also highlights more than twenty-five years of "best practices" learned by doing and shared to help others who may choose to follow a similar path.

TO BEGIN AT THE BEGINNING

I was inspired by the "hands-on" concept many years ago on museum visits with our four children when we accompanied my husband, a physics professor, to national laboratories and universities. Museum visits all over the United States and Europe were included in our itineraries. What stood out as the greatest fun for the children and for me were the museums where visitors were encouraged to play or experiment with exhibits, such as those at the Children's Museum and Museum of Science in Boston and the Exploratorium and Lawrence Hall of Science in California. I discovered interactive exhibits and the great fun and learning they afforded. I still remember vividly the glee of our three-year-old son, Mark, when he picked up a giant puzzle piece and fit it into the room-size puzzle at the Children's Museum in Jamaica Plain, Mass.

THE IDEA OF HANDS-ON EXHIBITS IN A MUSEUM

The idea began to take form while I was a student completing my graduate degree in museum practice at the University of Michigan. I interned at the Detroit Institute of Arts, and I had two wonderful opportunities to travel to study museum programs in depth. Two grants allowed me to visit thirteen art and science museums in Boston, New York, and Washington, D.C., and I also visited science and children's museums in California, Washington, and Oregon. Staff members at these museums were extremely helpful, providing advice and suggestions. These museum and science center visits were the seeds of inspiration for the idea of starting a participatory museum in the old fire station in Ann Arbor.

IN THE FALL OF 1978

Fate played a hand in helping the seed begin to take root. Ann Arbor's central fire station, built in 1882 and listed on the National Register of Historic Places, had sat empty for several years, since a new state-of-the-art fire station was built next door. An article in the *Ann Arbor News* said that the city was looking for ideas for a new use for this beloved landmark. I drove past the building daily, and although I had not been inside, I started to daydream about a use for it. I finally went to take a peek through the windows of the closed-up building. When I saw its open spaces, I realized that very few changes would need to be made. It was perfect. Another newspaper article appeared describing how the Ann Arbor Civic Theater had approached the mayor with the idea of a theater and he had seemed receptive. I realized that my daydreaming was just that, unless I shared it. My husband suggested I write a letter to city hall, as there was a deadline for proposals. Little did I know that they meant financially backed proposals and that my letter would be deemed merely an idea!

KIDS VERSUS CULTURE

There ensued a time of much drama and publicity under the headline "Kids versus Culture" when the idea of a museum and a theater were given front-page coverage. A group of children and adults went to Ann Arbor's City Hall to attend a city council meeting, and it became

Figure 1.1. Ann Arbor Hands-On Museum, Huron Street, original 1882 National Register historic firehouse and new addition of the major expansion in 1999 (left). Photographed by Fred Golden.

quite a memorable event when David Mouw, a father, took a horse's skull out of a pillowcase and passed it around to council members, something he often did for show-and-tell at the school where we both volunteered. He admonished council members that the museum idea was a "gift horse" they should look at more closely.

The city accepted our suggestion of joint use as a political solution. However, several weeks later, the theater group withdrew when another option became available to them. City officials were surprised at their withdrawal and imposed stiff conditions on us: we had to prove that the idea and plans were feasible; we had to raise the funds to renovate the building; we had to incorporate as a nonprofit organization; and we had to report to city officials regularly on our progress. Our reports to the city became our first newsletter.

On January 8, 1979, the Ann Arbor Hands-On Museum organization received permission from the city council to undertake planning to convert the old fire station into a new kind of museum unique to Washtenaw County—a participatory museum with hands-on exhibits for all ages. Its mission was to bring informal science education to the community. The group incorporated as a nonprofit entity and applied for tax-exempt status with pro bono help from attorney Phillip Bowen. From the beginning of the project the museum's name was very important, and we spent many hours discussing it. Joan Ross was particularly interested in its naming, as her husband, Marc, recalls:

> [Joan's] math education experiences using games, puzzles, toys and manipulatives had convinced her of the crucial importance of hands-on learning. Therefore she strongly urged calling the museum by its present name. A further advantage to this name would be that many different disciplines or interdisciplines could be appropriately featured in exhibits.

ORIGINAL MISSION

The museum's primary goal is to promote learning through a hands-on, minds-on approach—learning by doing, handling, taking apart, putting together, experimenting, and discovering: DOING! Central to this philosophy is the Chinese proverb, "I hear and I forget. I see and I remember. I do and I understand." In the type of environment conceived as a "learning playground, permanent and changing exhibits will be created to stimulate visitors to explore concepts in the sciences, technology and the arts." (Excerpt from the original Mission Statement)

Lucy Kirshner recalls the early discussions about how to accomplish this mission:

> Right from the start, the philosophical discussions were most engaging for me. I recall late-night conversations in various living rooms and kitchens around town as we planned the first exhibits for the museum. We debated how people would learn about science and what they should learn about art at our museum. What was science, anyway? Should exhibits be grouped or arranged in any particular order? Should we mix exhibits on physical science with exhibits on life science? How much was this to be a museum for children and how much a museum for adults? Should we give information about each exhibit or let experiences convey the information? What should we call the museum? I realize now that these questions about learning have been at the center of a powerful evolution of museums over the last fifty years. The answers are not obvious nor are they the same for each institution. Today, strong museums are built around a shared vision of the museum's educational mission, giving the staff an opportunity to discuss, debate, and contribute to the vision.

What followed were four and a half years filled with excitement, anxiety, frustration, satisfaction, exhilaration, laughter, and joy—emotions that ran the gamut. At first, the group grew erratically as publicity brought many people in, while many others left because they expected instant results. Meetings were scheduled weekly at people's homes and we literally passed a cup around to collect petty cash. It was difficult to play a leadership role when I had no experience knowing what steps needed to be taken. Many of us who stayed the course were educators who were totally innocent about fund-raising, but we were excited about the project and determined to try.

The city allowed us to set up a small office in city hall, which was directly across from the firehouse.

University Microfilms International adopted us as an employee community relations project to help get the group started. They assisted with marketing ideas, a logo, stationery, and the printing of a prospectus. We began to collect information on every hands-on museum we heard about, and we read the few books and articles written on the subject that were available at the time. We gathered photos taken on trips and visits to these types of museums. We created a slide show of what had been done in other communities, and used it as a basis for spreading the word.

SPREADING THE WORD

Our first major presentation was to the Ann Arbor Area Chamber of Commerce. It was a total disaster despite our careful planning and presentation of our newly printed prospectus. They were harshly critical of everything about the idea, and the *Ann Arbor News* reported on this event. We were very discouraged and had only a lone supporter, Mark Ouimet, who later became a trustee. However, we were determined to keep on trying.

That summer, Joan and Marc Ross and I went to the Exploratorium under a Fund for the Improvement of Postsecondary Education (FIPSE) grant program to learn about their exhibits and operations. During this period, we met Frank Oppenheimer, Rob Semper, and Mac Laetsch, director of Lawrence Hall of Science. The important lessons learned there formed the basis for our future successful exhibits and education programs (see Rob Semper, chapter 46).

Marc Ross recalls the experience:

In the summer of 1979, Cynthia, Joan and I spent four weeks as interns at the Exploratorium in San Francisco. The time there was a revelation: seeing the thinking behind exhibits, seeing them made in the shop, [and] seeing how people interacted with them. Another feature of the Exploratorium, which other museums have found hard to match, is that late teens and adults frequently come for themselves, without children.... [Our] vision was to aim in a very general way at a fourth-grade child, while hoping that exhibits and activities would appeal to a wide [age] range. Members of our start-up group were interested in children, but did not want a "children's" museum per se.

OTHER EARLY NOTEWORTHY ACTIVITY
- Mott Children's Hospital had a weeklong activity at the newly opened Briarwood Mall, where doctors and nurses set up a hands-on medical exhibition to alleviate children's fear of hospitals. Leo, the lion, was

the patient; children wore hospital gowns, took temperatures, and tried out wheelchairs and crutches. The Hands-On Museum set up panels of photos and materials of what the museum could become.
- Volunteers made T-shirts and set up tables with children's activities at the annual Ann Arbor Street Art Fair. (We had a booth there for about fifteen years.)
- A Chamber Pops Concert benefit, organized by Marguerite Oliver, who later became a trustee, was held in the firehouse.
- A reception for community leaders was held at the University of Michigan President's House at Vivian Shapiro's invitation. We brought several of our new exhibits from our traveling exhibition.

THE FIRST EXHIBITS—A TRAVELING EXHIBITION

Thanks to the efforts of Lorraine Nadelman, we received a small grant of $2,250 from the University of Michigan to celebrate the International Year of the Child (1980). This allowed us to create a prototype traveling exhibition to demonstrate the hands-on concept to the community. We built our first seven exhibits, adapting six that were easy-to-build favorites from the Exploratorium Cookbooks. We found talented folks in the university's marine engineering department to build the exhibits and recruited fifty volunteers to staff the exhibits and to create an instruction manual. Lorraine describes the experience:

In 1978, I was invited to join a small group of volunteers who were determined to turn Cynthia Yao's vision of a hands-on museum into reality. I had been teaching child development and doing research with children for years, and could be useful in planning a facility geared for fourth-graders through adulthood. At the least, I could, perhaps, help to translate the scientific, professional descriptions about what and how each project worked into everyday simple language. In 1980, I received an International Year of the Child grant from the University of Michigan, and used it for the first seven exhibits for our museum.

Since the museum was not yet open, these exhibits were first set up in Rackham Auditorium, Mott Children's Hospital, and the Exhibits Museum on the University of Michigan's campus. The exhibition, staffed by volunteers, was taken to several vacant storefronts on Main Street, downtown Ann Arbor and Kerrytown, a new marketplace. It then traveled to each public school for a week—educational fun for the students and fine public relations for us. That was the start for our later successful Outreach programs.

After moving these exhibits seven times in three months, we were relieved to find a temporary summer home with the dinosaurs at the university's Exhibits Museum of Natural History. It was a bold move, requiring concessions from the director, Professor Robert Butsch, as their exhibits were definitely "not touchable." High school students staffed the exhibits, and that fall, the Ann Arbor Public Schools agreed to take over the exhibition. They created a slide show and an unemployed teacher was hired and took the exhibits to twenty-six schools within the district. More than 15,000 students enjoyed them. Later, three outlying school districts, Ypsilanti, Dexter, and Whitmore Lake, took the exhibitions. An estimated 40,000 students were reached. The exhibits traveled for more than two years while fund-raising and renovation continued and the museum was ready to open. Response was very enthusiastic, and many children looked forward to the opening of the museum.

MORE EXHIBITS

A very active exhibits committee met often and planned which exhibits to build. Exhibits were funded by several small grants, were primarily about health and the human body, and were adaptations of Exploratorium Cookbook recipes. Local companies that had already given cash support, such as Sarns and University Microfilms, built many of these exhibits. Dick Crane became the exhibits liaison, and we formulated a plan for exhibits on both floors: the first floor was titled "The Subject Is You," and the second floor, "The World Around You." We incorporated the original seven traveling exhibits that came "home" and had twenty-five exhibits built by volunteers by opening day.

As work in the building continued, individuals and employees of local industries began building exhibits, and a small cadre of exhibit builders emerged. Dick Crane, retired chairman of the University of Michigan's physics department, became our chief exhibits person and was our liaison with industry. New exhibits were produced every few months. He and Bil Mundus, whose grandfather was the first fire chief, became a team and were joined by others. Crane's "research associates" came every morning to help maintain the exhibits, which were in working order nearly 100 percent of the time.

After a year and a half, we could be proud of several major accomplishments. Our first community advisory council was formed, which later evolved into our first board of trustees. Notable among them were Joe Fitzsimmons, president of University Microfilms, George Goodman, mayor of Ypsilanti, and Jerry Weisbach, president of Warner Lambert Parke-Davis Research Division. Other trustees were faculty and the wives of faculty—the beginning of a successful town-and-gown collaboration. This was important as our base became broader and we gathered more experienced and knowledgeable people who had the ability to help us raise funds. George Goodman describes the tireless efforts of turning the idea into reality:

In the late 1970s, my wife, Judith, and I were approached about coming to a meeting to participate in a discussion about the possibility of establishing [a] children's hands-on museum in Ann Arbor. At the time, I was mayor of Ypsilanti, and it was obvious that the formation committee was attempting to reach out beyond the boundaries of Ann Arbor. Among the items discussed was the issue of funding. I suggested that one group that should be approached was the building trades. It seemed that this type of a project would be one that the building trades would embrace.

Additional discussion included others who might be willing to lend their time, treasure, and talent. One of the names I suggested was Bob Lyons. Bob had been active in a number of community fund-raising activities and when he commits to a cause, he does so with total enthusiasm. After the meeting, Cynthia and I met with Bob for lunch at a local restaurant. The meeting resulted in Bob's enthusiastic agreement to help, and the rest is history. His enthusiasm, energy, and financial support were immediate. The other key supporter of the museum was Joe Fitzsimmons. Cynthia, Lorraine, and I called on Joe while he was still the head of University Microfilms. We made the pitch about the museum: its needs and its potential. After a fairly short discussion, it became clear that the hands-on museum bug had bitten Joe. He agreed to get involved and the result was a complete and total commitment of his time, talent, and treasure. There were many, many people who were responsible for the museum's success. Without the many hours of their time it would never have become a reality for the community.

NATIONAL REGISTER GRANT-IN-AID—FIRST-PHASE RENOVATION

The original goal was to raise $250,000 to bring the old central firehouse up to code and convert it into a museum. It was a formidable task. The group applied for its first major grant, a National Register Grant-in-Aid for $60,000, written by Jean M. Bollinger. The museum succeeded in obtaining the grant on a two-to-one match, good news, though it meant we now had to raise $120,000!

THE BRINK OF DESPAIR

After 12 months of fund-raising, the museum had raised only $40,000, and the grant deadline was upon us. The board decided to give up on the project, comforting us with the thought that we had tried our best. At the end of the meeting, three of us, Jean Bollinger, Lucy Kirshner, and I sat in stunned silence. We had become close allies and could hardly give up at this point. Fortunately, out of desperation to save the failing project, the Washtenaw County Local Building Trades agreed to donate labor and materials, an estimated value of $60,000 as in-kind match. This incredible contribution, made during a serious building recession, when many were out of work, was negotiated and orchestrated by trustees George Goodman, Dick Brunvand, and Bob Lyons. Joe Fitzsimmons recalls those early years:

> It was probably in 1979 that I received a call from George Goodman, for an appointment to talk about an exciting new project for the Ann Arbor community. Always interested in trying to help out and intrigued to learn more about the potential project I agreed to the meeting. In walked George along with Lorraine Nadelman and an energetic woman by the name of Cynthia Yao. She was an unbelievable advocate for what has now become the Ann Arbor Hands-On Museum. I was fascinated with the idea of putting a hands-on science museum for children in our old fire station, which was vacant and in need of a facelift. Cynthia explained the idea (new to me) of building exhibits that demonstrated principles of science and engineering (I was a graduate engineer from Cornell) that could be operated or run by children of all ages! What a way to learn and have some fun at the same time. What a way to bring families together—parents and grandparents taking their children to the old fire station!! I had five children of my own so the idea was very appealing to me. But it was Cynthia who sold the concept with her enthusiasm and the vision she wanted to accomplish. Needless to say I was hooked! . . . That was the beginning of a twenty-four-year relationship that continues today. . . . Looking back, other than my family and UMI, it is something of which I'm most proud to have played a part.

Architects Meneghini-Overhiser had just opened their offices and volunteered their services with the understanding that if we raised the money, we would pay them. Eventually, we were able to pay them and they also became construction managers when no one else would undertake the difficult task of overseeing donated labor during a recession.

RENOVATION BEGAN IN SUMMER 1981

More than 200 local tradespeople donated 6,000 hours of their time and labor to the construction. Businesses and individuals provided materials and dollars. Soon more donations came in, much of them in-kind materials and services, such as 600 gallons of paint, and all the painting labor contributed by Painters Local 514 and several paint companies. In addition, all the plumbing and much of the fire suppression system were donated. The dedication and cooperation of many people in this major community project totaled an estimated value of $300,000, at an actual cost of $100,000. The Hands-On Museum (HOM) was, and continues to be, a triumph of civic generosity!

The city gave the HOM a fifteen-year, renewable lease at $1 per year. This has now been extended to a longer-term fifty-year lease, contingent on viability. The HOM is responsible for all building and maintenance expenses. The city retains ownership of the historic building and has been cooperative with bus parking, snow removal, and other services.

MUSEUM OPENS

On October 13, 1982, the hundredth anniversary of the firehouse, the Ann Arbor Hands-On Museum quietly opened its doors for the first time with twenty-five exhibits built by the donated efforts of individuals and businesses, a few grants, and many cash donations.

Although we were open, we had limited hours, no elevator (we had obtained a three-year waiver), and a projection of 6,000 visitors in the first year. Well, we had 25,000 visitors! Everyone loved the HOM and it was an instant success.

FIRST STAFF

The board decided to hire an executive director in the summer before the museum opened and advertised the job in the *Ann Arbor News*. They asked me to apply for the job. There were five applicants, and I was hired as the executive director. Shortly after, Lucy Kirshner and Jean Bollinger became assistant directors. We had been volunteers for over four years and worked well together, so the transition to staff went very smoothly. We made a great team (with museum, early childhood, and psychology backgrounds) and learned how to create programs and work with visitors together. Our children were the first testers of our exhibits and programs. Lucy Kirshner describes the transition from being a volunteer to becoming a staff member:

> When I responded to a letter from Cynthia Yao in the Ann Arbor News, regarding a group of volunteers interested in turning the Old Firehouse into a new kind of interactive museum, I had no idea that I was taking my first step into a career. It started as intriguing fun and turned into a deeply interesting vocation, still intriguing and still fun. In retrospect, what I learned from the earliest experiences at the Ann Arbor Hands-On Museum has guided me for the last two decades. . . . Several volunteers and staff members still live vividly in my imagination and contribute to my inner conversations to this day. And the laughter. One day at the end of the first long winter after the museum first opened, Cynthia Yao and I were opening the doors for visitors. We saw that the bees in the indoor observation hive were frenzied, buzzing, and frantic, waiting anxiously for a warmer day so they could take flight. Many bees had died. We unscrewed the plastic tube at the entrance to the hive in order to remove several dead bees. Then the phone rang. The plastic tube was replaced but not screwed down while the two of us began the day of work elsewhere in the building. Later in the morning, a foreign, non-English-speaking visitor knocked shyly on the office door. She gestured with concern about the bees. It took us a while before we realized that the entire building was swarming with wild, if relieved, bees. They filled the air and defecated all over all of the signage we had just completed in the new museum. To this day, when the immediate problems of work drag at my heels, I have only to recall that day when Cynthia and I scrambled around waving giant sheets of cardboard trying to herd thousands of bees back into their hive and I laugh all over again.

I confess, I was the one who forgot to screw back the cover because I left hurriedly to answer a phone call and then got busy with the day's activities. (See Lucy Kirshner, chapter 21; and Kirshner is the focus in "Living Evolution: A Passion for Science Communication," *ASTC Dimensions* March/April 2006: 13.)

1982–1986

The first five years were spent developing and establishing programs. It was a very exciting time during which we learned about the day-to-day operations of a museum. We were open in the mornings to school groups only and in the afternoons to the public. This gave the small staff time to develop and experiment with new programs. We were on a learning curve for everything. Some issues that we had to work through were as follows:

- Making the transition from volunteers to staff who were running the organization
- For me personally, taking on a new role and becoming the boss
- Creating and maintaining exhibits, exhibits layout, and graphics
- Creating the Discovery Room with students from the School of Natural Resources
- Creating education programs, such as summer and fall classes and workshops and weekend demonstrations
- Starting the Volunteer Explainer Program and recruiting and training volunteers
- Creating a gift shop combined with admissions
- Accommodating school visits scheduled in the mornings
- Meeting annual budget projections every year through earned income
- Navigating the first audit
- Creating and sustaining a newsletter
- Learning how to deal with rapid attendance growth

Margaret Evans and Henry (Gus) Buchtel share their experiences, as volunteers, members, staff, and visitors to HOM:

> Our connection with the Ann Arbor Hands-On Museum began soon after we arrived in Ann Arbor from Canada in 1980. Margaret had a long-standing interest in science and science education and had written a chapter in a book for parents, encouraging preschool children's interest in science. When our children became old enough to enjoy museums, the Hands-On Museum figured prominently in our outing schedule. Our children, now graduated from Yale and Oberlin, enjoyed many happy hours at the museum. It was natural that Margaret would start to teach there: fossils and rocks were her specialty, reflecting a life-long interest in evolution and the evidence of change over the millennia. There were two kinds of teachers at the Hands-On Museum. One group developed courses that are offered to the public, often [on] a Saturday, lasting 1–2 hours. Usually children are the participants in these classes but sometimes the topics are of interest to young and old alike. Both of us taught these kinds of [classes] and enjoyed them immensely. The other group of teachers consists of "explainer guides" or docents, who spend many hours at the museum, answering questions and showing people things they might otherwise miss. Margaret did this

when the children were young.... The Hands-On Museum changed our lives in several very positive ways. It gave our children an invaluable introduction to science; it also afforded us the opportunity to share our interests and expertise with the community and our children, but, even more importantly, it has given us ideas that we are currently pursuing in our academic and research lives. Currently, Margaret is investigating the cultural and cognitive factors that influence the acquisition of beliefs about the origins of species, both in museum and school settings. She would not be doing this particular research if it were not for her years at the Hands-On Museum.

Note: Recently, Evans has been a learning researcher on Explore Evolution, an exhibition funded by the National Science Foundation (NSF), with Judy Diamond, PhD, as project director from Nebraska State Museum, which showcases cutting-edge and timely exhibits on biological evolution at five museums. This experience led directly to another NSF grant, Life Changes, on which she is coprincipal investigator with Martin Weiss at the New York Hall of Science. They will develop a traveling exhibition on the basis of Evans' research and use it to test whether informal museum-based interventions prepare children to accept the scientific basis of evolution by targeting their intuitive concepts. Evans' article "Intuition and Understanding: How Children Develop Their Concepts of Evolution" appears in *ASTC Dimensions* March/April 2006: 11–13.

In 1984, our first Institute of Museum Services–General Operating Support (IMS-GOS) grant was awarded!

1986–1992: SECOND-PHASE RENOVATION—SECOND CAPITAL CAMPAIGN

In 1985, after attendance had grown to 56,000 annually, the museum launched another capital campaign to raise $400,000 to add 50 percent more space by expanding into an unused attic. The new space would include an elevator, air conditioning, more exhibit areas, classrooms, new bathrooms, a new entrance, a greenhouse, offices, and storage.

Again, the community rose to the challenge. Despite some unexpected structural engineering problems that added to the construction costs, fund-raising this time was much easier. The museum received its first Kresge Challenge grant for $80,000, and Hobbs + Black, a local architectural firm, and O'Neal Construction both donated services valued at $25,000 each. Another great example of the giving spirit was when the head of a local fire suppression company found out that Lorraine

Nadelman was involved, he donated the installation of the fire-suppression system because he had taken a class from her. Construction began in August 1986. The museum had to be closed for several months of construction. Melissa Pletcher recalls this transitional period:

I came to the Ann Arbor Hands-On Museum in the fall of 1986, just after the museum closed for five months to complete its second renovation in less than 10 years in the 100-year-old firehouse. It was an exciting time. My first impressions were of a small, quirky but extremely talented and dedicated staff struggling to deal with the organization's inevitable growth. There were all kinds of growing pains. The staff was trying to continue to work in a building that was under major construction, five people jamming into a tiny space that had been the exhibits workshop. It seemed like new construction challenges popped up every day. Exhibits had to be built off-site and enrichment classes had to be held down the street in a church. Even the administrative framework was changing. Until then, there were no medical benefits for full-time staff members.

MARCH 17, 1987: THE MUSEUM OPENED ALL FOUR FLOORS TO THE PUBLIC

Museum attendance increased by 30,000 visitors in one year, and we had 170,000 visitors in just eighteen months. This was a period of continued rapid growth and maturity. We hired new staff with science center experience and science and mathematics backgrounds to ensure the continued success of our expanding programs.

Ensconced in the mini office on the darkened new Light and Optics gallery on the third floor, working away on our first computer, one could hear "Awesome!" "Mom, Dad, look what I did!" and the constant thumping of galloping feet going down the steps, as school groups descended from the new fourth floor. These sounds became music to our ears!

OUR FIRST CRISIS

We were named "Best of the Best," by a reporter who had visited ninety places with his child over a two-year period (*Detroit Free Press*, Martin Kohn, Feb. 1987). This attention precipitated our first major crisis. Detroit discovered us and it was spring break. We had lines around the block for several days. The police and firemen came out to help us. People were ten deep around each exhibit. We had to create several emergency measures to alleviate the crowding problems. We gave out tickets fifty at a time and scheduled them in half-hour intervals; we created a Math Walk around several blocks and gave away hundreds of free passes. After this near fiasco, we adjusted our schedule, opening the museum in the mornings for everyone and giving scheduled school groups a time limit. We also began to keep a schedule of spring breaks, major holidays, and snow days so that we could plan for more volunteers. Melissa Pletcher shares her perspective on these times:

> Throughout the fall of 1986 we worked diligently to get our ducks in a row and re-open as soon as possible with two new floors of exhibits. I can clearly remember working until quite late on Christmas Eve cleaning and preparing for the re-opening the day after Christmas. We re-opened the original two floors in December and opened two new floors during the busy Easter week. The new, expanded museum was an instant hit. We had people lined up around the block to get in. After a visit from the fire marshal, we began limiting admission by handing out timed tickets. Even after that week ended, we continued to have phenomenal attendance. It was really heartening to see so many people support us.

1987 HIGHLIGHTS
- Attendance grew to 170,000 visitors.
- The museum offered its first Summer Science Camp.
- The museum had its first Outreach Program with a van purchase.
- Camp-Ins for Girl Scouts and Boy Scouts began.
- More IMS-GOS grants were awarded.
- Howard Hughes Medical Institute grant for a biology program to underserved schools for five years was awarded.

1988: ASTC'S INSTITUTE FOR NEW SCIENCE CENTERS
The HOM achieved national and international attention among science centers, as Ann Arbor became the site for the Association of Science-Technology Centers' (ASTC's) first Institute for New Science Centers to assist communities in planning new science centers. The Ann Arbor Hands-On Museum was chosen as a model start-up science center. Enthusiasm and praise for the museum were resounding. More than thirty participants came from many states and other countries, including Ireland, New Zealand, Denmark, and Finland. They also visited Cranbrook Institute of Science, Detroit Science Center, and Impression 5 in Lansing.

Sally Duensing, Science and Museum liaison for the Exploratorium, came as faculty for ASTC's Institute for New Science Centers and sent us this note:

> After visiting your museum for several days last summer, I was extremely impressed by the entire operation. I have been able to see many of the new museums that have started in the past ten years and I am delighted to say that yours is one of the best, with only 15,000 sq. ft. that communicate ideas more effectively than some museums twice your size.

1990 HIGHLIGHTS
- We had 500,000 visitors! Annual attendance was 112,300.
- First NSF grant was awarded for Geometry in Our World.
- The museum received an invitation to the White House with over 300 directors of museums that received IMS grants to celebrate the White House becoming an official museum on May 18, National Museum Day.

1992 HIGHLIGHTS
- The HOM celebrates its tenth anniversary.
- Dick Crane's *Explore & Discover Exhibits Guide* was printed by University Microfilms International and supported by the Institute of Museum Services– General Operating Support grant (IMS-GOS) and became an ASTC bestseller.
- We began to have serious space problems again in our 10,000-square-foot space filled with 200 exhibits. The trustees hired a consultant and held a retreat. He recommended that we find a new location as we had outgrown our space. We began to look at various options, such as a nearby armory as an annex or an empty Kmart store on the other side of town. None of these options appealed to anyone.

1994 HIGHLIGHTS
- One million visitors!
- Museum purchased the buildings next door.

PLANS FOR A MAJOR EXPANSION
Fate played another hand when the Ann Arbor Area Chamber of Commerce contacted us and invited us to make an offer on their building, which directly adjoined the HOM on the west side. The museum purchased the building (actually five connecting buildings) with a $600,000 accumulated surplus from self-generated revenues. We renovated one section and added a prototype preschool gallery and a space for temporary traveling exhibitions. We rented the other half of the space to the Chamber of Commerce and to the Navy recruitment office as a temporary measure, while they looked for other sites. This provided us with income and time to plan and raise

funds as we launched our third and most important capital campaign—which would allow us to become four times larger.

1995–1999: PERIOD OF INTENSE PLANNING AND PREPARATION FOR THE BIG EXPANSION

Some key issues were the following:

- Board asking me to take on the responsibility of overseeing the capital campaign, and Leslie Kimmell, the longtime head of education, becoming the associate director, taking over the day-to-day management of the museum along with her other duties
- Planning long-term financial stability, to fund the bigger operating costs and to sustain a stable financial future
- Preparing a five-to-10-year budget based on projections and new revenues
- Starting an endowment fund
- Planning what the new museum would be like—created a comprehensive plan of exhibits and galleries
- Planning the new galleries, and refurbishment of original galleries
- Building exhibits to fill new galleries, making duplicates for traveling through ASTC, as well as upgrading our old exhibits
- Planning a large new Explore Store gift shop, hiring an experienced store manager, and using a new point-of-sale system
- Upgrading the internal infrastructure—purchasing new computers, integrating two platforms, Mac and PC, training staff with new accounting system, fundraising software (this was done with a $75,000 grant from the Chrysler Foundation)
- Reorganizing staff structure and organization
- Preparing a marketing plan—new logo, new identity, brochures, graphics, signage
- Planning the expansion
- Changing entrance from Fifth Avenue to Ann Street
- New entrance, bus parking, visitor, and group logistics
- Dealing with construction issues and running the museum at the same time

DIFFICULTIES ENCOUNTERED

The loss of key personnel and hiring of new less-experienced staff presented real difficulties—Leslie Kimmell retired in 1997 after fourteen years of service, just as the construction was about to begin. Also, we had a turnover of exhibits staff when we were in the middle of planning new exhibits and completing some NSF projects. Hiring new staff and getting them on track quickly was a challenge. Despite these problems, the construction and planning continued and we were able to accomplish all the goals that were set.

HIGHLIGHTS OF THE CAPITAL CAMPAIGN

- Launch of Ready, Set, Grow! campaign—original goal was to raise $4,200,000, then it became $5,200,000, and finally $6,500,000!
- The campaign was cochaired by Joe Fitzsimmons and Bob Lyons again and was well-planned and orchestrated with multiyear pledges.
- We sold named bricks for the Yellow Brick Road at the entrance walkway.
- We were awarded our second Kresge Challenge grant for $600,000.
- We received our largest gift of $1,250,000 from one individual donor!

GETTING THAT BIG GIFT

Phil Jenkins was a long-time, $1,000-per-year donor who generously gave the HOM $250,000 for the capital campaign. A few months later, after demolition began, we had to make some drastic changes to our plans for renovating the buildings. We discovered some serious structural problems and had to demolish all the buildings, except for a 1950s structure at the back, formerly, the gas company's offices. This precipitated raising the campaign goal to $5,200,000! A million more dollars! We decided to ask Phil Jenkins for this great sum. Norma and Dick Sarns, Bob Lyons, and I went to visit him and he said "Yes!" almost immediately; we were ecstatic! Then I handed him the carefully prepared letter and materials in a folder. He took one look at the cover page and said "Lyn will be awfully mad at the misspelling of her name!" In horror and embarrassment, I took the packet back with me, and returned a few hours later with a bouquet of "forgiveness" flowers and a floor plan highlighting the Lyn and Phil Jenkins Gallery. Sadly, Lyn passed away a few months after the grand opening, but she did cut the ribbon! We all owe thanks to a wonderful, generous man who enjoys giving money away to causes he believes in.

Phil Jenkins describes his personal interest in supporting the museum:

> Taking my eight grandkids to the Hands-On Museum got me interested and made me a believer in the concept—and then I met Cynthia and was sure the idea would fly. . . . My overall reason for backing the Hands-On was that the USA is constipated by the legal profession. In my opinion, new products—innovations and other market leading changes—are never brought to

market for fear of legal problems. If I could save "one" kid in Ann Arbor from becoming a lawyer, my million plus dollars were well spent. We have way too many lawyers in the U.S. (70% of the world!) and not enough good, well-educated engineers, and I think the Hands-On can interest kids in the engineering and science world.

Tom Beddow, whose firm 3M gave the largest corporate gift, describes the rationale behind the donation:

Under the leadership of Cynthia Yao, the Ann Arbor Hands-On Museum played a key role in the nurturing process in southeast Michigan by turning math and science into fun and wonder, often with the simplest of devices built by the hard work and innovation of museum staff and volunteers. When the museum reached the happy crossroads of being so well attended and utilized that it needed to expand, Cynthia led an impressive effort to raise the necessary funds. My company considered both cash and in-kind donations and finally settled on a combination. We wanted to give a gift that would "keep on giving" for many years and decided to provide high-bandwidth telecommunications wiring for both the existing structure and the expansion. This would enable the museum staff to "digitally innovate" for years to come without running into bandwidth constraints.

My company is very proud of this gift and based on its successful implementation at the Ann Arbor Hands-On Museum we recently committed to a similar gift at the National Museum of the American Indian, the newest addition to the Smithsonian Institution in Washington, D.C. The NMAI opened in September of 2004.

David Esau of Cornerstone Design Architects recalls his involvement in the major expansion of HOM:

In 1996, my architectural firm (Cornerstone Design Inc) was given what may be a once-in-a-lifetime opportunity: to design a major expansion of the Ann Arbor Hands-On Museum. For the next four years, through the expansion's opening in October of 1999, this project became a huge part of my life. It filled my workdays and many of my weekends, affected how I spent my free time (visiting similar museums), and even resulted in my being introduced to the woman who is now my wife. Through the process, I made many great friendships, and had a rare chance to make a significant impact on my community. The museum's needs for large, open, flexible spaces could not be met with minor renovations of the four small buildings,

and it [was] very difficult to meet life safety and accessibility requirements without major changes. In addition, the museum needed to move its main entrance from the crowded old firehouse to somewhere on the addition. That somewhere, we quickly realized, would have to be around the corner on what most people viewed as the back of the site. From the new entrance, the old firehouse would not even be visible. The new entrance would not only need to be attractive and eye-catching, but would have to become a new symbol of the museum, affecting everything from interior signage to letterhead.... In the end, the Hands-On Museum ended up with a very attractive building, at a cost far less than many similar civic structures around the country. It is always a source of pride to mention that we were the museum's architects, especially when people tell us how much they love the museum.

1997 HIGHLIGHTS
- Museum celebrates its fifteenth year.
- Attendance reached 1.5 million visitors!
- Construction and renovation of major expansion begins.
- Museum closed for several months; staff moved to temporary offices across the street; and $5,800,000 was raised, exceeding the original goal. A new goal of $6,500,000 was initiated to include an endowment fund and a healthy growing reserve for the future.

OCTOBER 19, 1999
- The museum reopens—grand opening.
- HOM became four times larger, growing from 10,000 square feet to 43,000 square feet.

Figure 1.2. Ann Arbor Hands-On Museum, new entrance on Ann Street. The museum grew from 10,000–43,000 square feet. Photographed by Fred Golden.

- Three hundred and fifty new and renewed exhibits in ten new and renovated galleries:
  - KidsWorks: a preschool gallery, dedicated to little ones under five years old
  - TechWorks: Lyn and Phil Jenkins Gallery: science, technology, How Things Work
  - WonderWorks: traveling exhibitions
  - MediaWorks: telecommunications, media, laser harp, computers
  - HealthWorks: health, body, perception
  - WorldWorks: waves and resonance, structures
  - NatureWorks: refurbished Discovery Room exhibits
  - LightWorks: Crane's Roost—light and optics in honor of Dick Crane
  - NogginWorks: math games and logic puzzles
  - Lyons' Old Country Store: in honor of Bob Lyons
  - Five ScienceWorks Labs: a library/board room, and a volunteer room

BEST PRACTICES THAT LED TO SUCCESS

1. Good Management and Fiscal Responsibility
- Met annual budget every year—growth from $50,000 to $1.75 million.
- Surplus of $600,000 enabled purchase of adjoining building.
- Successful capital campaigns raised $100,000, $400,000, and $6.5 million.
- Two Kresge Challenge grants were awarded: $80,000 and $600,000.
- Nine General Operating Support grants from the Institute of Museum and Library Services were awarded to support special operating projects.
- Six National Science Foundation grants were awarded.
- Hands-On Museum was left debt-free.
- Start of endowment fund to support general operations.

2. Grants for Exhibits and Programs
Grants for exhibits and programs are important to try to pursue, and we were fortunate to receive many.

- Six National Science Foundation grants were awarded: Geometry in Our World ($241,000), How Things Work ($374,000), Solve It! Puzzles and Challenges ($575,000), TEAMS-Clothing: Science from Head to Toe ($200,000), MidWest Wild Weather ($100,000), and Children's Object Centered Learning Conference ($50,000).
- A Howard Hughes Medical Institute (HHMI) grant of $150,000 was awarded for a five-year biology program for underserved schools.

- Ford Motor Company Fund, Ann Arbor Area Community Foundation, the Kiwanis, Galens Medical Society, and other businesses sponsored individual exhibits and programs.

IMPACT OF NSF GRANTS
- We learned to make professionally designed and travel-hardy exhibits.
- Dan Goldwater from the Franklin Institute was our national advisor for Geometry.
- Three copies of Geometry in Our World were built, instead of two.
- A copy of Geometry in Our World was sold for $40,000 to Discovery Science Center's Launch Pad in Santa Ana, Calif.
- NSF invited HOM to bring some Geometry in Our World exhibits to open its new gallery in Arlington, Va.
- Success with Geometry in Our World opened the doors for other NSF grants.
- Making two copies of exhibitions enabled HOM to fill new galleries with more than 100 exhibits.
- We shared and exchanged our exhibitions with other science centers around the country.
- ASTC circulated the exhibitions reaching 2 million more people who learned about HOM.
- Rental fees from ASTC provided income for the museum.
- Full-time exhibits staff was hired for the first time and equipped an exhibits workshop.
- Initiated first Traveling Exhibits at Museums of Science (TEAMS) grant by inviting three directors of small science centers to Ann Arbor to plan the first TEAMS request to NSF.

3. Developing High-Quality Programs and Services
We developed excellent programs initiated by staff and working with educators and volunteers to implement all our programs: weekend demonstrations, special events, teacher training workshops, overnight science camp-ins, labs, lectures, Family Math, Family Science, and so on.

Some good practices of our education programs include the following:

- Education programs earned income to sustain a balanced budget.
- We consulted with scientists and with educators in schools.
- We learned about new programs from other science centers.

- Our preschool programs for fifteen years promoted visitations from age two; it has been the best way to start the habit of weekly visits.
- Kidspace gallery was created and is a now a favorite.
- Family Science Outreach became a big money-maker: HOM staff go to schools and all families come to school gyms for Family Physics, Family Math, Family Biology, Family Chemistry.

Sarah Wasserman was influenced by a class at HOM.

I think it's great when adults treat children with respect—like people with the potential of contributing something to the community and to the world. Children get a real sense of self when they can experiment with new ideas in a supportive environment; it also starts them thinking enthusiastically about possible careers. The Ann Arbor Hands-On Museum did all of these things for me, when I was just nine years old.

I imagine that when my parents encouraged me to take some of the classes offered at the museum, it was to get me more involved with science. However, far more than this was accomplished. By valuing my ideas and fostering my creativity, the program set me on a path to learn more about the process of invention. It also spurred members of the community—at the public library and a private law firm—to get interested and involved. These indirect effects helped to shape both my and my brother's career paths.

I chose to take an invention workshop, and after some weeks of learning about both great inventions in history and whimsical mousetraps, we were instructed to invent something ourselves. We had no constraints—the inventions could be realistic and practical or they could be complex toys requiring technology that did not yet exist. However, at nine, I was learning that the creative process requires both work and inspiration. The inspiration came when I was sitting on the couch, trying desperately to invent something. My father was in the kitchen, making a racket as he looked for the right size pot top. It suddenly seemed obvious then that someone should make a top that fit many different-sized pots. Then came the work: How should the top look? How would different models affect ease of manufacturing and ease of cleaning? My older brother offered to help build a prototype, and soon we were both swept up in the process, cutting and taping cardboard into a mock-up of the Universal Pot Top and various-sized pots.

The museum held an invention fair at the end of the workshop, and my pot top was met with enthusiasm. People stopped and talked to me seriously about the invention, and it felt really good to be spoken to with respect, like a grown-up. It made me realize,

at nine, why grown-ups worked hard, and the satisfaction that could come from creating something new. What started at the museum spread outwards, as other people in the community became interested and showed their support. My father and I went to the local library, where they showed us how to do patent searches, and when we had exhausted those resources, a local lawyer took his own time to help us with the patent process.

The Hands-On Museum taught me about the process of creating something new, and the people there gave me the tools I needed to do it. But the program did more—it drew in those around me—the community supported what the museum had started. The workshop brought out strengths I didn't know I had, but which I have since developed. When I graduate [Yale] Law School this May [2004], I'm headed to a firm that specializes in patent law. I'll be on the east coast, and closer to my brother, the engineer.

The following is a recent note from Sarah Wasserman who is working for a law firm in Boston that specializes in patent law.

You made Ann Arbor a better place to grow up in!...And here's something you might be interested in—our firm goes to the local science fair every year and finds the most patentable invention and drafts a patent application for that child. I think it's a great program.

4. Staff

The staff of a small science center has a different work situation compared with a large center. Members of a small staff have to wear many hats, and they have to maintain a very close cooperative relationship as exhibits, programs, and daily management are all intertwined.

Here are some good practices:

- The executive director should have a strong work ethic, lead by example, be able to juggle many balls, and love the job.
- Staff should be qualified and experienced science- and mathematics-trained educators from science centers or schools.
- Despite its small size, as soon as it was able, HOM provided its staff with good medical and retirement benefits.
- Professional development of staff: we provided funds for staff to go to conferences and workshops.
- Visitors are the priority and the science center's clients. We try to have someone, either staff or volunteer, on the floors at all time. Staff and volunteers know that they are there to serve our visitors.

- Staff offices are placed on each gallery floor. The benefits are (1) it encourages staff to interact with visitors, (2) it makes staff aware of whom they are serving (visitors like this very much), and (3) it provides semi-security as the museum is spread out and staff numbers are small. This is an uncommon but highly effective practice.
- Manager for the Day: senior staff take turns as Manager for the Day who coordinates all activities for that day. (See Andrée Peek, Manager on Deck program, chapter 13.)
- Senior staff, including the executive director, work on weekends, as there are often more visitors on weekends than during the week.

5. Governance—Trustees

Trustees are the governing policy makers and standard bearers for the mission of the museum. Trustees are entrusted with the civic responsibility of the museum and should follow the highest ethical standards. I have been fortunate to have worked with 130 community leaders who were trustees of the Ann Arbor Hands-On Museum. Most have been wonderful benefactors to the museum and mentors to me.

Norma Sarns, longstanding trustee, donor, and supporter, writes the following:

In the summer of 1982, I was honored to be invited to join the Board of Trustees of the Ann Arbor Hands-On Museum prior to the opening of the Museum to the public in October of that year. It had been wisely decided that Cynthia Yao would be the first Director of the Museum. Not only was she highly qualified for the position, she was the founder and driving force behind the creation of the Museum.

The staff and volunteers were filled with creative ideas and enthusiasm for hands-on education. The board members were given a tour prior to the opening. I walked from one exhibit to another. Lorraine Nadelman, an early supporter of the Museum approached [me] and asked why I wasn't working the exhibits. So ingrained was the idea that one only looks when in a museum, I was actually not touching anything. How liberating it was. What fun! Just as the children, I could touch, experience and learn. In the days and years ahead the building throbbed with energy as the children and families entered to experience a museum where children could actually touch the exhibits. Every fun-filled exhibit taught a science lesson. The high caliber staff worked diligently to provide a quality education for the children who were just having so much fun learning. Fascinated, the children and their parents returned again and again. I was excited to be on the board of this wonderful community adventure.

Here are some good practices to consider for trustees:

- Provide a good orientation and overview of the history of the museum.
- Include Helmut Naumer's *Of Mutual Respect and Other Things* in board packets (available through American Association of Museums [AAM]).
- Clarify trustees' and director's roles: trustees set policies, director or CEO implements policies and manages the museum.
- Trustees of 501(c)(3) nonprofit organizations must conduct the business of the science center with the highest ethical standards and exemplary practices.
- Trustees should not use the museum for their own personal gain.
- Trustees have the fiscal responsibility to see that all financials are closely monitored beyond an annual audit. It is their obligation to investigate any question of impropriety, even if it maybe one of their own.
- Trustees should not usurp their power to create jobs for friends, family, or other board members.
- If a job is created, it should be advertised widely to select the best qualified candidate.

(For more on governance see Harold and Susan Skramstad's chapter 36 and Code of Ethics for Museums, chapter 37.)

6. Volunteers

Volunteers created the HOM and were the lifeblood of the HOM. There were over 300 volunteers who gave 12,000 hours annually.

VOLUNTEER EXTRAORDINAIRE H. Richard Crane, PhD, professor emeritus, retired chairman of the University of Michigan's physics department, member of the National Academy of Sciences, who received the Presidential Medal of Science, was our chief exhibits person and built many of our early exhibits. He is an inventor with a lifelong dedication and interest in teaching. He received the Oersted Medal and gave the prize money to the HOM. The museum received a grant from the NSF to build How Things Work, inspired by Dick Crane's columns in *The Physics Teacher*, a journal for high school physics teachers. Crane celebrated his ninety-eighth birthday on November 4, 2005.

FAVORITE DICK CRANE STORY Dick Crane installed the first in-house exhibit, the Hot Air Balloon. The base was made out of giant electrical cable spool from the construction, the balloon was powered by a toaster and a counter read "Your Flight Number is—." The whimsy and creativity and intent to teach science of this exhibit is a perfect example of the exhibits that filled the museum. One day, a mom called us about her son, a third grader, who had been to the museum and was inspired to build a similar hot air balloon for his science fair project. After spending many hours experimenting, he could not get the balloon to rise up very high. After many trials and tears, she called the museum for help. Crane had a conversation with him and he learned the secret—a counter balance! Several weeks later, we received a newspaper clipping with a photo of her son, grinning from ear to ear, with his hot air balloon behind him, as the first prize winner!

Here are some good volunteer practices:

- Having a volunteer coordinator on staff
- Creating a good volunteer recognition program
- Training, retaining, and appreciating volunteers as keys to success
- Having trustees, CEO, and staff participate in volunteer recognition

Wyatt Bardouille was a volunteer and staff at HOM:

When I was a teenager, about 3 years before I graduated from high school, I went to the Ann Arbor Hands-On Museum (AAHOM). I thought it was a really amazing place because all of the exhibits were ones you could touch and interact with; very different from other museums I had been to in the past. I found this most fascinating because the museum was full of exhibits about science and technology, two of my favorite subjects in school. . . . Knowing how much I enjoyed my visit there, my mother suggested I volunteer at the Ann Arbor Hands-On Museum. I thought that I would give it a few months and see if I liked it. So I started out volunteering at the museum and ended up spending 7 years there in various roles throughout the rest of my high school and college years. My volunteer experiences at the museum turned into part-time work while I was in college. I became a member of the staff and gained new responsibilities as a museum assistant.

One of the most gratifying events of my time spent at the AAHOM was when I interviewed for and was offered a full-time job at the museum after graduating from college. Although I went on to accept a position with another company (Microsoft!), I felt much honored by that experience.

The AAHOM also "gave back" to me in many ways. I gained valuable work experiences; I got an opportunity to explore career possibilities that I would not have considered otherwise. I learned leadership, teaching and presentation, planning and coordinating skills, which have served me well in various positions over the years. I grew in responsibility and accountability. I made friendships for life.

Our best and most reliable volunteers were retirees. Gene Kress was one of them:

My experience with the HOM began late in 1984. I had just retired from a career in secondary education. The museum was very new when I began my job as a guide. There was an orientation period and then a timed period on each of the two floors. A few years later, a third floor was added. I really enjoyed that aspect of the HOM. The staff was small in numbers but that's what made the HOM seem like "family." They were friendly and helpful. I actually looked forward to returning to the museum each week.

As the museum grew in stature, other events took shape. Science shows, Halloween night, and similar events required volunteer participation, which I was a part of, parties that were given for the volunteers and awards were given. We were recognized as important parts of the whole.

Over the eighteen plus years and approximately 6000 hours, I witnessed the HOM grow from that small two-floor museum into large enterprise that is today. However, the memories linger on. All of the changes were Cynthia Yao's ideas and she brought them to fruition. She is the one who created the HOM and should be given all the credit for its success!

Anthony Witcher is a special volunteer, whose story made it all worthwhile:

As a hearing impaired student at Eastern Michigan University, I struggled academically. Participating in team sports and social clubs didn't satisfy me and left me unfocused. The relaxed atmosphere and flexibility of the AAHOM to give me time and space to build small exhibits, props and static displays gave me satisfaction. Working with several people from the community to provide technical supports, I was able to refocus my life and plan my future. I met several people, not possible at other places, at the AAHOM over the time I was volunteering. These people ranged from beekeepers to stay-at-home mothers to business leaders to research physics professors, all dedicated to the education of general sciences. All of us worked together to build the AAHOM. This helped me to think positive and give something back to the community. After a few years of volunteering for AAHOM, the support of the staff helped me to enter into Rochester Institute of Technology in Rochester, New York. From there, I went on to graduate from the heavy structured academic program, and now hold a position as a mechanical engineer at a major automotive supplier and I am learning to fly an airplane.

7. Resources and Affiliations

The information provided here are not exactly practices, but are unique to HOM, which may help others with ideas that they can emulate similar relationships which will help them to be successful:

A. UNIVERSITY CONNECTIONS The museum has enjoyed a mutually beneficial affiliation with the universities and community colleges nearby, as well as the public and private schools.

- Faculty and staff of the University of Michigan's departments of physics and chemistry and schools of engineering, education, and natural resources, as well as the university's Museum of Art, Exhibits Museum of Natural History, and Kelsey Museum of Archaeology have collaborated on many HOM projects.
- Members of the University of Michigan's physics department have had a very special relationship as founders, trustees, builders of exhibits, lecturers, and advisors; students come to the museum as course requirements.
- Vivian Shapiro as trustee and Jean M. Bollinger as founder and assistant director have played key roles supporting the museum (UM presidents' wives)
- Thirty to fifty work-study students assist with education programs annually.
- Engineering students have built many prototype exhibits.
- Eastern Michigan University provided physics and education students as Explainers to our exhibits, and EMU education students received class and pre-student teaching credits. Several of their faculty members taught Summer Science Camp classes.
- Washtenaw Community College built an exhibit and allowed the museum to hold classes on robotics using their facilities and robots.

B. THE ANN ARBOR PUBLIC SCHOOLS (AAPS)

- AAPS contributed $5,000 annually in exchange for free visits of third graders to the museum.
- AAPS's Science Curriculum Coordinator Joe Riley was an HOM trustee. He created a position for a science teacher to spend a sabbatical year to create curriculum-related programs for field trips. This resulted in that all third graders visit the HOM for units on Electricity and Magnetism, Simple Machines, and Matter.

- HOM's senior staff became judges for the Southeastern Michigan Science Fair.
- HOM prize awarded for Best Potential Exhibit.

8. Collaborations and Other Relationships

Here are some more good practices:

- ASTC, AAM, Association of Children's Museums (ACM), National Science Teachers Association (NSTA) memberships are excellent investments for small museums.
- ASTC has been the most significant resource. Going to annual conferences is a must do. ASTC has the best and most supportive staff.
- Develop collaborations with other small science centers to create projects such as TEAMS for NSF and other grants.
- Develop relations and collaborations with other small and large science centers locally, regionally, nationally, and internationally.
- Collegiality: small science centers have many common issues. Promote collaborations, deep friendships, and camaraderie; share best practices; share and solve problems; and learn about new programs.
- The *Handbook for Small Science Centers* is an outcome of these relationships.

PERSONAL PERSPECTIVE

It has been a wonderful experience to spend some thirty years of my life enjoying every day of work/play. I looked forward to each day as an adventure and enjoyed the simple daily ritual of going to the museum. There was joy in walking through the galleries, greeting visitors, playing and demonstrating my prowess at making sand patterns on the Chladni Plates, picking up scraps of paper, checking if there was enough water in the Soap Bubble Capsule, and best of all, hearing the "gee whiz" exclamations of delight.

As founder of the Ann Arbor Hands-On Museum, I am proud, honored, and privileged to have had the opportunity to create a place where children of all ages can play and learn or PLEARN (a new word!). As its executive director for eighteen years, I did the best job I could and loved every minute of it. My hope is that the Ann Arbor Hands-On Museum will thrive and its leadership will continue the great mission of promoting public understanding and love of science.

Part II

CASE STUDIES OF OTHER SMALL SCIENCE CENTERS AND START-UPS

CHAPTER 2

Developing in Phases: A Case History

Charlie Trautmann

CREATING A SMALL MUSEUM IS A MESSY, ORGANIC process punctuated by agonizing disappointments and exhilarating successes. Every museum, just like every museum founder, is an individual case with its own personality. With small museums, there is no universal formula for success.

But despite the individuality of each project, several universal principles can contribute to the success of a small museum. In this essay, I describe the development of a small science center in rural upstate New York and tease out several lessons we learned about first-time museum building. Some of these lessons are generic and could apply to almost any museum project. Other ideas are more specific and may not apply elsewhere. However, I hope that through this narrative, others embarking on the profoundly satisfying adventure of creating a small museum will get a taste of the issues to be faced and some of the possibilities for creating solutions.

THE EARLY YEARS

Museums are commonly founded by a small number of individuals who fervently want to share their mission and vision with their community. The vision can be modest ("let's provide more science to the children in our community") or grand ("let's create a 'world-class museum' to improve science education, make a significant collection accessible, and save the regional economy"). Most of the time, the founders have little or no previous museum management experience.

In our case, Debbie Levin and Ilma Levine, two people with a background in science and interested in children, met while volunteering at an inner-city elementary school in the late 1960s. For the next fifteen years, they wheeled activities to classrooms on carts, staffed a science resource center, and ran an animal-lending library. During this gestation period, they solidified their love for doing science with children, espe-cially those with the fewest outside opportunities. They also developed a style of hands-on programming and community involvement that became the foundation for their development of the Sciencenter. These two women were eventually cited in a presidential award for their volunteer educational work and highlighted in Carl Sagan's final book *The Demon-Haunted World: Science as a Candle in the Dark* (1995).

In 1983, they began to consider expanding their offerings to the community and working toward sustainability by starting a small storefront discovery center. A blurb in the newspaper attracted fifty enthusiastic people to an organizational meeting, from which a founding board emerged. The elementary school where they had taught wished them well by presenting them with the Sciencenter's founding gift: a brown paper bag filled with $280 in nickels, dimes, and quarters donated by the children and their families. In the months that followed, they recruited volunteers to build exhibits.

A landlord donated storefront space. Children and parents flocked to the fledgling center. Many more volunteers came forward. As each donated storefront was in turn rented, the center moved, but always in a downtown location that was accessible to children of all socioeconomic backgrounds. Five years later, after more than twenty years of program delivery and hands-on activities and exhibits at the school and four subsequent storefront museum locations, the founders had built a sizable following of enthusiastic parents, scientists, community leaders, and politicians without ever having thought about a permanent museum building.

Idea 1: Small museums improve their chances for success when they each address a community need that is confirmed by programmatic demand, operating experience, and significant volunteer involvement. By 1998, the board had become enthusiastic about creating an attractive, community-built facility where children and accompanying adults could come repeatedly, have fun, and learn science together—using exhibits as the props.

Soon, the board was visiting other museums for ideas, reading museum literature, recruiting the advisory board, and building community support through public meetings, media articles, and events. A committee looked at some fifty building sites, and the unlikely winner was one-half of an abandoned sewage treatment plant located on the main state highway through the city. Although it sported high visibility and accessibility, its condition was grim, at best. After visiting a few more museums, however, the board soon came to appreciate the wisdom of the site selection committee.

Idea 2: A highly visible site that has adequate parking and is easy to reach by car is well worth the search. A second-rate location can haunt a museum forever. Most of this activity took place during a convergence of positive economic and educational climates. As a result, fund-raising took a backseat to the dreaming, researching, and conceiving that were needed to fit a structure on the site. A stunning model on a hardwood stand, built and donated by the architect, was frequently on display at public events. The architect convinced the board not to fixate on money initially, advocating instead that an inspiring concept, properly communicated, would garner support when the time came. Sure enough, one board member, a member of the site selection team, was able to secure a $50,000 gift from his company to both make a down payment on the land and to hire an executive director.

Idea 3: Financial commitment to a project often comes by way of direct personal involvement. Time taken to involve the right people is time well spent. The next step was to search for an executive director, and as is often the case with small start-up museums, the board went out on a limb by hiring someone with no museum experience when they hired me. I had a background in science, but my main connection with the science center field, however, was simply that I was an aficionado of science museums and had a lifelong interest in science, youth education, and construction.

Idea 4: When hiring a new director for a small museum, give weight to enthusiasm, tenacity, a vision for the organization's future, good references, an ability to communicate about science with board, staff, and sponsors, and enjoyment of talking with many different kinds of people. During my interviews, I was particularly comforted by the fact that fund-raising would not be an issue. The board had recruited an experienced

fund-raising volunteer and had completed a feasibility study with a budget. Campaign brochures were in press. All I needed to do was manage the construction of the project and expand the programming afterward. The budget seemed low, but I had no point of reference to judge whether the situation I had inherited was normal or if we were in for trouble before we began.

Soon, I was on a fact-finding trip. The first meeting confirmed my hunch that our plan was destined for trouble. Clearly identified were an inadequate operating budget and lack of groundwork for the proposed $2.6-million campaign. The next stop on my trip was to meet with Greg Prince, the visionary president of Hampshire College who had helped found the Montshire Museum. He provided a crash course in fund-raising, with coaching on how to approach the Kresge Foundation. Finally, I met with David Goudy, director of the Montshire Museum. This meeting produced a new perspective on what it would take to fund, build, and operate a museum.

Idea 5: Never take previous planning results for granted. Upon returning home, I recommended to our board that we (1) hire a fund-raising consultant to conduct a fund-raising feasibility study, and (2) hire a museum consultant to lead a board retreat and develop an operating plan. Calls to nearby Cornell University's development office and Bonnie VanDorn of the Association of Science-Technology Centers in Washington quickly pinpointed the right people for our needs. These consultants (Paula Sidle and Carol O'Brien of Carol O'Brien Associates, and Sheila Grinnel, then an independent consultant) prepared our board for the work ahead by leading us through the steps of developing a mission statement and creating fund-raising and operational plans.

Idea 6: Use networks of contacts and referrals to locate consultants who can provide the expertise needed by your museum's development and management teams. Look for consultants who have solved your problem many times before, and avoid low-priced assistance from less-experienced individuals. Interview the people who will actually do the work, not just the president or marketing person who makes the presentation to your board. Among other results, our consultants helped us reconfigure the project into phases that would promote financial and operational success and give us valuable experience at each stage before moving ahead. While initially unacceptable to several influential board members and our architect, the board retreat made the case for phasing a logical choice.

Idea 7: Developing a small museum in phases can enhance the chance for success by (1) reducing the consequences of wrong decisions, (2) lowering initial fundraising needs, and (3) providing opportunities for board, staff, and volunteers to get involved and gain experience at a small, manageable scale.

THE PLANNING PHASE

As we reviewed the few initial interviews from our fund-raising feasibility study, the feedback surprised us. Although we thought we were a highly visible and indispensable part of the community, many community leaders told our consultants that they knew little about us, our mission, or the programming we offered. Significantly, the dollars we had assumed were "out there" were most definitely not going to be found among our interviewees.

To address our problems, we stepped up efforts to build community awareness through frequent appearances at public festivals and other events. A public relations expert wrote science spots for us and recorded them for broadcast by a local radio station under our byline. We provided volunteer opportunities for hundreds of students from local colleges to assist with events, program delivery, exhibit development, and fund-raising.

Perhaps most significant, we committed to building the first phase of the museum with volunteers through a community barn-raising construction process. This process eventually involved 2,200 individuals over a period of ten months. About that time (fall 1991), the feasibility study had concluded that we could raise only $750,000 in cash—a far cry from the $2.6 million originally envisioned—and suggested widespread support for a phased development strategy. So we settled for a $1-million Phase I project in which in-kind donations of labor and building materials would fill in the gap.

Idea 8: Always, always, always perform a fund-raising feasibility study before a capital campaign, even if you are sure you know how much money is out there. If performed by a skilled consultant, the study will pay for itself many times over in terms of donor cultivation, increased gifts, and increased professional stature when applying for corporate gifts and foundation grants. We bought a textbook on business planning and began generating spreadsheets (Bangs 2002, 256). We read publications by Association of Science-Technology Centers (ASTC), Association of Children's Museums, American Association of Museums, and others and asked business members of our community advisory board to review

drafts of the plan. We found helpful rules of thumb for many of our questions ("estimate $0.75/SF for cleaning and $1.50/SF for utilities" or "figure on 8–10 visitors/SF of exhibit space, depending on location and exhibit quality" or "provide a parking space for every 1,000 annual visitors" [Anderson 1991, 93]).

We decided to apply for a challenge grant from the Kresge Foundation, the only national foundation supporting bricks-and-mortar projects without regard to location. What impressed our program officer most was the detailed business plan we had prepared. He studied the plan carefully and asked questions that forced us to tighten up several key assumptions and the associated financial projections.

Idea 9: A realistic, well-written business plan is the road map that convinces board members, donors, foundations, and community leaders to join in the project. Although well-done brochures and articulate presentations help foster interest, a convincing business plan instills confidence that the program will succeed once construction is completed and excitement from the campaign has faded.

BUSINESS PLANNING

Understandably, few board members or museum directors have written a business plan before setting out to create a new museum. However, embarking on a museum project without a business plan is akin to starting a long trip without a road map. Additional information is available in Trautmann (1996). A plan should include the following elements, as a minimum:

Part I: Description of Operations
This section should describe the following items:

- Project Goals: What demonstrated need will the museum address? What are its mission, vision, and value statements?
- The Organization: What makes it unique? Has it had any honors, awards, significant grants, or other recognition?
- Exhibits: Describe the exhibit philosophy. Who will design and build the exhibits and at what cost? How often will they change? How much exhibit space will you maintain? How will they excite the public?
- Programs: Describe the program philosophy. Who is the target audience? Who will create, fund, and deliver the programs? How will they serve a wide spectrum of the community, including those with the fewest outside opportunities?
- Location, Hours, Facilities: Describe the facility. What percentage of the space will be devoted to

exhibits? How will the offices be set up and how much space per person will there be? Will it be American with Disabilities Act (ADA) compliant? What special features will be present, such as theaters, a community room, volunteer room, exhibits shop, and so on?

- Audience and Market: What is the distribution of population around the museum for 100 miles? How accessible is the museum for automobiles and public transportation? How many visitors can you expect, anticipating a falloff of attendance in years two and three? Any estimate over eight visitors per square foot of exhibit space for a discovery-type center or children's museum is probably optimistic. For natural history museums, two to four visitors per square foot is common. (Note that once the first year's data are in, consultants often attribute lower attendance to inadequate marketing expenditures. Equally important: attendance per square foot will depend on the quality of the exhibits and the location of the museum.)
- Competition: What other young-serving programs vie for the discretionary time of the target audience (be broad; this is much more than just listing other nearby science museums)?
- Management, Staffing, and Governance Plans: Provide position descriptions for senior staff and an organizational chart. List the board and provide a chart cross-referencing their expertise in programming, fund-raising, and management. List the advisory board and their affiliations and qualifications. How are volunteers involved?
- Fund-Raising: For construction funding, some of this can be excerpted from your fund-raising feasibility study. How much money can you raise and where will it come from? How long will the campaign last and who will lead and staff it? What is the backup plan, should fund-raising fall short? For ongoing operations, annual fund-raising estimates may be more difficult. Be realistic: look at other similar organizations. And please don't bait campaign donors by saying that they will be making a one-time-only donation.
- Construction Plan: Will you use a construction manager or work with a general contractor? Who will provide oversight, and what is the timeline?

Part II: Financials

Contrary to human nature, when writing a business plan, it is best to estimate income on the low side and expenses on the high side. If you do, you will be rewarded. The financial section should outline the financial aspects of your plan and include the following:

- Sources and Applications of Funding for Three to Five Years: This is a listing of all the revenue and expense items and lists the three-to-five-year totals from the budget.
- Balance Sheet: This sheet provides comparative data for two to three years if you have it.
- List of Capital Equipment: This list has all the major capital items, including model numbers and prices, for everything over $1,000 that you think you need.
- Cash Flow Projections for Three to Five Years: This is the most important part of the entire plan. Create a master spreadsheet showing revenues and expenses by year, with totals for the entire period. These totals become the inputs used in the listing of Sources and Applications of Funding. Details and data can go in accompanying schedules.
- Breakeven analysis: Show a graph of operational revenues and expenses (vertical axis) as a function of attendance (horizontal axis). Show revenue at three levels (low, realistic, and best case) of unearned income (such as annual fund-raising). Where the revenue and expense lines cross is your break-even point.

Part III: Supplemental Data

This section provides the documentation to accompany the text of the business plan, such as résumés, job descriptions, lists of trustees and advisors, the 501(c)(3) letter, and so on.

While bound copies are necessary for external distribution, I think of a business plan as a living document and prefer the flexibility of a three-ring binder for internal use.

Idea 10: Researching these details adds to the value of doing a business plan, so don't farm this out to a consultant. The director and senior board members should be directly involved in researching the assumptions, creating the financial projections, and writing the plan.

EXPANSION

After opening with a skeleton staff in mid-1993, we were soon out of space and expanding into the basement, as well as leasing program space in an adjacent city-owned building. Several key advisors had told us that "a soufflé rises only once," but we quietly recruited volunteers to help acquire and rehabilitate this leased building and completed Phase II without fanfare three years after opening the initial phase. We wrote National Science Foundation (NSF) grants, collaborated with over a dozen departments at nearby Cornell University and Ithaca College, and continued to build our program. We borrowed or built three exhibitions per year to provide fresh experiences for our visitors. Some regular visitors complained about their favorite exhibits being gone but kept returning to see what was new.

Following a yearlong effort in which we brought two dozen members of city council and its relevant committees and staff to the site individually to discuss our plans and attended untold meetings, the city gifted us the other half of the block, including the building we had just rehabilitated. Grants from NSF led to four exhibition projects totaling 9,000 square feet. By this time, we had spilled out into two office trailers and six windowless basement offices.

We decided to proceed with the third and final phase of development. Another board-staff retreat, led this time by Washington consultant William Weary, identified changes in organizational structure needed to support the expansion. A fund-raising feasibility study conducted by Lancaster, Pa., consultant Andrea Kihlstedt, confirmed support for a $2.1-million campaign goal and also generated the plan for an educational donor-recognition exhibition called the Wall of Inspiration that eventually increased the campaign total to over $5.5 million

Our financial plan called for a four-pronged approach to the increased funds needed for a larger operation.

1. We anticipated a 25–50 percent increase in attendance income, based on a tripling of the exhibition space.
2. We planned to double our annual fund to $100,000.
3. We planned to resume writing more program grants after the capital campaign.
4. We would pursue a long-term planned-giving program emphasizing bequests.

The third phase of development included a community-built wood section, this time constructed by 1,200 individuals performing over 6,000 hours of volunteer carpentry over a four-month period (Trautmann and Gattine 2002). As in previous phases, volunteers had a major hand in prototyping and fabricating individual exhibits. A generous board member who owned a landscaping business donated most of the plant materials needed for landscaping, and another who owned a metal-working shop fabricated several key exhibits. Other volunteers created a large hardwood compass rose inlay for the carpet in the lobby, built more exhibits, installed the public-address system, and

Figure 2.1. Creative donor recognition: examples of one hundred 30-inch-by-30-inch plaques created to inspire visitors and honor donors to the science center. Rachel Carson and Hans Bethe plaques, Sciencenter, Ithaca, N.Y.

created colorful tile mosaics for the new early childhood area.

We have fostered an environment of creativity (Trautmann 2002) and invested in a program of events, programs, and exhibits with coordinated public relations that have brought our offerings widespread community awareness. Some examples include the following:

• Sagan Planet Walk: We raised $100,000 to build a public scale model of the solar system stretching 1,200 meters from the Sun station at the center of

Figure 2.2. Human Hand Battery for Dedication on February 28, 2003. Sciencenter, Ithaca, N.Y.

the city's pedestrian zone to the Pluto station at the Sciencenter (Trautmann 1997).

- Applied Jean-etics: To open an exhibition on Clothing Science by the Ann Arbor Hands-On Museum, we obtained the donation of a crane and fabricated special steel fixtures to "test" how many pairs of blue jeans it takes to lift a Volvo station wagon. The story was picked up by the Associated Press and appeared in 165 newspapers around the country.
- Egg Drop: This annual egg drop event, in its eighteenth year, now draws 300 entries, attracts an audience of 1,400, involves 100 volunteers, and has become a favorite event to emcee for local politicians and community leaders.
- Segway Raffle: We purchased a new Segway HT (Human Transporter) and sold over 250 raffle tickets for $20, turning a modest profit and delighting hundreds of people with rides and further information about the museum.
- Human Battery: To build community involvement, we gave all of those attending the opening of our final expansion a copper-zinc electrode and created a huge human battery to trigger the melting of the ribbon cable.

Idea 11: Creative events can help fund-raising by jump-starting the initial steps of generating awareness and providing information on the way to involving potential sponsors. Such events engage the media and help them become enthusiastic about a science center; reporters enjoy coming to events that are interactive and different from those they usually cover. Some of the best features of a small museum—one-of-a-kind exhibits, special building details, new programs, or events—are created by volunteers.

Idea 12: Volunteers are the lifeblood of small museums. Provide the best tools available and whatever materials they need. Give them responsibility and autonomy to develop their ideas. The rewards to the museum, the volunteers, and the community can be astounding.

SUMMARY

Small museums are in many ways fundamentally different from large museums. Their culture is often characterized by remarkable levels of volunteer involvement, staff enthusiasm, fleetness of foot, and a responsiveness to the surrounding community that is much more difficult for a large museum to achieve.

Small museums are commonly started and staffed by people with little museum background. Mistakes are made that might have been avoided by professional museum developers, but there is also a creative, pioneering spirit that can be absent in large projects dominated by professional museum planners. The combination of factors makes for a dynamic, exhilarating development process—one that can be navigated somewhat more readily by recognizing the common issues faced by other small start-up museums. In my opinion, these issues can best be addressed by (1) maintaining a willingness to build the organization slowly and enjoying the process of museum building as much as the eventual product of the museum itself, (2) developing the eventual physical facility in phases, (3) providing a detailed, realistic business plan, and (4) carrying out a fund-raising feasibility study before embarking on a capital campaign.

REFERENCES

Anderson, P. 1991. *Before the blueprint: Science center buildings.* Washington, D.C.: Association of Science-Technology Centers.

Bangs, D. H. 2002. *The business planning guide: Creating a winning plan for success.* 9th ed. Chicago: Dearborn Trade Publishing.

Sagan, Carl. 1995. *The demon-haunted world: Science as a candle in the dark.* New York: Random House.

Trautmann, C. H. 1996. Business planning 101: The foundation for a successful museum project. *ASTC Bulletin:* 17.

———. 1997. Sagan planet walk dedicated. *Informal Science Review* 26 (Sept–Oct): 2.

———. 2002. Innovation in informal science education. *Informal Learning Review* (March–April): 14–16.

Trautmann, C. H., and M. M. Gattine. 2002. Raising a museum. *ASTC Dimensions* (May–June): 7.

CHAPTER 3

Hockey, Nickel Mines, and the Pursuit of a Vision: Getting It Done at Science North

ALAN NURSALL

EXPERIENCING HOCKEY AND SCIENCE

In June 1984, a new science center opened in an unlikely spot, a Northern Ontario mining town better known for its blackened rocks and the tallest chimney in the world. The construction of the twin stainless steel snowflakes that comprise Science North represented a significant change to the landscape of Sudbury—not just the physical landscape but the cultural one as well.

No one has ever accused Sudbury of being a haven of intellectualism. It is a hardscrabble, blue-collar town of about 150,000 residents, built on a century of intensive nickel and copper mining. The metals refined from Sudbury's rich ores have played an essential role in the development of twentieth-century technology. The demand for nickel remains with us today, mainly owing to the ubiquitous use of stainless steel, made stainless because it is up to 30 percent nickel. The miners of Sudbury still extract vast mineral wealth from the ancient rocks of the Canadian Shield.

Building a science center in a town such as Sudbury encourages a different paradigm, a different mindset, maybe a new set of eyes. Over the next few pages, I will try to convey some of the central principles we have adopted to give our science center its own personality, one that reflects the community yet proclaims that there is more to Sudbury than mining.

THE SCIENCE ARENA

For the first ten years of its existence, Science North shared its site with one of the most commonplace of all Canadian institutions, a hockey arena. Here in Canada, we play and watch a lot of hockey. It is our national sport, and we follow it passionately. This hockey fervor leads us to build a lot of arenas. Travel to any Canadian town, and you will see the distinctive domed profile of the arena, usually near the center of town.

The presence of the arena is made possible by the efforts of the people in the community. They invest their time, money, and energy to build and maintain it. And its presence stands as an icon, a testament to the importance of hockey to the community surrounding it. It carries fundamental cultural messages, just as other major edifices convey messages about the social and cultural fabric of the community.

The presence of the science center is no different. A science center, especially one like Science North, which boasts a prominent location and striking architecture, conveys a powerful message to residents and visitors alike. It says that science and the intellectual exploration of the world around us is a valued activity in the community, so much so that they have built a major facility for science as a leisure activity. There's a behavioral comparison as well that we should not shy away from. Consider how people in the community actually use the hockey arena, and what their expectations are. Leagues play scheduled games, accommodating many age groups. Groups book occasional ice time for games of pick-up hockey, where the only objective is to work up a sweat, have a great time, and try to avoid taking a frozen puck in the shins (it really hurts). Many others come to the arena just to skate. Mothers and fathers come with their children to introduce them to the joy of skating, to be together, hold hands, laugh, and enjoy one another's company. It is a family place. Experienced skaters flash by teetering novices. Groups of youngsters chase each other around the rink.

Others come to the arena for figure skating. And some come just to watch. The arena is exciting and alive. In their own ways, all of these people use and enjoy the arena. Some young people go to the arena regularly, are instructed there, and work hard over many years to improve their skills as hockey players. A small minority may even reach the very highest echelon, playing hockey professionally. As for me, I love to play hockey. But I don't play to get better, and I don't aspire to league play. I play because of the pure joy.

Figure 3.1. Science North, Sudbury, Ontario, Canada.

There's a lesson in there about visitors to science centers. The vast majority don't come to become scientists, or even to become better at science. They come for the fun of engaging in science, of taking part in a playful shared activity with their family and friends. Our visitors come to explore science, to engage in science, to do science, and to have some fun doing so. It's science pursued not for revelations or cognition but for stimulation and enjoyment. The biggest difference between the group playing hockey and the group in the science center is that one sweats a lot more than the other.

I don't think there are a lot of differences between the way people use the arena and the way they use science centers. People generally don't go to science centers to become scientists or even necessarily to learn about science. Some do, but most don't. We are doing our job, when we are serving both these groups. A science center can illustrate to visitors that science is an energizing human activity and that great works of science are as passionate and inspirational as great music, art, and sport.

I think that science centers must aspire to something much more important than just to teach the public about science; they must provide an opportunity to enjoy science, to do science, to laugh at and about science, to be skeptical of science, and to be awed by science. We need places like that—science arenas—where we can play with our friends and let our minds work up a sweat.

DOCUMENTING OUR VISION

First Principles: The Big Picture

How do we go about trying to fulfill this vision at Science North? In order to succeed, we believe that the organization requires well-articulated principles and priorities. And not only must they be well articulated, but also they must be evangelized throughout the organization and well understood by all. And finally, they

must be focused, having a relatively fine point. This is not about mushy epigrams and platitudes; it is about defining the character and quality of the institution. In short, decide what is important, write it down, and train people to it. I will describe three documents in use at Science North that illustrate this point.

The keystone of Science North's principles is the mission statement. So what? Every organization has a mission statement. But not all organizations use their mission statement every day, in documents, on meeting agendas, in PowerPoint presentations, or on the wall.

The first mission statement was adopted in 1987. It read as follows:

> The mission of Science North is to provide a stimulating, bilingual program of formal and leisure learning involving people in the relationship between science and technology and everyday life, particularly in the North. Science North will be a significant attraction for tourists and visitors and will enhance community life.

It's not great literature, but it has meaning to the staff. All the key words, particularly the adjectives, were carefully chosen and defined. Most important, the mission statement had a point. It attempted to encompass what was really important to us—Northern Ontario, two languages, tourists and residents, a particular scientific emphasis.

Every two years, the entire staff reviews the mission statement as a group. As a result, the mission statement changes over the years, in subtle but important ways, to reflect the changes and growth in the organization. As in 1987, a deep meaning is embedded in each word of the mission statement. In 2004, the mission statement has slightly evolved to the following:

> Science North creates and markets high quality science education and entertainment experiences in English and French involving people in the relationship between science and everyday life.

How and why does one go about changing a mission statement? Economic circumstances change. New opportunities arise. Organizations mature. In 1987, Science North saw itself as a regional attraction with a focus squarely on Northern Ontario, in terms of content and clientele. Twenty years thence, we operate in a global community and the organization's efforts take place across Canada, throughout North America, and around the world. And thus the mission has evolved to reflect this.

While the central goal of the organization has always been science education, we work in multiple media—exhibits, programs, film, theatre, software—so we have adopted the word *experiences* as shorthand for all of these media. One of the more contentious changes was the inclusion of the word *entertainment*. Anyway, without apology, we now state that we are in the entertainment business, with an emphasis on science education. Some changes came about simply out of a desire to reduce the number of words; *science and technology* became just *science*, an expedient bit of shorthand as we recognize that the two are not the same.

There's always something a little jarring about reading other people's mission statements—they never quite seem to scan properly. I don't think ours is any different. It includes little bits of code that have to be interpreted through the culture of the organization. And thus, every change in the mission statement of Science North has involved the entire staff because without their ownership, the mission statement would be poorly understood, perhaps meaningless. And if that were the case, it would be a shame since the statement is the basis for all activities conducted by the organization.

First Principles: The Science Experience

In 1987, after the first somewhat rocky three years of operation, Science North staff created a document still in use today. We had sufficient experience to articulate what was important in our science center and what were the defining qualities of our visitor experience. That document is the Characteristics of Excellence for Exhibits at Science North.

Over time, lots of people have seen and used this document. Some think it's brilliant, some think it's dumb. But it has been an essential element for training, planning, decision making, and continuity. It defines the good stuff.

There are six characteristics. (Why six? Because the Science North logo is a hexagonal, stylized snowflake. So we do lots of things in sixes.) In outline, they are as follows:

- Real Science—A clear and accurate portrayal of science, backed up by exhibits and activities that engage visitors in the true experimental, observational, cause and effect world of science—simulations are used as a last resort, if at all!
- Tools—The stuff you get to play with to do Real Science: microscopes, computers, beakers, telescopes, stethoscopes, scales, and even pliers and wrenches.

- Fun—It had better be!
- Comfort—It has to work, be accessible in French and English, and make sense.
- Exhibit Tree—A good exhibit extends its learning opportunities like the branches of a tree, in many different directions and each equally fascinating.
- Learning—Learning is fun, but more important, fun is learning.

Each characteristic describes a quality of the science experiences found at Science North. The very best experiences have all six qualities in abundance. Some excel in some qualities but not others. In all discussions of the visitor experience, however, we are able to refer to these tenets as an indication of how true an experience remains to the personality of Science North.

We weren't creative enough to dream up all of them. We unabashedly "stole" the exhibit-tree concept from an article written by Frank Oppenheimer, founder of the Exploratorium, and we are indebted to his genius. The other five we can take credit for, and the first two are the keys to defining the spirit of Science North.

We have a twenty-page document that fleshes out these characteristics and provides context. Our characteristics are not really transferable. They are a product of, and in turn define, the institutional culture at Science North. They give us vocabulary and a framework for making decisions. When someone has an idea for an exhibit, a program, or some other type of science experience, the first thing we do is consider it in light of the Characteristics of Excellence and, of course, the Mission Statement.

First Principles: The People

There are few things, perhaps nothing, as important to the success of a science center as a well-trained, motivated, and empowered staff. At Science North, we have taken to calling our staff Bluecoats because the

Figure 3.2. Swap Shop and Bluecoats, Science North.

science staff wears blue lab coats. The term *Bluecoat* has been adopted by the whole organization to symbolize the qualities that each staff member must possess and present to visitors. Bluecoat has become a proper noun, a verb, and an adjective. The word has meaning and substance.

A while after we created the Characteristics of Excellence—fifteen years to be precise—we decided that we should create another document to clearly articulate the essential characteristics of a Bluecoat.

A Science North Bluecoat exhibits six characteristics. Every Bluecoat exercises all of the six qualities, but the emphasis may shift depending on function. The first three focus on the core visitor experience:

- Scientist—Be a scientist and use your knowledge and the tools around you to explore the world—and take visitors along for the journey. Do science in the science center!
- Entertainer—Have fun with science and make people smile.
- Ambassador—To every visitor you encounter, you are the face, voice, and ears of Science North.

The other three are a little more practical:

- Initiator—Don't just stand there, do something—better yet, do some science!
- Caretaker—Clean up after yourself
- Troubleshooter—When something goes wrong, do what you have to do to make it right. We're counting on you.

There is a very important operational strategy behind these behaviors—to create smart Bluecoats. Each Bluecoat is trained to understand unequivocally that they are to exhibit and exemplify these behaviors. For the very best, it gives them a structure in which to make decisions about how to spend their time wisely, adding value to the visitor experience. For others who may need a little more supervision, it provides the framework and vocabulary for that supervision. In fact, the appraisal system for Science North part-time staff is now structured to a large extent around these six be-

haviors. And so the Bluecoat characteristics become a tool to be used each day as a framework for training, motivation, appraisal, and development.

Another important function of the Bluecoat Standards is to keep staff focused on *now*. It is far too easy for staff to get wrapped up in the big projects, the new developments that take a lot of resources and planning and, frankly, keep the staff excited. However, there is a very large disconnect between all that developmental work and the daily execution of great science and service that is expected by the visitors who have just walked through the door. The Bluecoat Standards address this. They are all about the visitor. They describe how each Bluecoat can contribute to making each visit a success.

Best of all, the Bluecoat Standards apply to everyone in contact with visitors, including volunteers. One of the joys of incorporating volunteers into the experience at a science center is that they can devote all their energy to the visitor. They are unencumbered by all the process and drudgery that must be dealt with by staff in the course of operating a science center. They provide a face and personality to a science center that is literally priceless.

None of these documents offers a prescription for operating a successful science center. What they offer, however, is a context in which every staff person does his or her job and in which they make informed decisions. They empower staff by creating agreed upon principles and making sure everyone is aware of them. Backed up with practical training materials and logical examples of how these principles are manifested every day, they are the organization's vocabulary and identify what is important.

Within this well-articulated and documented framework of principles, staff can envision the future, feel connected, and make a difference to the science center's success. Create your vision, write it down, share it, and then run with it. Finally, do a Google search for "Big Nickel" and "Inco Superstack." Imagine how a science center can alter the future of a city that used to be known mainly for these two structures.

CHAPTER 4

One Small Center's Story: Grass Roots Thrive on Flexibility

ADELA "LADDIE" SKIPTON ELWELL

GETTING THE PICTURE

I am attending a high-powered meeting at a conference in a large city, hoping to learn how to select and manage the board members that our new little science center needs so badly. I want to learn how to most effectively tap human resources in our small city. The speaker has some great ideas, and I could implement them, too—if the available pool of board members could and would pay a few thousand dollars or more for the privilege of serving. The folks in our community either don't have that kind of money to give or they are already supporting a number of other social needs in the community. There are a few people whom I know have funds at their disposal, but they are not even remotely science-oriented folks and already support various sports or the arts. When they are approached, they kindly express disinterest and cannot grasp what that little science center could bring to their community.

The speaker describes how to raise wonderfully large amounts of money through board-organized events. That old sinking feeling begins to come back: Is it foolish to think that I can really pull this center together? My whole community has never raised that amount of money for anything because the money just isn't there. On the other hand, there is that solid core of supporters who really care about the mission and what this center could accomplish. This thing has really got to happen.

As I attend other meetings intended to provide support but succeeding only in making me better understand the real cost of exhibits and staff, I am worried about meeting a payroll that may total $300 that week, or trying to make a rent or a mortgage payment that may be bigger.

The meetings go on, and there aren't many sessions that speak to the problems, which revolve around the question "How can we provide an affordable science center for our community?" There are lots of consul-tants recommended, but their per hour or per diem charges are more than anyone on my staff, founding committee, or board has ever heard of being charged in our small community. Even if I could hire them, I know darn well what our supporters would say and what the community would think. . . . Furthermore, the consul-tants are used to working with organizations that have real money, and I'm not really sure that they would help any more than the wonderful core of volunteers we already have.

The science center we want to grow requires that exhibits and programs demonstrate basic concepts in science in an enjoyable way. Sports and the arts are well supported here, but there is little other than the public library to provide informal educational activities for kids and families who want to understand concepts and mechanics used in everyday life.

At Headwaters Science Center (HSC), the problems have to do with our community's size and nature. Bemidji, Minn., has the distinction of being the first city on the Mississippi River. It has a population of 12,400, though if one includes the area within 5 to 10 miles of town, the population is 30,000 to 40,000. The nearest science museums are in Winnipeg, Manitoba (220 miles north) and St. Paul, Minn., (230 miles south). The nearest larger cities are Grand Forks, N.Dak., and Hibbing, Minn., both over 100 miles to the west and east respectively.

Other regional characteristics are its population, its role as a regional center and tourist area, its poverty, its seasonal changes, and its potential for extreme cold. Three Ojibwe reservations, Leech Lake to the east, Red Lake to the north, and White Earth to the west, are within 15 to 30 miles of Bemidji, and Native Americans constitute from 10 to 90 percent of area communi-ties. Bemidji State University, regional Department of Natural Resources headquarters, and extensive retail, medical, and legal facilities are located in the city. U.S.

Forest Service headquarters for the Chippewa National Forest is nearby.

The local school district is the size of the state of Rhode Island. There are no locally owned large corporations. Employment varies widely, from logging and farming to law, medicine, university teaching, service occupations, and research. Most people live here to take advantage of the natural environment's lakes and forests and value their time outdoors. Fundraising is extremely difficult, and social needs are great. Minnesota has a large number of charitable foundations, but few serve this area. Headwaters Regional Development Commission provides sobering statistics regarding family income, births to unwed mothers, and other markers indicating significant unmet human needs.

Like many or most small communities, Bemidji has a short attention span. Community leaders are limited in number and are always in demand for some cause, and causes come and go with amazing frequency. We can't maintain needed attention for very long without something substantial to show for it. Situations are constantly changing, and in a small community changes are felt more quickly. Resources must be grabbed when they are available because they may quickly disappear.

THE BEGINNING

Four members of our original group had been involved in a program funded by Blandin Foundation through the Science Museum of Minnesota (SMM) entitled Opportunities in Science (OIS). This program made resources and programs available to northern Minnesota science teachers before the advent of the Internet. While spending time at SMM, we saw hundreds of northern Minnesota students, who enjoyed one or two hours there, then went on to other attractions in St. Paul and Minneapolis. These youngsters spent a minimum of ten hours on buses during their round trips from home.

In the late 1980s, Blandin Foundation funds dried up and SMM could not support OIS activities in northern Minnesota. We then incorporated OIS as a nonprofit, hoping to continue science support activities in other ways. The office base, with computers and copy machine, were critical for what was to follow. We began organizing in spring of 1992 by running a news article about wanting to start a science center. Amazingly, for a town in which educational meetings were very poorly attended, thirty people showed up! It is interesting to note that virtually all of these first meeting

Figure 4.1. Headwaters Science Center, Bemidji, Minn.

attendees were people who had lived in places where museums existed. This group soon boiled down to a smaller number and was joined by enthusiastic local folks to form a stalwart group of about twelve active people who believed strongly in making a science center materialize. A University of Wisconsin-Milwaukee physicist, Glenn Schmieg, who occasionally visited to provide programs, supported us from afar.

We looked longingly at an empty storefront, a former J. C. Penney building built in stages in the 1920s and 1930s and last renovated in the 1960s. It had been mostly unoccupied for several years, and its 75-foot expanse of unsightly, empty storefront was a big problem in the midst of downtown Bemidji—and no one ever shoveled the snow! With help from various community members, we were able to move in, rent free, for about six months. The building had 26,000 square feet of space, but a third was basement space, which couldn't be used for exhibits or programs. The store had originally been on the market for $500,000, but, with help from some wonderful community advocates, we were able to purchase it for $100,000 with $50,000 down and a five-year mortgage. We now had a home.

As one of the founding group members, I volunteered to serve as an unpaid director. With help from a Minnesota state employment program, we hired a staff person with professional acting skills and strong interest in science. He spent a lot of time that first summer building a bicycle gyroscope, which remains one of our most popular exhibits. We ran all kinds of programs to attract participation: Aero Prop™ contests, mystery festivals, informational programs in various fields of science, and even bubble blowing events. We had summer custody of an 8-foot long python named Monty, who became very popular, and an Amazon parrot, a pair of lovebirds, and a few puzzles rounded out the exhibits. We began to order very small numbers of field guides, science toys and books, thinking that a store might bring in more people. We couldn't afford to buy any ads, but we submitted news articles about all our activities to any media outlet that would run them. We printed a newsletter and disseminated it as widely as our very small financial resources would allow. A lot of very funny things happened that year!

People began to come in, though there were days that were pretty discouraging. We gradually began to attract more local citizens who also knew that Bemidji needed more constructive activities for kids and families. Our core group pulled off some real coups, such as helping arrange the purchase of two well-packed truckloads of science lab items from a soon-to-be-demolished science building on the University of Minnesota campus in Minneapolis. Other group members made colorful signs and worked hard to make our meager store look important.

I had attended an Association of Science-Technology Centers (ASTC) annual conference in St. Paul some years earlier and had been impressed by the nature of the science center/museum community. We knew that national resources exist for small science centers, and Jim, my husband, and I went as HSC representatives to attend an ASTC new centers session in Boston in 1992 at our own expense. It was at that session that we began to believe more in ourselves, encouraged by experienced science center and museum personnel who understood the magnitude of our dilemma. Their advice seemed sensible and reinforced some of the gut feelings we had shared. Our founding group had felt that a needs survey would be a waste of time, and the experts concurred. They also helped confirm our choice of the name Headwaters Science Center. ASTC staff and many members of the museum community were and have remained critical allies.

At the new centers session, we learned that Pacific Science Center (PSC) had received funding from the National Science Foundation for its Science Carnival program, and HSC was thrilled to be selected as a partner in that program. When eleven large and colorful Science Carnival exhibits and two PSC staff members arrived in Bemidji on a subzero day in March 1994, the future immediately looked brighter and local people began to pay much more attention. Along with the exhibits, two staff members visited Seattle for a week of training in presentations and exhibits. Coming up with the matching funds was another very difficult problem, but we did pay our debt. (For more on Science Carnival see Lynn Dierking, Foreword; Dennis Schatz, chapter 27; Stephen Pizzey, chapter 10; Karen Johnson, chapter 6; and Cassandra Henry, chapter 8.)

In less than two years Headwaters Science Center had become real, though it had a long way to go.

FLEXIBILITY HELPS HEADWATERS SCIENCE CENTER CONTINUE

The most outstanding positive characteristic of Headwaters Science Center is flexibility, which gives us the ability to roll with the punches, jump on opportunities, and evade disasters. We fly by the seat of our pants. We have not been able to consider Best Practices but have simply done what we could.

Herein is a major advantage of a small science center in a small community. The board is smaller and intimately involved. All have made significant contributions to HSC, either in kind or financially, and most have done both. There are fewer staff members, and being keenly aware of center needs, they often make valuable connections with community resources. Initial exhibits and programs can be fairly basic. Financial investments and facilities are smaller. Board and staff members can make rapid decisions and use opportunities that might be missed by having to run ideas through a large board and complex staff while maintaining a large facility.

Though initially its audience may be less sophisticated and more easily wowed, a small center must have a good base of available science-oriented folks or it is likely to lose its educational value. We want to appeal to all ages and provide challenges to the minds of all our users. Most wonderful aspects of our survival and well-being have to do with our dedicated staff and the talented and generous folks who come in and help develop and build new and truly unique exhibits, edit a newsletter, and help provide programs. University staff and professionals from other organizations generously give assistance.

Headwaters Science Center may work more closely with other community nonprofits than most other centers. We invite them to join us in providing Halloween events for the community and meet with them regularly to plan other community events. All local nonprofits are keenly aware of the limited local dollars that we share, but we choose to work together to make the best of a tough situation.

THE PRESENT

In 2002, Headwaters Science Center served over 24,000 visitors in many venues, including over 15,000 visits to its exhibit floor. Two full-time and eight part-time staff members work hard to meet all needs; an additional two full-time staff members are volunteers. A $465,000 grant from Howard Hughes Medical Institute permits our staff to educate over 1,000 seventh and eighth graders in five school districts about watersheds, water testing, invertebrates, and the uniqueness of water. The center has diversified, thanks to grants from local and state foundations and other generous benefactors. At this writing, HSC is ten years old and has never had continuing support from any governmental body other than a state after-school grant for our Science Club. Current economic conditions led to the withdrawal of those funds, but Science Club continues. Our annual budget ranges around $270,000. HSC owns its building, which we have renovated, and the center currently has no significant debts.

Headwaters Science Center has plans for its future. Several years ago, a grant from the Minnesota legislature allowed us to develop an outstanding architectural plan for a 52,000-square-foot center to be built near the Mississippi River about five blocks from its present location. The County Historical Society's museum would be a neighbor.

Problems remain. A major one is the need to find and pay for a new professional director and financial officer to replace the volunteers now in those positions. More space is needed for exhibits and programs. The upper level that we often use for programs is not accessible to people with mobility impairments. Staffing is thin, and we rely heavily on about seventy volunteers for many aspects of operations. Our major strength is the ability to meet, greet, and personally engage visitors.

Headwaters Science Center would not exist except for the many grass roots that sought out nutrients in the sandy soils of this region and brought energy, enthusiasm, joy, and life itself to the endeavor. Countless people worked to make some single thing happen, others have thrown awesome amounts of love and energy into ongoing projects. Granting agencies took chances on us, and badly needed funds came from some most unlikely sources. Our staff is enthusiastic and committed to the center's mission. Meanwhile, we know that we've promoted science education and made a difference for some kids and families, and that feels good.

CHAPTER 5

Focus and Balance for the Small Museum

Rebecca Schatz

The Works is a small fish in a large pond: a small, hands-on museum in a large metro area that already has a large, spectacular science museum, a renowned children's museum, two zoos, several fabulous art museums, and more than a dozen historical museums. Why do we exist at all? The Works survives (and thrives!) because of focus and balance. Focus has earned The Works enthusiastic visitors, volunteers, staff, and donors. Balance has brought us credibility, kudos, and hard-won financial viability.

The Works is a hands-on, minds-on museum of technology, open to the public since 1995. Our unique exhibits and acclaimed educational programs delight visitors ages five to fifteen, demystify technology, and inspire interest in the way things work. With a paid staff of three to eight people (varying seasonally), dozens of talented volunteers, 7,000 square feet of space, forty-seven exhibits and two classrooms, The Works serves more than 10,000 children every year.

Whatever your museum's subject and size, a consistent focus is critical to your success. Focus creates your identity and brings your institution to life. What is your theme and why is that theme important? Who is your audience? What will the visitor experience be like? Articulating a clear and compelling focus helps distinguish your institution as a worthy choice among the riot of contenders for people's time and donors' dollars. Focus also creates a culture to guide and nurture your organization. Think carefully about what makes or will make your museum distinctive, and then excel at doing just that.

The founding vision for The Works was to create a hands-on, minds-on environment for exploring technology. Technology—a blend of science, engineering, and mathematics—encompasses not just computers and robots, but the whole world of human creations: levers, lasers, bridges, bicycles, eyeglasses, telescopes, houses, helicopters, and on and on. Technology is also

a process: the process of invention, design, innovation, and manufacture. This view of technology was inspired by the American Association for the Advancement for Science (AAAS) Project 2061 Panel on Technology Education, chaired by Jim Johnson, and by the seminal work of the International Technology Education Association.

In the early 1990s, there were very few hands-on exhibits about engineering or technology in any museum. Locally, the Science Museum of Minnesota was pulling away from early efforts with technology exhibits, and their president encouraged me to go ahead independently. It seemed that a new museum with a focus on engineering and technology, done well, would have a place in the community of Twin Cities museums.

Technology was for me a personal passion; I delight in looking inside the box, in seeing how something fits together and in trying to piece together what makes things work. As an engineering executive in the fast-paced telecommunications industry, I also understood technology as a critical factor in any nation's economy and was concerned about the lack of hands-on engineering in primary and secondary education in America. In addition, I knew that exploring and creating things is a thrilling, empowering experience for young children, and I wanted make the joy of engineering accessible to children from all walks of life.

Inspired by the Ann Arbor Hands-On Museum and unreasonably optimistic about how long it would take, I initiated the effort, which ultimately produced a new hands-on museum of technology. We called it The Works because we wanted to show people how things work.

Originally, we planned to create exhibit areas based on the place each technology is used. One exhibition would explore the technology in a home, from blenders to television. Another exhibition would show the technology of medicine, another, the technology

35

of the office, and so forth. We soon realized that this would result in a hodgepodge of unrelated ideas and decided to group the exhibits by concept, rather than by the place that each technology was used.

We selected image machines as our first topic and created The Interactive Image, an exhibition that premiered in 1995 and remains popular at The Works to this day. Imaging technology was and is an exciting field with growing applications ranging from scientific simulation to consumer electronics. It is based on simple, powerful ideas and lends itself to appealing, engaging hands-on exhibits. (Not every topic does!) A few relevant exhibits could be adapted from other museums, giving us a small but significant jump start.

Most technologies, including imaging technology, are based on a combination of concepts. In contrast, most of the best museum exhibits explore a single concept. To design a good exhibition about technology, we decided to separate out each concept and to create wonderful exhibits exploring that one concept, then group exhibits together so that our visitors could understand a complete machine. The concepts, which became exhibit sections within The Interactive Image, include optical sensors, digitization, animation, color mixing, and basic optics. These concepts are the foundation of technologies ranging from digital cameras to television to medical imaging. Once we had the sections outlined, the next step was to create wonderful exhibits for each section.

A wonderful exhibit attracts and engages a child and allows this child to discover something important or make something magnificent. A wonderful exhibit provokes curiosity then satisfies it. "How does that work? Oh, I see!" "What can I build? Come look at what I made!" With a wonderful exhibit, a child *becomes* a scientist, *becomes* an engineer for a time. A wonderful exhibit creates the environment, the context, and the tools to allow this to happen. Wonderful exhibits are also grounded in reality and are doable, durable, maintainable, and affordable.

The ideas for The Works' exhibits originated from many sources: from reading about technology, tinkering with machines, watching children play, visiting other museums, and talking to engineering and other professionals. Articulating the concepts was straightforward; much time and imagination went into finding opportunities for hands-on fun that would illuminate each concept.

Creating wonderful exhibits is an art, a discipline based on love of ideas and love of people. It involves building prototypes and watching, learning

what works and what doesn't, what inspires, and what confuses. It is a slow but vital process of rethinking, rebuilding, retesting, throwing out lovely ideas that just don't work, and digging into unexpected ideas that do. There are guidelines now and guidebooks to help you learn to create wonderful exhibits or to copy others. But there remains a kind of magic, a gestalt behind the truly wonderful exhibit.

The Works managed to create a few wonderful exhibits and a dozen or so good ones for The Interactive Image. We aimed for a balance of exhibits, deliberately including exhibits that would involve different senses, skills, or learning styles. We blended whimsy (a harp with strings of laser light) with real world applications (a street light). We made some exhibits that could be understood quickly (an optical window that measures your height) and some that offered a more in-depth experience (robots that the visitor can program using bar codes). This balance prevents museum fatigue and builds meaningful, memorable visitor experiences. We put our first seventeen exhibits together surrounded by vibrant color and tucked into odd shaped

Figure 5.1. Light Harp at The Works, Minn.

crannies, in a space designed to be welcoming and comfortable.

It worked! By focusing on wonder, on meaning, and on comfort, we achieved a highly successful exhibition. The Interactive Image attracted 11,000 people in the first three months and launched The Works. Over the years, we added an exhibit area about structures called Build! and one about simple machines and mechanisms called Gears and Gizmos. The Works now has forty-seven exhibits and more underway.

Another key factor in The Works's success is our focus on the experience of each individual child. We can do this, in part, because The Works is a very small museum. This focus pervades daily, fundamental choices on everything from exhibit selection to program design to hiring philosophy. We strive to create an environment where *each* child feels welcome and can personally experience the thrill of *doing* engineering. We hire people who radiate love and high expectations for each child and ooze knowledge and enthusiasm about science and technology. These exceptional people create the exceptionally good programs—workshops, scout events, summer camps, and more—that attract and delight a growing audience.

While a strong focus can bring an institution to life, balance is what keeps you there. For The Works, a crucial issue throughout the start-up years was balancing our need for stability with our need to minimize costs. The two biggest expenses any museum faces are space and people. The Works tried to minimize both by using donated space and by relying heavily on volunteers. Perhaps you can learn from our experience.

Donated space seemed reasonable initially but proved increasingly problematic. The Works's first exhibition, The Interactive Image, premiered in 1995 in a 3,000-square-foot gallery in the Bell Museum of Natural History at the University of Minnesota. A hugely successful thirteen-week run produced rave reviews and slightly more revenue than expenses. This inspired us to begin year-round public operations. We wanted clean, accessible space with heating, air-conditioning, parking, elevators, and good bathrooms at rock bottom rates. After touring some cheap but questionable real estate, we brainstormed sites that might offer us free space and came up with the idea of shopping malls. After walking through many malls to get a feel for ambience and vacancy rates, we asked four different malls to consider donating space for our museum. To our delight all four offered free space.

The issue was stability: we wanted at least a two-year commitment and the longest initial offer was twelve weeks. After considerable negotiation we persuaded one center to guarantee twelve months, and to give us six months notice if we had to move. Exuberant, we began our life in the mall. There were benefits: the center gave us the chance to test our exhibits early on, we enjoyed great visibility, rent was free, and it even paid our utilities. We started a small retail store, which did extremely well, and we enjoyed a continual steam of teenaged "mall rats" in addition to our other museum visitors. The significant problem was the continual worry about when we'd have to move.

And move we did, four times in six years. It became a crazy cycle with the frantic search for new space at a different mall, a frenzied installation, and a precarious first year as we fought to pay off the expenses of moving and establishing the new site. In the second year at each site, we'd come into our own, attendance would double and we'd be flying high. We always looked forward to the wonders that the third year would bring, but never experienced them. We moved twice after barely two years. Then in 2001 we had to move unexpectedly after sixteen months with only three weeks notice. We closed abruptly, canceling thousands of scheduled school groups. The loss of space, momentum, and credibility nearly killed The Works.

It was clear that The Works had to secure a more stable site, dramatically change its focus, or close down. In a series of difficult board meetings, The Works revisited the original vision, worked and reworked the numbers, and resolved to move ahead. After an extended effort to locate and finance a site, we signed a five-year lease on excellent space in a nearby community center. The Works reopened there on January 11, 2003, to capacity crowds of delighted visitors.

Would I recommend the free-space-in-shopping-malls approach? Yes and no. If you negotiate hard, pay as little as you can, and get a long-term commitment in writing (aim for five years), then you'll be okay. Otherwise, view the site solely as a temporary location for portable exhibitions or fixed-duration programs and maintain a headquarters office with a consistent address somewhere else. The Works used the free-space-in-malls approach too long, creating great frustration for our staff and volunteers, plus damage to our professional image for donors, visitors, and the media.

The other primary expense for any museum is the people. The Works tried, with reasonable success, to balance the expense of staff salaries with the vicissitudes of volunteers. The pleasure of working with volunteers is that you see the best people in the world, at their best. The reality is that (with delightful exceptions) volunteers are difficult to find, inspire, schedule, train, and retain. Only one out of fifty people you ask will volunteer, and only one out of twenty volunteers will consistently produce results for your organization.

The Works is blessed by a terrific core of volunteers, distilled over the years, and active in specific areas. Volunteers have built and maintained most of The Works exhibits, provided the muscle for our many moves, painted every site, designed our gallery, produced our website, updated our computers, and helped with accounting and publicity. The Works's programs use volunteers more sparingly—for scout events, public celebrations (like National Engineers Week), and weekend hands-on activities. We have not used volunteers to staff the welcome-admissions desk or as explainers for groups, relying instead on the consistency of trained staff with dependable schedules. While our staff has always been tiny, it is vital to have a reliable core that you can extend with volunteers or work study students.

While balancing the cost of people and space is essential, there are many other aspects of your museum that also benefit from balance, including programs, the calendar, and your own time.

To balance programs, decide who your key audiences are and what you want to offer them. Consider your mission and the strengths and weaknesses of your institution. Start with a few programs and build on as you see what is successful and what is not. Think about what is important to your mission and what is important to your budget. For instance, The Works has always strongly subsidized school groups and Girl Scout events, which are central to our mission. We also do a lot of birthday parties and summer camps, which subsidize themselves.

The Works, like most museums, balances the calendar so each part of the day, week, and year has an audience (and hence revenue): school groups visit on winter and spring weekdays, scout groups and birthday parties after school and on weekends, homeschoolers in the autumn, day camps during the summer. Almost nobody visits in September, which becomes a key month for planning and publicity.

Do balance your own time and priorities. Starting and running a small museum is thrilling and deeply satisfying; it is also exhausting and can be brutally difficult. There are always a gazillion things that you should do as you strive to create something wonderful or struggle to make it through the day. The art of running any small business is the art of balance. Read those books on setting priorities and read your own signals. Burnout will sap enthusiasm. Take care of yourself.

Building The Works has not been easy. We have been frequently on the edge financially and with one move more or one major contribution less we could well have sunk. Several generous, visionary members of The Works board of trustees and committed patrons outside the board have rescued The Works more than once from critical points of financial stress. I have sweated payroll more often and more recently than I like to admit. The unexpected and abrupt loss of space in 2001 nearly killed us before it catalyzed our move to the beautiful, functional site we now enjoy.

The future looks good. Settled in a stable site, The Works's board of trustees has completed our first strategic plan. Over the next three years, The Works will expand public programs, attract new audiences, build a sustaining base of annual support, and invest in new and improved exhibits that will bring hands-on, minds-on learning about technology to a new generation. Focus, balance, and persistence have brought The Works survival, stability, and success.

CHAPTER 6

Discovery Science Center Start-Up: Launch Pad in a Mall

Karen Johnson

THERE IS NOTHING AS CONTAGIOUS AS A WONDERFUL new idea, as all of us who want to start science centers have experienced. My beginning was in 1987 when two board members hired me to help build a science center. One of them had been to Seattle with his thirteen-year-old son and stopped in at the Pacific Science Center; they nearly missed the flight home because the experience of the center was so interesting. He decided that Orange County, Calif., should have a science center.

For anyone who has been to a science center, the experience is compelling and he or she understands the need for and value of the learning that occurs. The first obstacle to overcome for those wanting a science center in their community is to introduce the exciting science center experience to others who have the capacity to help with funding or designing the project. The first rule for starting a science center is getting adequate funding.

In our case, from 1989 through 1996, Orange County was in a recession, and the majority of the philanthropy was coming from the land development community, which was experiencing the severest repercussions of the recession. To further exacerbate the funding problem, the county declared bankruptcy in 1994, further shaking the confidence of donors.

From 1989 when the board voted to move forward with the science center project, the board members struggled with the need to raise significant capital to build a building, fabricate exhibits, and hire competent staff. By 1991, we had identified $2 million toward a goal of $24 million, and the prospects for raising many more dollars looked bleak.

It was about this time that the employees of a major aerospace company committed the largest amount in their history to the science center project: $300,000 to be used immediately on something but not used for capital that is, not for a building. What was not said was that this employee group felt that the permanent science center might never be built because of the adverse economic conditions. Location is always a crucial decision, so when a shopping mall was suggested for a preview facility, it seemed to be a great solution. Retail malls were having trouble keeping their spaces filled, so the idea of opening a small preview center resonated with the mall owners as well as both the aerospace employee group and our board of directors.

Through contacts made at several Association of Science-Technology Centers (ASTC) conferences, I heard about a project, being submitted to the National Science Foundation by Pacific Science Center (PSC). PSC would mentor several small community organizations. Successful applicants would receive a package of new but tested science center exhibits, a week-long management course, scripts for demonstrations, materials and curriculum for a teacher-training session, and advice from PSC experienced personnel for several years during and after the start-up phase. We were one of the seven centers selected for the project and planned to open a preview center in the shopping mall. We had approximately $200,000 in the budget, two staff members (myself and a fund-raiser), and a great vision.

After securing a site of approximately 5,000 square feet from the mall management for little more than $4,000 per month (no rent, but our share of the Common Area Maintenance [CAM]), we selected thirty exhibits from the Pacific Science Center's Science Carnival list. A week in Seattle, being tutored in the arts of exhibit design, fabrication, installation and repair, school field trip curriculum development, budgeting, marketing, teacher training, employment practices, and staff job descriptions, our heads were whirling. However, we had two very important decisions behind us. We knew the location *and* we knew what exhibits we would have.

The benefit was obvious. We would have a safe and secure location to which we could bring prospective donors (always in the morning to see the school children exploring), we could start educational programs, and we could recruit and train a staff, learn about meeting the public daily, and test our budget parameters. We named the preview center Launch Pad, and thereafter referred to it as Launch Pad, the Preview Facility for Discovery Science Center.

The benefit to the shopping mall was less tangible. They needed to create scarcity; showing prospective tenants that there was a demand for space. Furthermore, they believed we would generate traffic of moms and families, which might shop on the way to or from the science center. On the basis of our performance, the following year the mall owners moved us to 10,000 square feet of new space, paid for all the tenant improvements, and after the first year, donated both the rent and CAM for the entire five-year lease.

Our vision was to create a comprehensive science center, with exhibits, a demonstration area, classrooms, preschool area, and retail space. We also had to have an office, storage, and employee break room. Clearly, in 10,300 square feet, each of the spaces needed to be sized appropriately for the overall space. They were as follows:

Exhibits	4,300 sq ft
Demonstration space	750 sq ft
Classrooms (2)	(total) 750 sq ft
Preschool area	400 sq ft
Retail space	1,000 sq ft
Office	800 sq ft
Employee break room	600 sq ft
Circulation and storage	1,700 sq ft

The operations of Launch Pad mimicked any permanent science center. Schools scheduled field trips on weekday mornings, which included a science concept demonstration by a staff member. Afternoons had few visitors, in spite of youth group programs and birthday parties. Weekends were busy with those same programs and more family visits. Over the five and a half years of operations, Launch Pad averaged 60,000 visitors per year. On the whole, Launch Pad, the preview science center succeeded as a demonstration site, both from the fund-raising aspect and from the education programs and staff training perspective. It did have problems, however.

FINANCES

Although the capital to purchase exhibits and to do tenant improvements was in hand, the operations lost money. Folks in Orange County wanted to see

Disney-quality four-color brochures, bright, clean, attractive spaces, and easy parking. To attract visitors, we developed a full marketing program, more costly and detailed than was appropriate for our size. The media costs in our area (50 miles south of Los Angeles) are extremely high, and there are no commercial television or large radio stations located in Orange County. Print media costs are also prohibitive. We secured some free page ad space and a deal from the *Los Angeles Times* as part of their capital gift to the permanent science center.

A significant cost of operating in a shopping mall is staffing. Malls in Orange County are open 10 a.m. to 9 p.m., Monday through Friday, 10 a.m. to 7 p.m. Saturdays and 11 a.m. to 6 p.m. Sundays. Launch Pad was expected to be open and operating fully (that is, the exhibits and demonstrations available) during all hours of mall operations even though the vast majority of visitors came weekday mornings and weekends. Furthermore, employees had little to do during the low visit times, creating low morale. That held true all year except during the holiday season, which had longer hours of operation but more walk-up visitors.

The retail store could be reached through a separate door into the mall and had the highest revenue days during the holidays. Also, we regularly routed visitors through the store when they exited Launch Pad, which had positive and negative results. There probably were more impulse sales; however, there was more pilferage and breakage as well. The store was never a huge source of revenue.

Another issue surfaced when the mall management suggested that we offer babysitting as an inducement for moms to shop. Because of state licensing requirements, it wasn't possible. Two other competitors in the area, Fundazzle and Discovery Zone, both for-profit operations, offered babysitting, and interestingly, both have gone out of business.

As the permanent science center began to gather momentum, owing to improved economic conditions and a surprise gift from a major corporation in the area, (see below for the sponsorship from Taco Bell), excitement and anticipation caused yet another problem. At that time, in 1997, Launch Pad was operating fairly smoothly (although at a negative cash flow of about $180,000 per year, more about the budget later), and we had twenty-five employees there. The administration staff (ultimately eighteen persons) operated out of a separate office, 2 miles away, creating a "we and they" situation. Regardless of the efforts made by the organization as a whole, this two-camp mentality continued until Launch Pad closed in August 1998, prior to the December 1998 opening of the permanent science center.

OPERATIONS

Operating Launch Pad permitted us to try various education and membership programs, special events, and activities. Although we began with good teacher packets including pre- and post-visit activities, the children's visits were random pinball-type experiences. Students would run from one exhibit to another, with very little or no learning. Within two years, we changed to structured visits, with staff members giving groups tours and asking open-ended questions. The educators of the twenty-eight local school districts overwhelmingly approved this approach, which continued into the permanent science center. We tried birthday parties, which were a financial success, but which had our employees spending time setting up the room, giving extra demonstrations, and cleaning up the mess. For a price, we provided the room, (decorated with fun paper products and helium filled balloons), a demonstration at which the birthday child was featured, and pizza from Pizza Hut (it could be ordered and delivered). We would even include birthday bags (favors) from our retail store, for an additional charge. I still question whether the birthday party concept is viable as an earned income source.

There is a temptation to either ignore completely or design programs and exhibits wholly for preschool children in a shopping mall site. They do represent the majority of the children in the mall during school hours. We identified several exhibits, which were lower in scale and simpler and placed them in an area separated from the rest of the exhibits. We designed several special preschool programs and tried to attract toddlers to the venue in the afternoons. While it was somewhat successful, the overall increase in attendance income never materialized.

To promote the science center on the weekends, we created special events, which drew an amazing number of visitors. One book signing by Sally Ride generated more than 300 persons who stood in line for two hours to buy her book and have it personalized. In all, we had three or four book signings (including one by Buzz Aldrin) each of which attracted a large attendance. Other weekend specials featured a puppeteer, bubblologist (these performers continue to bring sell-out crowds, even at the permanent science center), nature talks (with live animals), and a sensational juggler. All of these programs were tied to some science principle and were advertised as heavily as we could afford. Most were hugely successful, although depending on the cost of the performer, attendance and memberships increased but not the bottom line.

LOGISTICS IN A MALL SETTING

You are constrained by the mall management in design and appearance of your facility. Your storefront has to fit in with the other stores in the mall. The level of construction and finishes of your site must be consistent with the building and conform to the Uniform Building Code; we opened with six public restrooms (three male, three female) and two for staff, instead of the required twenty-four. Installing exhibits had to be done between 10 p.m. and 7 a.m., hours when the mall wasn't open to the public. Further complicating matters was the size of the freight elevators. They are sized for retail, and measure about 10 by 8 by 8 feet. Many of our exhibits were 10 to 12 feet high. Each exhibit either had to be taken apart and brought to the third floor in pieces (with the hope that our inexperienced staff could put them back together), or brought in diagonally and assume once in the elevator they could be removed, as the elevator doors didn't extend the full width of the elevator. We used the existing electrical service in the site where possible; however, we had to bring in many more feeds, using the interior columns as locations for outlets or curly wires from the ceiling. The payoff of this work, however, was at the Grand Opening at which board members, members of the media, and donors were dazzled by the appearance of Launch Pad.

Storage at science centers is always in short supply, and in a mall setting, there is even less. The store fronts are built to display merchandise for sale, leaving very little "back of the house" to store crates, broken exhibits, restroom supplies, tools, retail inventory, cleaning supplies, or marketing materials. At times, we used nearby storefronts that weren't leased for exhibit storage. We built shelves above the employees' break room, stored things in the offices, used the exit corridors and crammed as much as possible in the two small storage areas designed into the center.

STAFFING

One of our goals was to recruit and train a staff that could deal with the public, seven days per week. Because there were two locations, Launch Pad and the office where the senior staff worked, we relied on the Launch Pad staff to handle all visitor-related situations that arose. We mainly hired students who wanted

Figure 6.1. Discovery Science Center's Launch Pad in a Mall, Santa Ana, Calif.

part-time work (more cost effective) and who had enthusiasm for the students, helping to ensure a positive experience for the visitors. A former employee of Pacific Science Center, who had worked with the exhibits we had purchased, supervised the staff. He was able to train new employees regarding the science concepts and oversee the operations. We found that actors did the best demonstrations, as we could teach them the science, but they could dramatize the presentation and had better poise on stage. We taught employees the basics of customer service, such as smiling when greeting a visitor and showing enthusiasm for the exhibits and programs. The staff was also certified in first aid and CPR, although in the mall, there are trained personnel available. On the whole, the Launch Pad staff worked well together and maintained positive evaluations from teachers and family visitors. Evaluations were requested from every teacher.

The last six to eight months prior to opening the permanent science center were frantic. In addition to the need to raise more and more funds (is there ever enough money?), we were balancing the construction and renovation of the building and site; ordering or fabricating and installing exhibits; planning and executing the opening marketing launch; hiring and training new staff; and developing additional personnel policies, job descriptions, security measures with the skills of a small staff, few of whom had done any of this before. We knew it was important to become a team . . . one that could work together and that could trust one another to make decisions and to follow through on action plans. We hired a former science center executive to meet with twenty staff members (mostly senior management and supervisors from Launch Pad) to develop a core culture. It was helpful, although tempers flared from time to time. The job required nearly everyone to work significantly more than the usual forty to fifty hours per week, with the result of short tempers, haggard faces, and general exhaustion. Combine that with the incredibly slow pace of the building construction and renovation, and we all were extremely anxious about meeting the December 19th opening date. It was the holiday season too, adding one more layer of responsibility to most employees' already busy schedules.

In order to try to mitigate these problems, I tried a couple of things learned from colleagues at ASTC. I gave (and announced about four months prior to opening) everyone working on the science center an extra day of paid vacation for every month they worked, up to a maximum of five days. We celebrated the holidays with a catered prime rib dinner at the new building in early November, complete with tablecloths, waiters, and a special gift. All those present received windbreaker jackets with the new science center logo and "Launch Team Member 1998" embroidered below.

TACO BELL GIFT

What permitted Launch Pad to operate at a deficit was the ongoing capital campaign. We could justify some of the expense of the senior staff and marketing as a fund-raising expense. And the gamble paid off with an early capital gift from Taco Bell. Although the gift gave credence and substance to the concept of building, equipping, and opening a permanent science center, it raised huge issues for our staff to resolve. The employees at headquarters wanted to help children and improve education so a permanent science center fit their goals for the community. Taco Bell offered $2 million to the capital campaign and assistance with acquiring funds from local, state, and federal governments. We raised more than $7 million from these sources. Taco Bell's gift was catalytic. However, the question of naming the building for Taco Bell was difficult. No science center or major museum that we knew of had ever been named for a public corporation. So began the internal debate about whether to accept this gift. On one hand, we knew it made the permanent science center real, but could we be trendsetting enough to accept the permanent name Taco Bell Discovery Science Center?

Questions arose immediately. Would prospective donors give if that were the name? Would the public expect to find the history of the burrito instead of a science center at the building? What would the reaction of school districts be? And would they permit and pay for their students to visit? The resolution of all these issues and more took an inordinate amount of staff time, focus groups, soul searching, and difficult decisions. Initially, while it was an enormous risk, our board agreed to move forward and accept the gift. The final agreement was detailed and evaluated many times before there was concurrence on the intent and practice for operating the permanent science center.

We both agreed that the gift named the building, not the organization, so letterhead, business cards, and organization-driven literature such as membership or education brochures were printed without the Taco Bell name. But, all marketing materials, press releases, ads, and invitations used Taco Bell Discovery Science Center. Any reference to the building or location included the Taco Bell name. What made this concept work was the extreme patience and good will of the Taco Bell Corporation and their representatives. From the onset of discussions, they understood the problems with other donors and with the education community. At present, the Taco Bell Discovery Science Center has become an educational and entertainment venue in Orange County, and the name is a non-issue. This was the prediction of the senior Taco Bell officer who originally had the idea, and worked tirelessly on the final agreement. I credit his vision that made this idea feasible for both parties.

This has been a case history of one way to start a permanent science center. It worked for us. Is it the only or best way, definitely not? At the discussion of whether to commit the money and resources to opening Launch Pad in the shopping mall, one of our board members said, "you have to walk before you can run." That certainly was true. We learned much from the experience of opening and operating Launch Pad. Our original goals to raise capital campaign funds, to train staff, and to start education programs were met. It was an advantage to have a preview center in a shopping mall, despite the constraints. We made the best we could of every unusual situation, got the most from dedicated board members and employees, and it worked.

CHAPTER 7

Curious Kids' Museum

MARY BASKE AND PAT ADAMS

IT WAS CURIOSITY THAT GOT US STARTED HERE AT THE Curious Kids' Museum. While on Christmas vacation in Ft. Lauderdale, Fla., in 1997, my husband and I took our three children, ages eight, six, and four, to the Discovery Center. It was our first time visiting a children's museum, and we honestly had no idea what to expect. The museum was housed in a small but charming two-story historic building and contained an infinite variety of wonderful hands-on exhibits, some scientific, others cultural, and all very interactive. I still remember how vibrant and exciting our visit was, petting their resident iguana, playing in their kid-size thatch-roofed Seminole Indian village house, crawling into their portable planetarium, cranking their bicycle generator, attending the impromptu workshop on fingerprints, and so much more. The museum kept our entire family totally enthralled for hours. As a special education teacher and mom, this children's museum concept was absolutely incredible to me, a kid-friendly place where non-traditional methods were used to arouse a child's natural curiosity—and done in such a simple and playful style. From that visit on, I had what I like to call an obsessive-compulsive museum disorder, museum on the brain, or a passionate desire to start one in my hometown. I vividly remember saying to my husband as we left the museum that day, "It doesn't look like it would be too hard to start a museum like that, does it?" Little did I know!

After we returned to our home in St. Joseph, Mich., I made the most important call I have ever made and that was to Cynthia Yao, the director and founder of the Ann Arbor Hands-On Museum. I told her I had just visited an extraordinary place in Ft. Lauderdale called the Discovery Center and wanted to start a museum, too. I desperately needed her advice. I remember that Cynthia didn't even laugh. We talked for over three hours that day. She outlined for me what this process would entail, what resources there were on the subject,

from literature about museums available through the Exploratorium, to articles and books available through the Association of Science-Technology Centers (ASTC) and the AYM, and finally, people to contact and museums to see. She warned me from her first-hand experience that I would need a lot of persistence and a patient husband. Luckily, I had both.

Soon after that call, I contacted a number of my good friends who I thought would be interested in this project, and told them what I was thinking. We later had a dinner party and talked about starting a museum. We all wondered whether our area could financially support such an entity. After all, St. Joseph already had a small art museum, a symphony, many community projects that needed support, and our population was a mere 12,000 people. After much conversation (and a couple of bottles of wine), all of their input was positive and supportive. The verdict was to ask the community college, Lake Michigan College, to conduct a feasibility study. Happily, they did one and their encouraging results propelled us onward. From that point on, the idea snowballed, thanks in large part to those original people and friends who had come to dinner at our house that night. Many of them later became board members, volunteered in some way, and one friend even became assistant director after we opened. Many remind me of that night sixteen years ago when founding the museum was just an innocent idea floating around a dinner party.

The first thing to do, according to Cynthia, was to visit as many children's museums as possible and talk to museum professionals. One of the first visits was of course to Cynthia Yao's Ann Arbor Hands-On Museum where she and her staff generously met with us for hours, fielding questions and then giving us a tour of their fantastic science-based hands-on museum. The quality of the exhibits and the depth of their programming and staff expertise were impressive.

Next, we visited Impression Five in Lansing, Mich., again meeting with their director, Bob Russell, and absorbing all that he had to share with us. Later visits included the Kohl's Children's Museum in Wilmette, Ill., and Chicago Children's Museum in Lincoln Park, the Detroit Science Center, Indianapolis Children's Museum, Los Angeles Children's Museum, the Magic House in St. Louis, and many more. At each destination, we visited with the directors or education directors, spoke to staff, volunteers, and board members, viewed their exhibitions, and observed their educational programs. We wanted to see what was working and what was not. The museum community generously shared horror and success stories that they had experienced along the way of founding and growing their museums. We found that each museum had its own story, its own history, its own emphasis, its own strengths and personalities. Often we would do these visits on weekends, and take our children, afterwards asking them what their favorite experiences were while there. After a number of months investigating, probing, and analyzing, it was almost time to form our own museum personality.

One of the most beneficial experiences we had during those formative months was to attend Boston Children's Museum's conference on How to Start/Not to Start a Children's Museum, a three-day workshop that focused on issues facing start-up museums. There, cofounder and friend, Liz Garey and I met others who were starting or at least contemplating starting children's museums. We were introduced to exhibit builders, museum association people, museum directors, educators, architects, board members, professional fund-raisers, grant writers, and public relations professionals and attended their workshops addressing some of the many issues that would face our future Curious Kids' Museum. After we returned home, although we knew that we were definitely on the right track, we also knew there was much to do.

On October 28, 1987, the Curious Kids' Museum was born. The museum incorporated and became a nonprofit 501(c)(3) organization. We were official now, with an official board, a mission statement, a name, a set of bylaws, and a seal. After leaving our lawyer's office that day, my fellow cofounders and I all looked at each other with a mixture of elation, excitement, and dread knowing that this was just the beginning of our long journey.

Next on our full agenda was to find a home for the museum. Our growing board formed a site committee and after only a couple of months located two potential sites, one being the lower level of the local YWCA, which had 5,000 square feet of space available to rent. The other choice was an unoccupied and rundown neglected memorial hall, built in 1918 by the federal government for the veterans, who now used its basement once a month for their meetings. At this point we invited museum angel Cynthia Yao over to look at our choices and make recommendations. She graciously reviewed the sites and advised the board to go for the site that gave us autonomy, and a possibility for expansion, that being the veteran's memorial hall building, a three-story building of 6,000 square feet that needed extensive renovation after years of neglect. After meeting with the veterans' group and getting their approval to share the building, we approached the city, which owned the building. The city manager and the city council didn't know quite what to make of us or our idea of a children's museum. Not only did we have to sell them the idea of the museum, but also talk them into letting us lease their building. Many meetings later, "the ladies" as we were affectionately named, were given approval by the city of St. Joseph, to lease their building for $400 per month, as long as we renovated it ourselves and made it handicapped accessible (including adding an elevator and ramp). One of our board members' husbands volunteered to spearhead the project, getting bids for the building renovations, finding an architect to design the improvements, and overseeing the construction. We were finally off and running.

Now all we needed was the money. The next stage of our plan involved public relations, getting the word out. Our core group of six women devoted hundreds of hours to spreading the word at lunches and giving video presentations. From all the travels we had taken to various children's museums, we had made a composite video of our favorite museums and interviews, so that this new concept of forming a children's museum could be better understood and visualized. We visited every organization, group, or club that would take us, including all the civic clubs, foundations, corporations, banks, women's clubs, schools, radio and television stations, art fairs, and festivals, going on everywhere and anywhere people would listen to us about the future Curious Kids' Museum. Making the community leaders aware of our plans, sharing our financial projections and needs with them, and finally asking them for their support were all part of the mission and the pitch.

I will never forget the very first important meeting we had with the corporate vice president of Whirlpool

Corporation, the largest employer in the area and a company we really needed support from. My friend and I decided to be creative in our approach, so I wore our dinosaur costume as we entered his office. To my great relief, he loved our enthusiasm and humor; he respected the people that we had enlisted to be on our board and supported the concept and plans for the children's museum. Soon we received our first large grant of $10,000. We were thrilled to have their backing. Later, the Community Foundation, the larger banks and corporations, and other private foundations began to join in with their grants and donations. Within six months of this campaign, we had raised over $120,000 to be used for the renovations (including the new elevator) of the memorial hall, exhibits, operations, and salaries. Everything was falling into place. The construction started at the end of June 1989 and we were ready and raring to go by September.

On September 19, 1989, after almost two years of conceptualizing, planning, raising dollars, renovating the memorial hall, and building exhibits, the very first of many classroom groups arrived at the Curious Kids' Museum's front door. They were preschool children from the migrant school in Dowagiac. The excitement of their day was captured on the city newspaper's front page that evening. The photo reflected the children's faces lit with excitement as their little hands were digging into the sand of the dino-dig, exploring the giant sandbox for fossils. During its first year of operation the museum attracted almost 40,000 young and curious visitors, school groups, and families. Within ten months of opening its doors, museum membership reached over three hundred families. The generosity of the community and its obvious value for the family and education could not be denied. We were dumbfounded by its instant success, and we were constantly trying to keep up with all of the demands that the success brought with it.

In spite of our shoestring budget, and thanks to many generous volunteers and organizations, we were able to construct and assemble a variety of wonderful hands-on exhibits when we opened our doors. Educators from local universities and schools, relatives, friends, local craftsmen, doctors, dentists, business owners, and experts in various fields devoted time and energy into making new exhibits for the museum. For example, the local Potowatomi Native American group designed and constructed a child-size longhouse and canoe exhibit. The Southwestern Michigan Beekeepers Association constructed and installed a live transparent beehive for our Discovery Room. Our local

Figure 7.1. The Curious Kids' Museum hosts a children's sand sculpture workshop for the Community Sand Sculpture Festival.

doctors donated an articulated skeleton and Take-Apart Torso for our Body Works exhibit. United Way funded construction of our first Wheelchair Obstacle Course and Braille exhibit called What If? The local chapter of Sigma Xi designed a transparent washing machine. A local Boy Scout group took on building the Dripping Water exhibit as a father-son activity for the troop. My father-in-law, using an exhibit design from the Exploratorium Cookbook, made the Pipes of Pan Sound exhibit. Our Shadow Room, the Bubble Chamber, and Pin Table were built by local craftsmen, teachers, and volunteers and financed by the local Rotary and Lions Clubs. The Discovery Room was funded by a local insurance company, complete with microscopes, animal track activity boxes, animal skins, and a fish tank with animals and fish native to Michigan. Our two-level boat exhibit "Kids Port" was donated by our area's largest family foundation. These first exhibits were simple, bright, well conceived, interactive, and fun.

Our original staff included many volunteers, including friends, retired teachers, and moms. I decided I would take on the responsibilities of the director as a volunteer at least until we could afford a "real" director (which turned out to be ten years later; not that we couldn't afford one until then, I just couldn't leave). My good friend and fellow founder took on the full-time paid position of assistant director. The first floor manager was hired from a nearby college. The gift shop and admissions area and the rest of the floor staff were run entirely by volunteers for almost a full year, until we realized that the popularity of the place demanded full-time paid staffing for managerial positions. Volunteers were still (and continue to be) very key to our success, but growth demanded more consistency and professionalism in order to serve our increasing public demand.

The museum has been an integral part of St. Joseph and our twin city, Benton Harbor, for over fourteen years now. We have become a popular tourist destination for Chicago, Indiana, and outlying areas of Michigan. The annual budget has grown from its original $90,000 to $450,000. In spite of this phenomenal growth the Curious Kids' Museum still maintains its original warmth, friendliness, accessibility, hands-on emphasis of exhibits, quality programs, and commitment to providing to its children and their families and schools quality hands-on educational experiences. The Curious Kids' Museum presently "edutains" over 72,000 visitors every year and has over 850 families as members. We have had one major building addition of 2,700 square feet, adding exhibit space, a birthday party room, office, workshop, and storage space. We now have fifteen part-time and five full-time employees, including a new executive director, a building/exhibits person, admissions and gift shop manager, and two museum managers/educators.

Our programs have expanded as well. The museum offers a variety of public programming including weekly family programs on various topics; a very active outreach program, which serves over 11,000 kids in schools each year with ten different programs; birthday and corporate parties; Scout Badge Quests and Overnights; Holiday and Summer Camps; Science Start—a program designed for the Tri-County area Head Start preschools; participation in many community annual events; and many special educational programs for the underserved community.

The Curious Kids' Museum's awards have been numerous, the most prestigious being the Institute of Museum Services grant, which we have received numerous times. The others include the AAA Tourist GEM award, the *Herald Palladium*'s award for the past ten years for the Best Place to Take Kids, and the *Herald Palladium*'s award for the Best Museum in Southwestern Michigan. The most gratifying gift, however, is the award we get at the museum every single day, when the children protest, and sometimes even cry, when their parents say, "It's time to leave!" We know what we set out to do many years ago around a dining room table has indeed happened, and as of this year, a million curious kids will have been a part of this hands-on children's museum dream come true. It was, and still is, an unforgettable journey.

Curious Kids' Museum is entering its teenage years and is planning for the future with recommendations from a recent MAP Assessment and Long-Range Vision and Plan. For the past two years, the board, staff, and over 1,000 members, visitors, community leaders, and children in classrooms have worked together to create and focus on a Long-Range Vision and Plan for Curious Kids' Museum. How do we maintain the intimate, cozy small-town space and feeling people tell us they love and still meet the needs of our community and visitors who continually ask for more space, more exhibits, faster exhibit turnover, a place "to feed my child," snacks "so we can stay longer," something for older kids, and more groups?

Our first goal is to sustain the museum and ensure that there will always be a Curious Kids' location here on the shores of Lake Michigan. It is vital that we raise an endowment of about $2 million, which would generate about one fourth of our annual operating income, ensuring a building and staff.

The second goal is to expand the facility to incorporate many of the above-mentioned needs plus add vital on-site exhibit workshop and storage space. Board members on the Long-Range Vision, Dollars, and Facilities Team are working with the City of St. Joseph, area foundations, and residents to find solutions, which provide a win-win situation for the museum and the community. Will Curious Kids' Museum expand on the bluff overlooking Silver Beach, a location loved for its views, scenic picnic and trails, and anchoring the downtown businesses? Or, will it be more advantageous to build a new site next to Silver Beach and combine efforts with other attractions such as the future Silver Beach Carousel, Water Parks, Cultural World Food Museum? These are the decisions and discussions facing Curious Kids' Museum as the future unfolds a whole new world and economic climate. The next five years will tell the story as we seek the wisdom to make the best decisions possible for Curious Kids' Museum to maintain and increase its destiny as a premier "edutainment location" where learning is fun.

CHAPTER 8

Science Spectrum: A Science Center for the South Plains of Texas

CASSANDRA L. HENRY

BUILDING FROM THE GROUND UP

The first meeting to establish a science center in Lubbock, Tex., was held underground—a remarkable feat considering that Lubbock is located high on the southern edge of the flat prairie expanse called the *Llano Estacado*, in the lower panhandle of Texas. As land is plentiful, few reasons exist to go underground, except to drill for oil and observe prairie dog communities. Phrases such as "getting in on the ground floor" or "building from the ground up" have amusing significance for the founders of this science center because this meeting to form a founding science center board and to begin incorporation was held in the basement of my home in 1986. From that moment on, the proverbial "snowball" gained momentum, resulting today in our three-floor, nearly 100,000-square-foot facility.

Top business executives and community opinion leaders were solicited to join the Science Spectrum's formative board. Looking back, one can attribute its start-up to this influential leadership assembled at the right time, driven by one person, the keeper of the vision and flame. There are not many cultural facilities in the city. The only other similar facility that exists to the one proposed at that time is the Museum of Texas Tech University. It contains themes of southwestern culture, art, and natural history. It is free with a traditional format and not interactive. The city was ripe for this new kind of exciting, family-oriented, hands-on, and science-focused educational experience.

SCIENCE CENTER CONCEPT

My first introduction to the science center concept came through conventions and tourism, not through my education background as a teacher. Through my husband's involvement in city government, first as a councilman, then as a mayor, I attended conventions that held special, usually family, events at these places called *science centers*. These social events blended fun, food, and people in a unique environment filled with activity and exhibit exploration. Mayors of all size cities with their families were having a wonderful time absorbed in discovery. The Franklin Institute was the first science center I experienced. From then on, I searched out science centers wherever we would travel. At this point, our story becomes like that of many others. I began to collect information and pictures, to talk to friends, and to assess the interest of targeted segments of community leaders. I found that science centers reflected their regions. They had many audiences, came in many sizes, and had variations in mission. One could work in our city. Some good advice given to me by science center professionals that I interviewed was that we should design our institution to a scale that reflected the needs, desires, character, and capacity of our regional area.

Roy Shafer, then president of Center of Science & Industry (COSI), suggested that we bring a group to Columbus to experience a science center firsthand. Community and regional opinion molders flew to Columbus. This trip proved to be pivotal. His staff answered questions, and we returned with resolve to form a board, incorporate, and begin fund-raising. We retained an attorney to handle the process of incorporation because we felt that it would be faster and would be done correctly. Our strategy in forming the first board of directors was to have the board composed of the top executives of the major entities in town whose interest, position, and sphere of influence would be needed to benefit the formation and operation of the new science center. We knew that they would not be able to attend all meetings, but they would be available for assignment. Therefore, we set no regular attendance requirement. There would be quorums needed for decisions. There would also be no board terms of office. Board members would serve until they resigned. This decision also proved to be critical to our long-term success. Continuity and longevity produced consistency and allowed us to move quickly. The board was a policy board. There were no standing committees, only short-term ones for special projects. Consistency has given us credibility and wisdom through these ensuing years.

The next advice we received was to bring to Lubbock a traveling, hands-on science exhibition to assess the community response. We secured an empty storefront for two weeks in our shopping mall and brought in Light and Sight from the Franklin Institute. Local response to the concept was very enthusiastic. At this point, a local grocery store chain represented on our newly formed board made space available in an empty former store for another trial exhibition. We obtained COSI's Science of Sports and set up the space as a miniscience center complete with admissions area, gift shop, and party space. During this project, Dinamation called to ask if we were interested in displaying dinosaurs as well. As we were already set up, we extended the project another few weeks to host the robotic dinosaurs, which had never been seen in this region. The two exhibitions were timely and huge successes. The utility companies, hospitals, newspaper, and city government were sponsors. These efforts were our practical feasibility studies. The concept was so groundbreaking here that we were not sure that a formal feasibility study would be productive. So we never did one.

A BUILDING IS DONATED

Our next important step came as the result of a contribution of space for a beginning science center by the same grocery store chain, Furr's Inc. Furr's loaned 30,000 square feet of space in the same empty grocery store near the regional shopping mall that had hosted our previous, recent exhibitions. The fund-raising began to open our first real science center. The city council was approached to provide funds from the city hotel-motel tax revenues for exhibits. COSI was contracted to construct the exhibits. In addition to the space, Furr's donated money for a director's salary for one year. Excitement was high. On February 11, 1989, the Science Spectrum was officially opened to the public as a science center for West Texas and Eastern New Mexico. We felt that we had really accomplished something. Our first science center programs included demonstrations and outreach presentations. The Junior League gave funds for the StarLab, which we took to local and surrounding schools. We remained at this location for

four and a half years until Furr's announced that it was moving its headquarters to Albuquerque, N.Mex. Furr's did not own the building that they had "sublet" to us; so we had either to pay rent, offer to buy this building, or find a new facility. We were beginning to experience problems in sharing the building and parking lot with the other tenants, notably a restaurant next door. At times there was not enough parking for museum patrons and restaurant customers during the same business periods. The dumpsters were by our outside wall. The rotting food smelled foul and attracted flies. We did not want to begin renting. We considered buying the present building but decided against it because it was a marginally well constructed building and held little future expansion and parking potential.

A board member knew of another building on the south Loop that the owner might consider donating to the right nonprofit institution. We submitted a proposal. In the meantime, another nonprofit project was being proposed in a redevelopment area in central Lubbock. We were strongly urged to join forces with this project at that location. But the growth of the city was southwest. Our proposal was accepted; so, we decided to locate on the Loop, where the traffic was. This was a reverse strategy from the redevelopment concept. This decision proved to be another wise one. The other project never materialized in the central location. Sometimes incentives for redevelopment projects are not in the long-term best interests of some community projects, especially museums that depend upon traffic. It is better to locate where people are than to try to entice them into redevelopment areas where safety may be a concern and the convenience of distance or proximity to other frequented places is an issue. Time and ease are factors in public decision making.

AT HOME ON THE LOOP

We now had a permanent home; so, one might say, "and the rest is history." However, our Loop location did not come without challenges. The building had asbestos and a boiler heating and cooling system. The asbestos had to be abated, and the boiler required quite a bit of rehabilitation after being dormant for over ten years. The roof also leaked, and the parking lot needed resurfacing. We needed to raise money for renovations and to plan a facility. Several wonderful grants helped us realize our dream, a federal grant for tourism development and a challenge grant from the Mabee Foundation matched by the Meadows Foundation. As the City

Figure 8.1. A beautiful day in the atrium of the Science Spectrum, Lubbock, Tex.

of Lubbock at that time was entertaining proposals for joint ventures to promote visitation and tourism to the area, we submitted a proposal to add an IMAX Theatre to the Science Spectrum. The proposal was approved, and we were commissioned to build the theatre and to operate the two as a single destination facility. We had begun to refer to the science center alternately as a *science center* or a *science museum* although the term *science center* was often confused with and thought to be a division of Texas Tech University, as they have many academic study centers.

Our new facility was three times larger than our former one. Our small set of starter science exhibits were four years old and only covered a small range of science topics. We needed to acquire more exhibits and develop new demonstrations. Our larger facility with a theater also meant improving our operational practices, including acquiring a ticket-software computer system. In one sense, we were a beginning, new science center all over again, as we were advancing to the next larger level of operation. We learned of the Pacific Science Center's new exhibit program for developing science centers. Help with operations, demonstrations, workshops, and a set of exhibits was offered to successful applicants. We were accepted. This opportunity for mentoring, training, and exhibits from a mature science center gave us the professional development around these wonderful, well-designed exhibits

that we so critically needed. Also, we were later accepted into a similar program of the Exploratorium. We had four years of experience on a very small scale but did not know how to advance to the next level. It is at these moments that we realize our isolation in West Texas. The Science Spectrum was no longer a child. We were ready for adolescence. But how could we get help? We knew enough to ask the questions, but more important, we could recognize and implement the advice. These experienced and successful world-renowned museums helped prepare us to move forward. For the future development of small science centers, it is critical that a program of mentoring exists to enable them to progress to mature ones in program and administration, no matter the physical size of the facility. We are indeed grateful to the Pacific Science Center, the Exploratorium, COSI, and National Science Foundation for helping us in this process.

Now, with the involvement of the city, the science center became a target for opposition by a political taxpayers' watchdog group. Though the members expressed that they were not against science centers, they indicated that we would be used to further their cause for leaner and better-managed city government. Articles and letters appeared in the newspaper, faxes circulated about town, and guests appeared on morning talk radio shows. A small cloud hung over our opening, but families responded with praise and enthusiasm.

TEN YEARS LATER

Ten years now after that exciting moment on October 23, 1993, the Science Spectrum is entering its next stage of development. It is now an established hands-on science center with an excellent performance record and an annual attendance of over 180,000. It is installing its first permanent, original exhibition, The Brazos River Journey, and it is evaluating programs for the next ten years. Renovations and expansion of exhibition and program areas are underway. We are gratified for our accomplishment and acceptance. The Science Spectrum is vital, relevant, and poised to be a community leader now in decidedly harder economic times. Funding remains a huge challenge. It has a lean operation with entrepreneurial instincts. It relies largely upon earned income with the help of grants from generous foundations for capital and special projects.

The annual budget of the Science Spectrum averages $1.4 million. Presently there are seven salaried full-time employees, six full-time hourly employees, and fourteen part-time employees. There is no organized volunteer support group; however, we use volunteers for many programs and special events as needed. Our facility and exhibition renovation is the present priority. We must continue to develop fund-raising activities. We will seek more partnering opportunities for programming with the university and other community organizations, such as science laboratories for home-schooling and private school organizations. Lubbock

Figure 8.2. Outside view of the Science Spectrum Museum, Lubbock, Tex.

is rich in state educational resources. Science centers do not have an exclusive on interactive science experiences in an entertaining environment. There are many commercial and other nonprofit competing programs. We must determine what the public looks to us to provide and what we do best. We must execute our programs exceptionally well in an inviting facility in order to compete for leisure and educational time. We will address what we need to do to be a better tourist attraction.

Dedication, tenacity, consistency, a sense of ownership, and pride characterize the community leaders who stepped forward to create this important regional asset and the staff who continues to oversee its daily operations. We must continue to inspire and to enrich the quality of life for our family of regional visitors.

CHAPTER 9

Science Center in a College: An Unabashedly Autobiographical Account

ALBERT J. READ

WHAT FOLLOWS IS THE STORY OF THE CREATION AND operation of a very small science center on a medium-size state college campus in a small city in rural upstate New York. The story is simply told as a personal narrative.

First, I should introduce myself. Until my retirement I had been happily engaged and engrossed in teaching physics, especially to students with no significant initial understanding of the subject. In my teaching, I emphasized the laboratory as the principal place of learning, with lectures being mainly commentaries based on the laboratory experiences. I found that in addition to devising original laboratory experiences that were really educational, I often had to design and construct the appropriate laboratory and demonstration apparatus. I mention this background because it was excellent preparation for my postretirement career in the science center field.

Shortly after I retired, a middle school teacher was remarking to me that class field trips to a science center were highly desirable educational experiences, but the nearest science center was too far. So, she added, why not set up one in our locality?

It turned out that we had all the necessary ingredients at the college I had recently retired from and with which I had still maintained active and cordial connections. I discussed the possibility with the college president, who agreed that it would be a good thing for the college and the community but that the college didn't have space and it couldn't provide any funding. I pointed out that there was a large unused space in the basement of the building housing the college's physics department, from which I had retired about a year earlier.

The room was a rather large unfinished basement space, 40 by 80 feet with an 11-foot-high ceiling. It had a dirt floor, poured concrete and concrete block walls and ceiling, and a large number of pipes running the length of the space a foot or two below the ceiling. The space was made more attractive by the fact that the physics department with all its apparatus was just upstairs, and that the machine and carpenter shop, which I had designed when the building was constructed, was a short distance down the basement corridor.

I also informed the college president that I thought I could raise enough money from the community to make the space usable and to set up and operate the center with no funding at all from the college budget, a condition about which the president was emphatic. Furthermore, I agreed to serve as director of the proposed center on a volunteer basis. If I had known then how much the room rehabilitation and initial setup would cost, I would not have had the courage to make such a rash commitment. However, it got done, as related below.

The president, after consultation with various authorities in the state university central administration and others in the college, gave me a letter officially appointing me to be director of our college's Science Discovery Center. One important aspect of this appointment is that it gives the director an official, if unpaid, position in the college and therefore the protection of the Public Officers Law against liability claims.

From this point it took just under three years to open the center. Fund-raising was the first main concern. I hate to ask others, whether friends or strangers, for money. It took a fundamental change of attitude for me to enter into it with the necessary enthusiasm. A major step in achieving this change of attitude was when I adopted the rationalization that I was not asking them for a gift but that I was offering them the opportunity to have a part in creating something good, an educational resource for the college and the community. Within a year and a half, I had raised over $80,000. About half of this came from individual small gifts and the other half came from some local foundations.

It took about a year to accumulate enough funds to allow me to proceed with the room preparation. The college assigned a liaison person from its facilities staff to work with me and to guide me through the requirements established by the state, since we were modifying a state-owned building. In order to pour a concrete floor, we had to hire an architect to produce professionally acceptable plans. Meeting all the state regulations required the architect to produce a project manual, with about a hundred pages stating organizational and job conditions and with the specific drawings and specifications pertaining to our room taking up only the last few pages. These had to be approved in the state capital both by the state comptroller and by the state attorney general. I became aware that a large organization such as a state government, when requested to take an action, looks to how it handled a previous situation and then follows the same procedure. This makes it a real challenge to get the state to do something it had never done before and for which there was no precedent. Fortunately, everyone I had contact with in the state's official circles was sympathetic to approving our request and were just being cautious in seeking to do it in a way that completely conformed to the laws. Eventually, the approval came through; we advertised the job for bids, let the contract, and got a floor and finished walls.

Conforming to the fire codes was also a major consideration in preparing the space for occupancy by the center. The 8-inch cement block wall between the room and the adjacent basement corridor did not meet the code for fire resistance, so we had to supplement the wall with two layers of fire-rated gypsum board. We also found that since we were a place of public assembly below grade, the code required that a sprinkler be installed.

The total expense of preparing a 3,200-square-foot area for occupancy by the center in the late 1980s was very close to $80,000, or about $25 per square foot.

Exhibits are the heart of the center, the reason for its existence, and where the personality of the institution is revealed. For better or for worse, I have taken advantage of my position as director to exercise absolute and sole authority concerning what is offered to our visitors. I have designed and made almost all of our exhibits, drawing not only from my decades of experience in teaching physics but also from many visits to other science centers, large and small. Like many others, I acquired early the Exploratorium's marvelous three-volume set of Cookbooks and pored over them with admiration and appreciation. Also, I resolved to avoid copying them for our own center, instead taking inspiration from their quality and attention to promoting understanding while concentrating on developing our center's own character.

An Association of Science-Technology Centers (ASTC) publication on exhibit design that I find especially useful is Jeff Kennedy's *User Friendly: Hands-on Exhibits That Work.* Along with practical information on such topics as exhibit dimensions for comfortable use by both adults and children, he also shows examples

Figure 9.1. General view of the Science Discovery Center of Oneonta, N.Y.

of exhibits that are flawed, explains why, and suggests corrections.

One pitfall exhibit designer-builders should avoid is being overly clever. As an exhibit design evolves in the mind and hands of the designer-builder there is the temptation to improve it by adding just a little more complexity here and a bit more there. Its functioning is still perfectly clear to the designer, but somewhere in the development a line was crossed, so a naive user coming up to the exhibit for the first time is baffled and quickly abandons it. Achieving simplicity and clarity often requires great effort, ingenuity, and self-discipline, but the result is well worth it.

A number of our exhibit plans have been included in ASTC's Cheapbooks 1 and 2. Brief descriptions of most of our exhibits may be found on our website, located under "academics" in our college's website: oneonta.edu. In our 3,000-square-foot center, we have over seventy exhibits, mostly tabletop size, and find that a school group of forty children can be completely occupied for a full hour or more. Family groups with children of various ages often have trouble getting the children to leave after two or three hours.

The total expense for our exhibits, including tables, was under $5,000 except for one item, a $5,000 video microscope setup, which was provided through a grant from a local hospital. I designed and made most of the exhibits in the physics department shop, just down the basement corridor from the center. Most of the tables were made there, too, with occasional selected volunteer carpenters and machinists assisting me. The tables were designed with an open front of adequate height to permit easy wheelchair access.

For the very small children, we have provided a corner with selected toys. The basis for making the selection was that the toy should promote some aspect of development of the child. Many of these toys involve assembling structures from numerous small parts to promote spatial visualization. Other selected toys involve fitting shaped objects into the properly shaped holes in a box, assembling arrays of intermeshing gears, balancing rubber stoppers of various sizes (larger than choking hazard) on a simple balance, and some wheeled small (push-around) toys.

Staffing the center is a different sort of challenge, but an interesting one since it keeps me in contact with many old friends and associates. For about ten years, we operated with a completely volunteer staff. It is just in the past few years that we have taken on a part-time paid staff person to supplement the volunteer staff. Even so, the majority of our staffing hours

are still filled by volunteers. The most likely volunteers are retired elementary and middle school teachers and retired college faculty, especially those who were in the college's various science and education departments. The retired college faculty have the advantage of possessing campus parking permits. These people enjoy working with children, are very good at it, and are glad for the opportunity the center offers to interact with children without the responsibilities that formal teaching entails. Each volunteer receives a letter of appointment from the college president, which, as mentioned above, gives the volunteer an official status in the college, and the liability protection of the Public Officers Law.

We have settled on Thursday through Saturday afternoon hours because of a requirement by the American Automobile Association (AAA) that to maintain a listing in their Tour Book, an attraction must be open to the public for a minimum of 12 hours per week. Since our AAA Tour Book listing is probably our most effective publicity, bringing in a rather steady stream of families and touring groups year-round, we make sure we meet their requirements. Also, the Tour Book listing is free, but the center must pass an occasional AAA on-site inspection to retain its listing.

Financially, the center operates without serious stress. The college provides the space and utilities. The major expenses the center must pay for include the usual office expenses, the equipment and supplies for making and maintaining exhibits, membership in professional associations such as ASTC, advertising expenses, and the wages paid for staffing assistance. Our total annual operating expenses are typically about $5,000.

The center's income is mainly through donations and membership payments with some supplementation from interest earned by the center's endowment fund. The center does not charge for admission or group usage.

Visitor numbers are rather low, typically about 5,000 annually. A crude statistical outcome is that our operating expenses are about $1 per visitor, which is one-tenth or less of the operating expense per visitor in a typical science center. About half of our visitors are elementary school children and their accompanying adults on class visits. Well over one hundred class groups visit us each year.

A good number of college classes in a variety of departments avail themselves of the convenience of having hands-on science setups on campus. Physics classes frequently come downstairs for part of a class or laboratory period to work with particular setups. A psychology professor regularly brings a class in to work on aspects of perceptions. A music class in sound recording comes in during the first week to get basic concepts with sound waves. Classes in composition from the English Department have been brought to the center to have unique experiences to write about. But for me the most satisfying use of the center is by students in the Methods in Elementary Science course. They have an assignment to prepare and present a science activity to the center's young visitors, working one-on-one with children, guiding them through doing the activity, and then talking with them about the science involved. This has been a popular program, attracting greater than usual numbers of visitors. The college students making the activity presentations are also very positively impressed by how eager the children are to do the activities and to pick up the ideas behind what they are doing. Many spontaneously come to the conclusion that their science teaching will definitely include as much hands-on activity as possible. It gives me much hope for the next generation of teachers.

To summarize, a science center can be located on a college campus with reasonable success and considerable economy. It can be a valuable service to the college as well as to the schools in the region. It can also be a cultural resource for the community and an attraction for tourists, bringing them to the college campus for a pleasant and memorable experience with intellectual content as well as entertainment value. Retired college faculty and schoolteachers will often enjoy the opportunity to help staff the center. A science center and a college can be mutually helpful, and can accomplish a good thing together, which neither could do alone.

Visitors' reactions are often our best reward. A girl about six years old signed our guest register as she came in. An hour or so later when she was leaving, she deliberately signed it again. She explained to her mother that she had enough fun for two people.

CHAPTER 10

Science Projects: The Observatory Science Centre, Herstmonceux, and Inspire Discovery Centre, Norwich

STEPHEN PIZZEY

THE TWO SCIENCE DISCOVERY CENTERS DESCRIBED here have their origins in a traveling discovery center called the Discovery Dome, which took to the road in the United Kingdom in five interlocking tents in 1988. The tour spanned seven years and resulted in the establishment of our two science centers and influenced the setting up of others at sites visited by the Discovery Dome. Science Projects Ltd. was incorporated in 1986 as a company with charitable status to receive grants and set up the exhibit production workshop to develop the exhibits and operate the exhibition. It is the parent charity that now operates both the Inspire Discovery Centre in Norwich, U.K., and The Observatory in Herstmonceux, U.K.

SETTING UP THE MODEL—THE ADVENTURE BEGINS

The original exhibit development and operational model was adopted partly from my own experience in industrial research and development. I was involved in getting prototypes from the laboratory into the first steps of production. This has always seemed the best way of regarding an exhibit as a product of integrated design-development and production for use in a demanding battlefront environment.

Secondly, the Indian science center model of a central resource feeding local regional centers seemed eminently well founded. It enables local centers to adopt their own individual quirks and character while drawing on a shared resource. A well-equipped workshop can also produce revenue through manufacturing for outside customers, to generate a financial reserve and broaden its own expertise and stay current. This was a central plank in setting up Science Projects. The workshop had the potential of freeing the organization of dependency on grants or imposed restrictions, which has since proved true, by building up an uncommitted financial reserve. In other words Science Projects can

take considered risks and venture into uncharted territory without the bane of political restrictions and other outside agendas and mind-numbing paperwork, subject to approval of the board of course. I also like producing, creating something that did not exist before. Creating exhibits, particularly hands-on exhibits, serves this personal view.

In order to receive initial funding, Science Projects was set up as a not-for-profit corporation with an educational mission as its banner. I was advised that the best way to set this up was as a company with charitable status. In order to survive we would have to run it as a tight, financially stable business. Regarding the workshop, the products of our workshop furthered our educational mission of taking science to the public and, it was assumed, would produce a financial surplus within the charity for our own development work. We were advised against setting up an outside subsidiary as this would introduce another level of management and risk losing the sense of mission and spontaneity, which was central if the project was to endure. The country's best charity lawyers were employed on our behalf by one of the foundations that wanted to support the venture.

And so it began. The model was put in place. A grouping of activities was supported by a central design, development, and fabrication resource. This was also capable of generating an income for sustainable expansion for the future, The sense of mission has endured ever since and is the glue that binds together all our activities. This has been enhanced by the chance to work on new projects—mainly overseas these days.

Another point is the importance of being located where there is some action. London is good for this: noisy, expensive, and always on the move, and it provides the edge. Rents are so high the exhibits have to flow through the workshop fast. The contrast of this with the rural calm of the Sussex countryside around

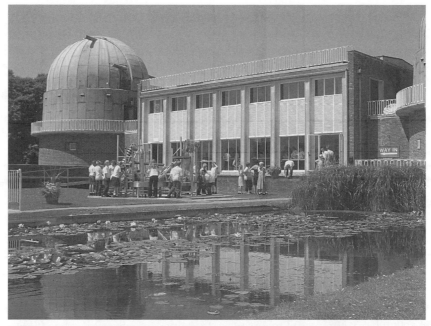

Figure 10.1.   The Observatory Science Centre, Herstmonceux, East Sussex, U.K.

our observatory with its historical connotations is all part of how we work.

SETTING UP THE SCIENCE CENTERS

The two centers we have succeeded in establishing benefit from the resources and commercial activities of the exhibit production workshop in London and share traveling exhibitions and expertise. The centers are self-funding in revenue terms by keeping the costs down through shared resources and management. Capital funding is mainly provided by the activities of the workshop. Adding more centers would be potentially cost effective, and it remains to be seen if new locations present themselves. This model of building up a consortium with a central resource seems to be effective so far, and it supports traveling exhibitions and circulation of school curriculum materials in addition to the two science centers.

THE OBSERVATORY SCIENCE
CENTRE, HERSTMONCEUX

This science center occupies the former telescope domes and buildings of the Royal Greenwich Observatory (RGO) in East Sussex in the south of England and lies in the grounds of a castle near the coast. The telescopes were moved there from Greenwich, London, in the 1950s to escape the enveloping lights of the city. The observatory closed in 1990. The venture to set up a science center and restore the telescopes to full working

order was described in one national newspaper as a labor of love unfolding in the Sussex countryside. This is perhaps the best tribute the center could have, as it recognizes the sense of mission.

The idyllic rural setting away from the city hardly seems an ideal location for a science centre in terms of catchment. But, the mix of hands-on exhibits and the presence of six large telescopes in their exotic domes set around a lily pond make it special and unique. This atmospheric setting contributes in no small part to the fond regard the visitors have for the center. A recent document from English Heritage described the buildings as "a conscious expression of modern science in traditional local materials, with the guise of eighteenth century garden temples never deflecting from their scientific purpose." The architect, Brian O'Rorke, made his reputation designing interiors for ships and flying boats, and the new observatory buildings, commissioned by the admiralty, is regarded as his leading work. By 1990, the site was no longer required by the RGO, and it was left to its fate. The RGO itself was closed down as an institution in 1998 after 323 years as the Royal Observatory.

The site was purchased by Queens University, Kingston, Ont., with funds from a benefactor, Alfred Bader, and a U.K. trust was established to operate it. A long lease on the observatory buildings was taken by Science Projects in 1993 to set up the science center. The tents and the exhibits from the Discovery Dome enabled the center to open at just a few days notice while work was carried out on the building to comply with various regulations. The center has since evolved as funds allow, and now it boasts an outdoor discovery park as well as the more familiar hands-on exhibits.

An essential element is an ambitious program of events and activities. These include science experience days for children, family events, and themed activities for schools. Viewing evenings for the public use the magnificent telescopes that have now been restored along with the associated buildings. An award of funds from the Heritage Lottery with matching funding from Science Projects's reserves made the restorations possible.

The center has several layers of activity to interest the visitor. First there is the visit to the site itself. Wandering around the domes and telescopes is an evocative experience and the grounds make a good picnic site, weather permitting. Then there are the interactive exhibits and the inevitable buzz that surrounds them. Visitors can take a guided tour of the telescopes to learn something of their history and their purpose and something of astronomy. The oldest telescope on the site was constructed in 1896 and in use until the 1970s. The telescopes on the site span across the time when photography became dominant in astronomy until the age of electronic data collection and imaging. The history of the observatory is recorded in a new permanent exhibition.

The Outdoor Discovery Park provides another experience entirely with much physical participation. The intention is to expand this to include environmental sculptures and trails. Then there are supervised workshops, the most popular being the construction of various types of bridges tough enough to walk over. For those stimulated to learn more there are evening classes and observing evenings and for those socially inclined there are events such as the ever popular Leonid Curry Night and Shooting Star BBQs. Staff members are encouraged to try out new activities with a view to enriching the visitor experience and bringing in extra revenue. The whole enterprise is mission led and cash conscious—it has to be.

EXHIBITS AND STAFF

The indoor exhibition space covers 12,000 square feet and is distributed throughout the various telescope domes and main building and a 2,000-square-foot geodesic dome. The exhibits are predominately hands-on and cover such topics as forces and gravity (twenty-one exhibits), time (eleven exhibits), the Earth (twelve exhibits), and light and optics (thirteen exhibits). The temporary exhibition gallery features one of our traveling Exploring Science exhibitions (sixteen exhibits). The permanent exhibition about the history of the site contains historic objects; some hands-on astronomy exhibits as well as graphics and text panels.

The outdoor exhibits occupy 20,000 square feet at present, although there are plans to extend farther. These exhibits constitute fifteen large structures such as the energy track and spiral, balance board, sound tubes, and the water circuit, which has six water-lifting and water-powered devices. Descriptions can be found on the website.

The Observatory has nine full-time (or equivalent) staff plus volunteers, and students are hired during the summer holidays. The attendance has been rising, and the center now attracts around 50,000 visits per year. At the time of writing, the observatory has just been awarded a grant from the national ReDiscover fund.

INSPIRE DISCOVERY CENTRE, NORWICH

Inspire is a small center that occupies St. Michaels Coslany, one of the finest medieval churches in the city of Norwich in the East of England. Alas, the church has a tower rather than a spire but that small detail has not prevented the name of Inspire being adopted. The center was the brainchild of Ian Simmons who until recently was its director. He had noted the success of the Discovery Dome when it visited Norwich and approached Science Projects with a scheme to set up a permanent center. It was agreed that Ian should be employed by Science Projects and that the center become one of the company's activities along with the Observatory and traveling exhibitions.

The center was established in 1995 and has since acquired a reputation and expertise in developing science shows that are performed both at the center and as an outreach activity. Although small in area compared with most science centers, and especially in terms of the prestigious new lottery-funded centers in the U.K., Inspire is a living example of how a small center can serve a local community and respond to local situations.

A recent project entitled the Study Support Scheme is a government-funded pilot where children from local schools in disadvantaged parts of the city use the center as a focus for practical activities and projects. These include developing a new exhibit or designing clothes from low-cost unusual materials. The project is spread over several weeks and includes a visit by the children to Science Projects's workshop in London. There they can use the production machines under close supervision

Figure 10.2. The Science Squad in action at INSPIRE, which specializes in shows.

to demonstrate how real exhibits are made and get some experience of the work environment. They also get advice on how to realize their various projects. The group took the opportunity of visiting the Science Museum and seeing some of the sights in London. This year, Paul Orselli visited Norwich to take part in the project and work with the children at their school—he was a real hit, "decidedly cool," they say.

Inspire has certainly become part of the community and provides a much appreciated educational resource and visitor attraction. This does not free it from problems, however: the church, though atmospheric, is limited in space, and the lease is for a limited period. Increases in rent have put a financial strain on the operation. Further grants are required for the outfitting of the center.

EXHIBITS, SHOWS, AND STAFF

The public exhibition area occupies the main body of the church and includes a soft play area for under five-year-olds and a seated area for children to explore discovery boxes. The total area is 3,500 square feet. There are thirty-five exhibits plus sixteen from whichever traveling exhibition is on display. The shows and demonstrations take place on a balcony at one end of the church. These shows are a speciality of Inspire and are linked to the national curriculum for science. Some of the titles are Animal Magic, Senses and Sensibility, The Really Gross Show (featuring George), The H_2O Show, The Air Show, E = Amazing Stuff, Play With Your Food, and Science Tricks. In all there are sixteen shows including a stardom planetarium. The center attracts around 30,000 visitors a year and is operated by four full-time, or equivalent, staff plus volunteers.

SHARED RESOURCES—THE LONDON WORKSHOP

Exhibits for the centers were developed at the central workshop in London. In addition to these, the workshop has designed and built traveling exhibitions that serve the general public. The exhibitions for the public are placed in our centers for periods of up to nine months at a time and at other times bring in revenue through loaning to other centers and museums both in the U.K. and overseas. The traveling exhibitions are under the banner of Exploring Science and titled Risk, Medieval Machines, Senses, Feel the Force, Good Vibrations, and Puzzles and Illusions. Other topics are in the planning.

A curriculum-based exhibition service for schools is operated from the workshop to serve the London area and is also available to travel into local schools around the centers as part of an outreach program. The program is entitled ScienceWorks and the individual exhibition topics cover Force, Sound, Light, Life, Materials, Electricity, and Energy. Each exhibition constitutes twelve table-top exhibits and is designed in a modular form for ease of transport on wheeled pallets. There are two sets of each topic. A school will rent the exhibition for two days at a time. These traveling exhibitions attract in excess of 150,000 visitors per year.

EXHIBIT FABRICATION

New exhibitions and individual exhibits are designed and built at the workshop in London. Workshop staff members carry out any major repairs while local staff members at the centers carry out the day-to-day maintenance. Special events are devised by the local staff and graphics are either produced locally or through the London workshop, whichever is most cost effective. In addition to this activity, the workshop designs and builds exhibits and exhibitions for science centers and children's museums worldwide. This activity generates capital for the two centers and for the development of new traveling exhibitions.

MANAGEMENT AND FINANCE

Day-to-day management and accounting are carried out locally for each center with payroll, tax, records, and analysis carried out for both centers at the London base. The workshop employs fifteen full-time staff including design and technical staff and senior management for the whole organization.

CONNECTIONS

We are making good links with smaller science centers that have grown up much the way we have, particularly

Teknikens Hus in Sweden, which has an exhibit development workshop like ours. They have also organized tours of our traveling exhibitions for the smaller centers in Norway and Sweden.

George Moynihan, former CEO of the Pacific Science Center in Seattle, came on a surprise visit to see the Discovery Dome in action when it was in Glasgow, Scotland. Following that visit, we got together to discuss working on a version for the United States and joined his team in Seattle for three months, making many new friends there. What a luxury that was: a decent budget, the ground already broken with our pioneering venture, big trucks, interesting locations, and good resources. Out of this Science Carnival was born and was apparently a great success. (For more on Science Carnival, see Lynn D. Dierking, Foreword; Dennis Schatz, chapter 27; Adela "Laddie" Skipton Ellwell, chapter 4, Karen Johnson, chapter 6; and Cassandra L. Henry, chapter 8.)

Our international travels did not stop there, and Science Projects went to help set up exhibits and to train staff in India. We made a sized-down version of the Discovery Dome exhibition called Exploring Science, which spent three years traveling around the Middle East and North Africa. It finally came to rest in Addis Ababa, Ethiopia. With the support of the British Council Office in Ethiopia, the exhibits became the inaugural exhibition that inspired the setting up of a science center there. It was really good to see the way teachers used the exhibition. The schools were poor and had few books, but teachers were resourceful and made versions of the exhibits. During our travels and staff-training workshops in some of these locations, we could always track down a workshop and find people who made things. This surely is the power of simple hands-on exhibits as a medium—it is accessible even to under-resourced communities.

WEBSITES FOR FURTHER INFORMATION ON SCIENCE CENTERS IN THE U.K.

Science Projects: www.science-projects.org
Observatory Science Centre: www.the-observatory.org
Inspire Discovery Centre: www.inspirediscoverycentre.com
Ecsite-uk, UK Network of Science Centers and Museums: www.ecsite-uk.net
BIG, British Interactive Group: www.big.uk.com

CHAPTER 11

Experimentarium: A Fairy Tale from Hans Christian Andersen's Denmark

ASGER HØEG

1. A VISION: EVERY CITY WITH A POPULATION OF MORE THAN 100,000 SHOULD HAVE ITS OWN SCIENCE CENTER

We have seen it in more than 500 towns around the world. Some dynamic people get together, and they start a process that ends with the opening of a large or small science center. There is always too little money to create the center of your dreams, but the pioneer spirit makes the money go a longer way than you might expect. The capital money comes from government, the city, the university, private companies, and foundations. They all want to support the center because they are worried about the lack of interest in science among the younger generation.

From day one, problems arise over how to develop and run a science center because a science center is certainly not a science museum. There is no permanent exhibition. On the contrary, the exhibitions must change over time because a science center survives on the revisits of the people living in its town.

But a town gains many advantages from a science center:

- The center stimulates in visitors a broad interest in science and technology.
- The center inspires teachers to improve their teaching by a wider use of experiments.
- The center fills out the role as a tourist attraction, which brings money to the city.
- The center creates jobs—all with challenging content.
- The center inspires museums and other activity centers to improve their visitor services.
- The center represents an economy with a turnover at the level of US\$1 to \$5 million, which benefits the city.

So there are a lot of good reasons to establish a science center. Actually, there seems to be no disadvantages to a city having a science center. Nevertheless, we are still confronted with the question: Why should Copenhagen have a science center? Is it really necessary? On the other hand, every small town takes it for granted that it must have an art museum. Art museums attract only a few visitors, but you find them in every small town in the Western world!

The vision is simple. Every town with more than 100,000 people will, eventually, over the next twenty years, acquire a science center. The cities and the world will gain from these 10,000 science centers. The synergy among the centers will be enormous, and science and technology will gain new territory in the minds of the younger generation, and the world will prosper.

2. THE DANISH FAIRY TALE

The Danish fairy tale began in 1984 when the chairman of the Egmont Foundation, Esben Dragsted, visited the Exploratorium and the Ontario Science Centre. Among its activities, the Egmont Foundation supports and initiates education projects. Inspired by his visits in North America, Esben Dragsted decided to examine the possibility of establishing a science center in Denmark. The impetus for this initiative was the lack of interest on the part of the younger generation in science and technology.

In early 1985, the Egmont Foundation commissioned a feasibility study for the establishment of a science center in Denmark. A fourteen-member project group was established, and Jannik Johansen was asked to be chairman. Tage Høyer Hansen and Nils Hornstrup (director of development at the Experimentarium) were hired to carry out the feasibility study, which was completed in January 1986. Their final report, called "A Danish Science Centre?" was submitted to the Egmont Foundation.

In connection with the feasibility study, the author visited several science centers throughout the world.

Figure 11.1. Climbing wall at the Experimentarium in Copenhagen, Denmark. What does it take to climb well and safely? Photograph by Per Arnesen.

The reasons offered in favor of establishing a Danish science center were as follows:

- The science center creates interest in science and technology. Especially among the younger generation. The science center arouses curiosity.
- The science center inspires the formal education system.
- The science center contributes to the scientific culture in the society.
- The science center creates a platform where the community, industry, commerce, and research can meet.

A budget was prepared and several funding options proposed for both the establishment and the ongoing operation of a Danish science center.

The report estimated that the capital needed to establish a science center with an exhibition area of 53,800 square feet would be about US$15 million and that such a center would attract 350,000 visitors per year. Also, the report expected the annual deficit to be US$.8 to $1.6 million, which should be covered by donations and grants from private and public sources. To introduce the concept of a science center to the public, the foundations, and the Danish government, the report suggested that a pilot exhibition be carried out.

To finance the pilot exhibition, the three Danish foundations—Egmont Foundation, Augustinus Foundation, and Thomas B. Thriges Foundation—created a new nonprofit foundation with the long name Center for the Promotion of Science and Technology Foundation, which was established on November 6, 1986,

with Jannik Johansen appointed chairman. After some months, the foundation adopted the working name Experimentarium.

An initial capital infusion of US$600,000 was spent in its entirety on developing and producing the pilot exhibition, "Man—Here Is Your Body!"

The exhibition was inaugurated by the minister of education, Bertel Haarder, on April 15, 1988. Two weeks before, I had been appointed as the new executive director of the Experimentarium. My first assignment was to design the new permanent science center, raise the funds needed for the project, and manage the marketing and the public relations connected with the three-month pilot exhibition period.

The problems began on day one. The criterion of success for the pilot exhibition was defined as 70,000 visitors in ninety days, from April 15 to July 14, 1988. On the two first days, the number of visitors averaged 280. For the staff, 280 visitors in an area of 4,300 square feet seemed like a lot. But if you only attract 280 people per day, in ninety days, your visitorship will only be 25,000—not the 70,000 as promised in the press releases.

The new director called a crisis meeting first thing Monday, April 18, 1988—the first time ever for the new director to meet the staff, a total of eight people. The question asked was, How do we raise the number of visitors and do so very quickly? The money for marketing was long since spent on newspaper advertising. From then on marketing should be executed by using no money. A well-known situation for many science centers. The answer was—television. The Experimentarium contacted the national television, DR TV (at that time, the monopolist of Danish television). A journalist came Tuesday morning and filmed just over two minutes from the exhibition to be shown during evening prime time. Approximately 40 percent of the Danish population would see the film from the exhibition. You could not wish for better marketing!

On Tuesday afternoon the prime minister took the chair in the Parliament and called for general elections. Of course the evening news told nothing about the Experimentarium's exhibition—everything was about the election.

I called DR TV each day and proposed that they show the film of the exhibition. The concept of a science center exhibition was a novel idea in Denmark and it would still be interesting news to the Danish people. One week later, in the morning, DR TV said they would show the film that evening. But, in the afternoon, a train accident happened! Three people were killed. No film from Man—Here Is Your Body! on the evening news.

In grim humor, I and staff all decided that God had no plans for a science center in Denmark.

The number of visitors to the pilot exhibition rose steadily. Word of mouth spread the good news—Man—Here Is Your Body! was very entertaining. And at the same time you learned a lot about your own body that you did not know. The number of visitors rose to some 500 to 600 per day. But when all was said and done, one would need a miracle to reach 70,000 visitors. The miracle came on May 25, 1988, more than a month after the opening of the pilot exhibition. The film of the exhibition was finally shown on the evening news. That evening the exhibition was open until 10 p.m. A few seconds after the news item, many people phoned in to the renovated garage where the exhibition was hosted. The next day, the number of visitors doubled and on July 14, 1988, no less than 1,994 people visited the exhibition. The number of visitors in ninety days: 71,900! Finally, success was reached!

In the same days of May when the visitor numbers doubled, the CEO of Carlsberg, Poul J. Svanholm, contacted the chairman of the Experimentarium's board, Jannik Johansen, and asked if the Experimentarium would be interested in creating the new science center in the Tuborg Breweries' old bottling hall, Tappehal Nord. This 193,700-square-foot building was seventy years old but still very solid. The building appeared to have been cut out to host a science center as the bottling hall could be transformed into a large exhibition hall. The canteen could be the science center cafeteria. The workshops could be transformed into the science center workshop. And so on.

Based on the donation from Carlsberg, the Egmont Foundation decided to donate US$3.8 million, not as capital money, but as money to fund an endowment that could be used as a buffer to cover deficits during the first years of operation when an annual grant from the government was not yet secured. This huge donation from the Egmont Foundation was a guarantee that the new science center would survive at least the first five to seven years.

A fund-raising meeting was held on September 28, 1988, with approximately 100 foundations, companies, and institutions present to learn about the science center project. The budget was not US$15 million, as the feasibility study suggested. But because of the donation of the building, the budget was reduced to US$8.5 million.

The fund-raising meeting was inspired by our visits, Nils Hornstrup and myself, in July 1988 to the Association of Science-Technology Centers (ASTC) Institute for New Science Centers, which was held in Ann Arbor, Mich., hosted by Cynthia Yao, executive director and founder of the Ann Arbor science center—Ann Arbor Hands-On Museum.

During the fund-raising meeting, no foundations or other potential contributors said a word. No questions. No comments. The Experimentarium's director was interviewed on the local radio, Copenhagen Radio. After thirty seconds of silence when Jannik Johansen asked for questions or comments, I was asked: "Director, do you have anything in the future to be the director of?" And I said, "Yes, I guess that almost all the foundations will contribute, with small or large donations." And so they did. And on January 26, 1989, the Experimentarium's board could declare the fund-raising successful.

3. CREATING A SCIENCE CENTER IN TWENTY MONTHS AND NINE DAYS

The staff that would create the Experimentarium was in place by May 1, 1989. So the science center was actually created in twenty months and nine days!

Today, the name Experimentarium is a brand. Not the strongest brand in Denmark, but the word is synonymous with quality experience. In those early days, the name Experimentarium was not known to anyone. Also, the concept of a science center was unknown to most people. When you asked for help or cooperation from someone, the telephone conversation normally started, "You are talking with Asger Høeg from the Experimentarium, the Danish science center." Response: "No, we are not interested in Scientology!"

Eventually, the reporters came to the bottling plant and saw the charming transformation of the building from an industrial building housing 600 working people to a science center, with the hope of housing 100,000 visitors.

The first exhibition was developed in two years. We had help from Daniel Goldwater, Stephen Pizzey, and Rogow and Bernstein; the overall design and content of the exhibition was decided during 1989. From 1990 until January 1991, 200 exhibits were designed and produced. Approximately 95 percent of the exhibitions were designed and produced in-house. The philosophy was that those who had produced the exhibitions should also be responsible for their maintenance.

Marketing for the new science center was easy. The press was very interested in the new type of interactive exhibition. Her Majesty the Queen of Denmark agreed to inaugurate the Experimentarium. It was a very important decision and a kind of knighthood for the Experimentarium.

On January 9, 1991, at 11 a.m., Her Majesty the Queen, Margrethe II, and her husband, the Prince of Denmark, Henrik, came through the main entrance. Eighty exhibition guides flanked the Queen and the Prince on their walk to the arena where the inauguration would take place. The Queen started a flame that burned for twelve seconds. The flame heated a pipe that made a clear sound and that was the official opening of the new Danish science center. After that, Her Majesty the Queen amused herself in the exhibition for more than an hour and when the Queen and the Prince left the Experimentarium, they were almost one hour behind schedule!

The Experimentarium had spent almost US$ 150,000 on different marketing activities. All that money was wasted because the science center was overwhelmed with visitors from day one. In 1991, 531,000 people visited the Experimentarium.

4. TEN THINGS TO REMEMBER WHEN YOU DEVELOP AND RUN A SCIENCE CENTER

The first and most important rule is that "Science center business is like all other kinds of business." When you establish a science center, everybody is inspired by the soul of pioneering. Salaries are no issue. Working late is no problem. But eventually, the staff wants to know what their rights are and what their responsibilities are—on paper. There is a cash flow in a science center. You have lots of issues with tax, VAT (value added tax), and related items. Summa summarum: You need an efficient human resource manager and a chief financial officer. After the first few month of honeymooning, you need to market your science center, as all other products need to be communicated to the public. Therefore, you also need a professional director of sales and marketing.

The second rule is that "Science centers must change all the time to give visitors a reason and chance for a revisit." It is well known that the running and operation of a science center is costly. But it is not well known that the presentation of new exhibitions all the time is a part of the maintenance. Such is the case! If you want to maintain a science center properly, you must present one or two new exhibitions every year. Having 360,000 visitors per year, the Experimentarium survives by revisits from its core group of visitors. In 2002, 70 percent of the visitors (426,300 people!) made a revisit to the Experimentarium.

The third rule is that *all* science center directors tend to overestimate the income and underestimate the costs. In the real world of businesses, a CEO is not accepted when he or she continuously makes budget errors for the forthcoming years. This is not true in the science center world. Since we live in a world where it is very difficult to secure balance in the accumulated net income, we all have a tendency to be too optimistic concerning the number of visitors and too optimistic concerning the level of expenditures. A science center director must be just as conservative and accurate in the budget planning as the CEO of any business.

The fourth rule is "The board of directors must be a mirror image of the users of the science center." The decision on who should be represented on the board is a vital question. The science center has a lot of users: the community, government, schools, universities, tourist organizations, companies in the area, and foundations. The board must reflect the many stakeholders of the science center. The board should take the responsibility and be the guarantor of the long-term survival of the science center.

The fifth rule is "Consider the visit as a chain of experiences for the customer and measure your service performance at all the points on the chain." When a visitor leaves the Experimentarium their mind is full of what has happened during the last two to five hours. But it is not always the exhibits that the mind may be full of. It could be the smell in the restrooms, the price of the food in the cafeteria, or the exhibits that did not work as planned. It could be the fact that there was no facility to change the baby's diaper, or the exhibition guide did not smile and could not answer a simple question. The fact is that you must have a fair score on all these services to make sure that a visitor is satisfied with the Experimentarium and will recommend it to families, colleagues, and friends. If your score is poor then the visitor will do the opposite and will recommend that friends, family, and colleagues *not* visit the Experimentarium. And, it takes ten good experiences to forget one bad. Every day, the Experimentarium asks twelve visitors eighteen questions that are designed to reflect the overall performance of the Experimentarium. This survey is the main tool used by the CEO to keep up service at the right level for the science center.

The sixth rule is "You need one kind of staff to develop a science center and another kind of staff to run a science center." When you develop a new science center, you need people who are creative, yet they are not always the best people at following budgets and time schedules. After the center opens, you need people to administer and perform the daily operations without problems. This is a new kind of staff. They might be less creative but more oriented toward budget constraints and time limits. When you turn from development to operation, you must reduce your staff in development and add new people in administration and operations.

The seventh rule is "The science center must define the schools as the primary target group. But when you do so, you also will have fulfilled the needs of your other target groups: families and tourists." The raison d'être for a science center is the schools' need for inspiration in informal learning and the need among the general population for lifelong learning. Experience tells us that if you adapt the message in the exhibition to a schoolchild, you will have ensured at the same time that the family seeking weekend entertainment and the tourist seeking a break from the beach holiday will be both satisfied with the level of communication in the exhibitions. Also, the school must be viewed as the number one target group to define the center as a serious communicator of science and technology and not just a Disney-style exhibition where the purpose is only fun.

The eighth rule is "The primary marketing person is a member of the staff." Half of all visitors are convinced on their own to revisit the Experimentarium. This means that any one of these visitors is ready to make a second visit because the first one was a success. And this can only be true if the exhibits were working, the guides were smiling, the toilets were clean, and everything else was working as expected. Twenty-five percent of all visitors are convinced by others to make a first visit. This means that 25 percent of all visitors are making the decision to visit the Experimentarium on the basis of recommendations from their families, colleagues, or friends. And this is only possible if those families, colleagues, or friends had successful first visits to the Experimentarium. Again, the overall performance of the Experimentarium is closely tied to recommendations from visitors. Recommendations in the press to visit the Experimentarium result in 12.5 percent of all visits. These people have been convinced by the press, on the sole basis that the reporter had a successful first visit to the Experimentarium. In the end, only a very few visitors are convinced to visit the Experimentarium on the basis of the center's marketing material, posters, leaflets, outdoor advertising such as billboards, or other similar endeavors. However, these things *are* still necessary to remind the people of Copenhagen that now is the time to revisit the Experimentarium.

The ninth rule is "Take the highest possible entrance fee. The fact that you dare to charge that much is a sign of quality." The sensitivity of visitors to entrance fees is much less than you expect. Therefore, you should start the entrance fee at a high point and then let the fee rise a little bit more than inflation. One day, you will hit the ceiling and you will not be able to raise the fees any higher. But the Experimentarium has not yet hit this ceiling and will continue to raise its entrance fees slightly more than inflation.

The tenth rule is "The cheapest way to create news in your science center is to rent a temporary exhibition by a colleague's sister science center in a neighboring country." The main exhibition must not be a permanent exhibition. It is very costly to change your main exhibition every few years. And the marketing effect of renewing your main exhibition will not be as strong as bringing a temporary exhibition to the science center. Therefore you must reserve an area in your exhibition galleries for a temporary exhibition or exhibitions. You can produce one temporary exhibition every two or three years. You rent out these exhibitions to your colleague's sister science centers and rent their temporary exhibitions on equal terms. In fact, you could make a barter trade where you swap exhibitions in the same period.

5. A FINAL WORD

The most beautiful job in the world is to be the executive director of a science center!

GETTING STARTED AND RUNNING A SCIENCE CENTER

CHAPTER 12

Planning the Building

PETER A. ANDERSON

SO, YOU ARE GOING TO BUILD A NEW BUILDING OR adapt an existing one as a science center? You are braver than most people! But you are following in the footsteps of many excellent and able leaders who have done what you are about to do. The first rule is to visit as many science centers and related institutions as you can, talk to the people operating them, and, if possible, the people who presided over the development. Here is a compendium of thoughts, spaces, and considerations to help you in your deliberations.

BASIC FACTORS

The building for a science center must be in the right place and it must suit the intended operations and exhibitions. To get the site and building right, it is essential first to answer some key questions.

What kinds of exhibitions and activities are planned?
What is the potential attendance, who are they, and how will they get there?
How long will an average visit last?
How large is the realistic funding target?
Is future on-site expansion likely?

Suiting the Exhibitions and Activities

Will there be large artifacts—such as an automobile—that require a large access door to get them in the building? Heavy things require the proper flooring support. Delicate items or live things need proper climate control. A discovery area for young children and other program spaces, such as classrooms or labs, need a lot of space. Will public and private events be important parts of the revenue stream, so that attractive spaces and provision for catering are needed? Will outdoor space be needed for outdoor activities?

Accommodating the Attendance

The potential attendance is influenced by the size of the center. For small science centers, attendance is also strongly influenced by the size of the regional population, by the center's neighbors, and especially by how many school classes can afford to bring students there. So, it is also necessary to fit the center to the expected attendance. Who will come, in socioeconomic terms, is governed by the design of the building, the exhibitions and programs, and by marketing. But the center's site can also be a very important factor, and parking is vital.

The Length of an Average Visit

If the center will be very small, with an average visit length of about an hour, then it needs suitable neighbors such as a shopping mall or a popular civic park. It cannot stand alone as a single destination. Ten thousand square feet of exhibition space typically will satisfy an hour's visit, and, of course, live programming can lengthen the stay considerably. A two-and-one-half hour average stay can support a stand-alone site, other things being equal. Somewhere in between one hour and two-and-one-half hours is the turnover point— and there are some in-between centers that manage with a modicum of neighbor support.

The Size of the Funding Target, and Future Expansion

Obviously the coat must fit the cloth. The location, style of building, whether recycled or not, and size are determined by the available funding. It is vital here, however, to include operating costs and future expansion in the equation. Almost all small science centers grow, and it may be unwise to accept a small site. Some small centers have moved to other sites as they grew— though this is not an ideal situation.

Visitors Come in Clumps

Most days see light visitation; most visitors come at crowded times. This is a crucial factor in all science center planning and operations. It is imperative to plan for peaks and valleys, not for averages. The crowded days must go very well, for this is when most of the visitors come. Typically, the top fifteen days of the year see 10 percent of the annual attendance. And, most people come in the early afternoon, so that about half the day's attendance is in-house at about 1:30 p.m. The building must be able to accommodate this pattern, and the operating plans must handle it well. It is just as important to economize at the quiet times.

The building helps at the peaks by having adequate space, not only in the exhibition halls, but in all the public spaces. When visitors in open space (not seated or queuing) have only 25 square feet of floor space each, they start leaving,[1] and they later tell their friends that the center is uncomfortable and too crowded. Taking into account typical peaking patterns, the maximum accommodation should be about twenty visitors per square foot of exhibition space per year. Thus, 10,000 square feet of exhibition space will not support more than 200,000 visitors annually[2]—and that may be optimistic. At best, this is operating in the discomfort zone, so planners really should target double the accommodation—20,000 square feet of exhibition space for 200,000 visitors yearly. Outdoor space helps too, but only partly, because attendance tends to peak on rainy days—or swelteringly hot ones, when air conditioning is critical.

Operational programming is also key to handling peak days. Demonstrations, theater programs, outside programs, and sit-down, make it–take it activities all help to bring other spaces into use and to concentrate people in acceptable ways.

A SHORT STORY

We opened the Ontario Science Centre in 1969, but it was not until a few years after that that we first plotted a year's day-by-day attendance on a graph. Seeing it that way was a real shock. High days were up to ten times heavier than low days. Ten percent of the annual attendance came in the top fifteen days, and the overwhelming bulk of the visitors came on crowded to overcrowded days. But, most of the full-time, senior staff saw the high days only during the summer. For the greater part of the year, we saw the relatively light weekday attendance, and that conditioned our thinking.

It wasn't that we hadn't noticed that days were heavy and light. We just hadn't built a clear model of the process and appropriate responses, so our responses had been ad hoc. This new insight explained why the food service attracted the most complaints (no normal food service can handle such a day-to-day variation). It

told us clearly that we were in the peak attendance business and that we should structure our activities to serve those peaks really well and learn to economize in the valleys. It made us see that more live staff on the floor is the best antidote to crowding—they facilitate; they give live performances and demonstrations; and they concentrate people in acceptable ways, such as watching demonstrations or theater programs. In addition, they help find lost children, keep exhibits working, and talk pleasantly and understandingly with stressed visitors.

THE SITE

The site can make or break the project. Key questions to ask are:

> Is the site large enough for the program? Will it accommodate expansion—or will moving to another site for expansion be acceptable?
>
> Is it easy to get there? Is there access to a convenient parking space for every 1,000 annual visitors?
>
> How will people get to know that the center is there? How visible will it be?
>
> Is it a place to where visitors will be happy to go?
>
> It should be a pleasant place for both middle-income visitors (on whom the center will depend for its revenues) and lower-income and minority visitors (toward whom much of the center's efforts will be directed through grant-funded efforts).

Site Development Costs and Planning Restrictions

It is essential to ensure that there are no unacceptable planning restrictions on the site, such as large setbacks or height restrictions, and that it won't have high site-development costs. The latter can arise from ground pollution, being situated on a flood plain, or lack of adequate nearby services.

THE BUILDING

Size of the Building

The building should be as large as the capital and operating budgets reasonably allow, and there should be planning for early expansion. Building space is like money in the bank—like an endowment. Size allows for a rich offering that builds attendance; it makes visits less crowded and more pleasant. Also, size can provide for more attractive spaces that bring in social or business functions and their attendant revenues.

The exhibition space should be at least 1 square foot per ten anticipated yearly visitors. Note, however, that one of the models used to predict attendance suggests that the exhibition area is the main determinant of attendance, and that therefore it should be as large as possible to attract more attendance.

Style and Cost of the Building

Signature buildings are great. They find their way easily into city tourist flyers; they quickly become destinations—it's free marketing. When problems with signature buildings arise, they typically are cost overruns and loss of functionality. In fact, much of a science center—the exhibition spaces—can be essentially warehouse structure. Some architects have used simple structures and added striking features that give the building more profile. Science centers do not have to be expensive structures and should be toward the lower end of the scale of cost per square foot in their regions.

Using Recycled Buildings

Recycled buildings are just fine, and the same general considerations hold for them as for new buildings. Sometimes an older building—like a fire hall—brings instant site recognition with it, and its continued use preserves some of the tangible heritage of its town.

Parts of the Building

APPROACH AND ENTRANCE The building should be attractive and pleasant to approach and enter.

RECEPTION AREA This area accommodates ticketing, orientation, and gift store access. It should be attractive and, if well designed, it can be a prime rental space. It should be able to accommodate peak in-and-out traffic of 300 visitors per hour (for 100,000 annually). In northern climates, a coat check is needed—or at least lockers—especially for school groups. Allow about 250 square feet for 100,000 annual visitors.

MAIN EXHIBITION SPACES

Lighting: A mixture is best. Natural light is advantageous for growing plants and seeing natural colors, and it has a pleasant feel. Controlled lighting is essential for many exhibits, especially exhibits of light phenomena and other science exhibits. Direct sunlight is almost always completely unacceptable, because it is so much more intense than almost any artificial light. For example, bright light on a computer screen washes out the image. If the back of the computer screen is to the light, then the reflection of the brightly lit visitor obscures the image.

Height: Fourteen feet clearance is ideal for exhibits, but 12 feet and clearances as low as 10 feet over short ranges are acceptable—for example, under mezzanines. Great height (say, 20 to 25 feet) over small areas permits some dramatic items.

Area: Overall exhibition areas (including temporary ones): For 100,000 annual visitors allow 10,000 square feet, as noted above.

Services: For electricity, allow three watts per square foot minimum. Provide water, drains, and gas as needed. For information technology and audio visual (A/V), run category five (or better) cables to all exhibits (80 to 100 cables for 10,000 square feet). An under-floor cable grid is useful in temporary exhibition areas, but it is probably not good value elsewhere. One school of thought advocates highly changeable exhibition areas with under-floor crawl spaces for quick installation. Before making this decision, it is important to ask whether the center is likely to have the personnel and financial resources to sustain frequent changing of the exhibitions. Most don't.

Floor loadings: 100 pounds per square foot is usually enough.

Density: Allow an average floor area of 100 to 120 square feet per exhibit unit.

TEMPORARY EXHIBITION SPACE Most marketing directors consider temporary exhibition space to be very important—at least 2,000 square feet for 100,000 annual visitors. Temporary exhibitions vary in size and many will fit into 2,000 square feet. But, many of the best now take up 4,000 square feet or more. If the reception and other areas are near, the exhibition can spill out somewhat to accommodate the larger ones. Forklift access from the loading bay is highly desirable.

EXHIBITION SUPPORTS

Workshop: For maintaining 10,000 square feet of exhibition, allow 2,000 square feet for workshops and parts and material storage. For fabrication, an additional 2,000 square feet. Provide 3-phase electric supply, water, gas, drains, and extra ventilation. The ceiling height should be 12 feet.

Storage: For 10,000 square feet of exhibition space, 3,000 square feet of storage space is an absolute minimum. Storage is usually seriously underprovided. Outreach activities need additional storage. Almost every science center activity needs storage. For instance, the crates for a temporary exhibition occupy about one third of the area of the exhibition. (That is, a 2,000-square-foot exhibition requires 700 square feet of crate storage.) As building budgets threaten to overrun, one of the first things to be cut is storage—but the center pays dearly for this decision after opening.

Loading bay: Provide a minimum of one loading platform accessible to eighteen-wheelers. Also provide forklift access to the workshop and temporary exhibition area, at least. The loading area should have space for waste storage—including chilled storage for food service waste in hot climates.

EDUCATION PROGRAMS Typically, out of an annual attendance of 100,000, 15,000 to 20,000 will be students in school groups. Again typically, 80 percent of these will be elementary or middle school students, the rest will be high school students. Class sizes vary, but forty is not uncommon. Demonstration spaces in the exhibition areas can accommodate many school programs, but laboratory programs require laboratories, which should accommodate classes of forty; at a minimum, half classes of twenty. Allow 1,500 square feet for a class laboratory with associated storage.

ASSEMBLY SPACE A flat-floored, general-purpose assembly space can be a great asset, not only for lectures and demonstrations but also for banquets and other revenue-generating functions. It should have a raised stage with chair (and maybe table) storage underneath. Projection and stage lighting equipment is an asset, as is a catering pantry nearby.

Revenue Generators

Before deciding on the specific facilities, it is worth considering carefully the ways in which the building can support revenue-generating activities. Food and gift shop sales are common even in the smallest science centers—and they enhance the quality of a visit. Lectures, movies (not great fund-raisers unless you are talking about large-screen theaters), science shows, dinner-dances, banquets, auctions, conferences, fund-raisers, product introductions, skills classes, camp-ins, scout programs—all these are common revenue-generating activities in science centers, and all have needs from the building. Outreach activities need office, storage, and parking spaces, whether they yield net revenue or not.

FOOD SERVICE Food service is a must for any science center. Some small centers rely on a fast food service in a nearby plaza, but there is nothing like having it inside, where the children need it mid-visit. Well-run, often using volunteer help, food services yield net income as well as being important to the visit itself. Simple service—even coffee, pop, sandwiches, and doughnuts from a well-equipped cart—is better than no service. Where there is space for a simple cafeteria, provide about thirty-five seats for every 100,000 annual visitors. Even then, it is wise to provide a food cart for the peaks, to shorten the cafeteria line-ups. The kitchen and other behind-the-scenes areas take up about one-third of the total cafeteria area. The real profit in food service is in catering, but a small center is unlikely to be able to

carry a catering kitchen. It is important, though, to provide facilities for catering if the center has good rental spaces—ideally a catering pantry, but at least a good supply of power. If possible, add a sink where the caterers can set up their curtains and bring their equipment.

GIFT SHOP Like the food service, the gift shop is an important adjunct to the visit. In a small center, it will make a profit only if staffed largely with volunteers, or placed and designed so that the admissions ticketing person can take sales in quieter times. Depending on the stock and layout, there may be extra eyes needed in the shop when school groups are visiting. Allow about 350 square feet for 100,000 annual visitors, and equal space behind the scenes for office and storage.

THEATERS Allow 15 square feet per person for a theater. The size varies with the proposed program. If in doubt, provide two seats for every 5,000 visitors expected per year.

A planetarium can be a major asset. Planetariums are coming back again as powerful features rivaling large-screen theaters. The new electronic systems can show anything from a star show to a virtual reality theater to a full large-screen theater movie, and the projectors are getting better and brighter all the time. But keep in mind the ongoing operating costs that will be required.

RENTAL SPACES These can be the entrance foyer, meeting rooms, assembly spaces, outdoor areas, or a theater. A well-located and attractive science center is a community resource, and some imaginative marketing can turn this into income.

Staff Spaces

For a science center with 20,000 square feet of exhibit space and 100,000 visitors per year, allow accommodation for up to 30 FTEs (full time equivalents). One-half to two-thirds of these will be full-time staff, and the rest, made up by part-time staff. This may be on the generous side, but peak times will surely fill it up. About five staff members will be based in the workshop. For the full timers, allow on average 200 square feet each—gross space, including passages, copy room, and storage. Designers require 400 square feet each. For the part timers, provide a staff room with lockers and a pantry—perhaps 200 square feet. This space may be extended to accommodate volunteers, too.

MEETING ROOMS Provide a meeting room for every twenty-five full-time staff members. If these are suitably situated and equipped, they can also serve as revenue generators.

Building Services

Building services include washrooms, heating, ventilating and air conditioning, together with other mechanical, electrical, water supply, and drainage facilities, janitor closets, and so on. There must be generous provision of washrooms, not just code requirements, and women require substantially larger numbers of facilities than do men. Consider providing escalators for height changes of more than 12 feet because it is tiring for families with small children to climb long staircases. For more than one story, a freight elevator is needed. It and its door should be 8 feet wide and 8 feet high. The depth should be as much as possible, and certainly not less than 12 feet.

Information technology and A/V: A server room should be at least 200 square feet with air conditioning and UPS (Uninterruptible Power Supply) and trays to run category five (or better) cables to all exhibits, office telephones, cash registers, and so on.

Volunteers' office/lounge: Provide a lounge but don't separate volunteers from the staff too much.

Members' lounge: The lounge also can be used for guests.

Accessibility

The whole vital issue of universal access has been greatly developed since 1990 and is too large to tackle here. Excellent sources of information are available in the United States from the Association of Science-Technology Centers and the American Association of Museums. Most other nations have comparable resources.

NOTES

1. Fruin, John. 1987. *Pedestrians*. Mobile, Ala.: Elevator World.
2. These numbers, and others like them in the text, are amalgams of data from many sources, discussions with colleagues, and the writer's own experiences. They are not infallible and should be taken as approximations. Detailing the derivations of them would make this chapter excessively long.

CHAPTER 13

Sci-Port Discovery Center: How We Operate and Position Our Science Center for Ongoing Success

Andrée Peek

OUR IDENTITY

Knowing who we are and what we want our science center to provide to the community is fundamental to our long-term success. A strong sense of identity enables us to position the science center well in the hearts and minds of the audiences whom we seek to serve, whether starting a new center or evolving an existing one. The public's ongoing participation, through attendance, donations, and rallying of government support, and our effective use of the resources enable continued progress with our mission and public service. The ultimate measure of our success is how well we perpetuate this cycle of public participation and resource reinvestment into our mission.

As background, when I joined Sci-Port Discovery Center as president in 1997, construction had already begun on a 67,000-square-foot new facility on the Shreveport downtown riverfront. Sci-Port Discovery Center had been successfully operating a 5,000-square-foot facility nearby for three years, serving 40,000 school students and families annually. Although this attendance was good for our community's population of 200,000, it didn't represent the diversity of the residents nor capture many tourists. Interestingly, despite this attendance success, there was little name recognition among the local public. It was common to hear suppliers, business contacts, and even news media call us Ski Port or Sky Port. The name Sci-Port was distinctive but clearly not intuitive, nor did it convey "science center." Those familiar with Sci-Port referred to us as the children's museum downtown.

We answered the "Who are we?" question during a board workshop designed to get everyone on the same page about the expansion project and the work ahead. The board reaffirmed our mission: To provide an educational and entertaining environment for people of all ages to explore and actively participate in the world of science and technology. Our mission's

key messages, discussed extensively at subsequent staff meetings, became the driving force defining our focus for developing and operating the new facility. Staff to this day is celebrated for learning the mission and having fun rapping it in meetings to handclaps and table beats.

MAKING OUR ACTIONS SPEAK LOUDER THAN WORDS

The mission's key messages that we wanted the public to embrace were that Sci-Port is a fun, educational, ever-changing environment rather than just a beautiful new building; the experiences appeal to all age groups; the focus is an exploration of science and technology; and the method is active participation by all visitors. We believe it is important to our long-term success for the public to voice this description of Sci-Port from their own experiences rather than having us tell them what our center is.

To put these goals into action, the programs and visitor experiences that we develop are self-tested to identify where the science is and where the fun is before implementation. To achieve active visitor participation, we provide make it–take it activities and workshops free with general admission. The activities change daily and periodically include featured demonstrations by guest speakers from the region. Examples of the visitor activities include glider golf, kite making, rocket launches, cockroach races, and egg drops. Featured guest demonstrators have included pottery makers, marine biologists, sky-diving instructors, and storytellers.

We also deliberately started using our full name, Sci-Port Discovery Center, at all times, when answering the phone, talking to the media, speaking to groups, or in any conversation, to transmit a more tangible image of Sci-Port as a discovery center. We consciously refer to ourselves as a science center rather than a museum

Figure 13.1. Sci-Port Discovery Center, Shreveport, La.

in press releases and public situations, and visitors buy tickets to explore our discovery areas rather than exhibit halls.

Soon after opening, we began hearing reports from newscasters, the public, and even conversations in Wal-Mart about Sci-Port Discovery Center being so much fun for all ages. In the first year of opening, our annual pass memberships rose to 4,000 households, up from 500 memberships when the center was at the old site. Today, four years later, Sci-Port continues to enjoy 4,200 annual pass memberships and a 67 percent renewal rate, among the highest renewal rates of U.S. science centers.

SUSTAINING OPERATIONS WITH A FOCUS ON PEOPLE

Our plan for sustaining operations is rooted in our long-term outlook for providing a community place of informal learning that evolves in alignment with our region's educational priorities and cultural changes.

Basically, our approach for sustaining Sci-Port Discovery Center is to stay focused on people. Our sustenance methods are to promote multiple visits annually by providing ever-changing visitor experiences and services that are so regionally unique and personable to also attract tourists. We believe serving the public well and addressing critical education needs will generate the earned income and contributed support vital to our successful implementation of our mission.

We take great care in developing what the ever-changing visitor experiences will be for the next eighteen months to three years. By March each year, our schedule of traveling exhibitions and IMAX films for

the public is firm through August of the following year. We also contract well in advance for our summertime traveling exhibitions, recognizing the opportunities for peak attendance from the vacationing public. For example, we contracted four years in advance to feature the popular Grossology exhibition for a summertime slot at our center.

Exhibitions

Our interest in offering high impact, educationally fun exhibitions, and engaging IMAX films naturally has to be balanced with our limited funding resources. In years when we scheduled a very expensive summer exhibition, such as Dinosaurs, we provided lower-cost, effective exhibitions for other periods that year, such as brain teaser exhibits during part of the school year. We've even created our own educational experiences in the exhibition gallery such as glider golf and large scale math and science activities for families during the months prior to standardized tests. Building an eighteen- to twenty-four-month schedule of visitor experiences at one time allows us the ability to visualize and work within our annual resource picture. As a concentrated, once-a-year decision process, we're free to dedicate our time during the rest of the year to implementing the program and marketing activities for maximum attendance and optimum visitor experience.

Booking the exhibition also establishes that period's theme for the variety of visitor experiences we offer rather than relying solely on the exhibition to sustain the center's attendance. Our IMAX films are then selected to deepen the content and widen the visitor's perspective of the theme, to appeal to a complementary subject, or sometimes to address a distinctly different targeted audience. By June each year, the focus and content of our school program offerings for the next school year are developed after meeting with our education advisors from the school systems for an update of current critical needs. We don't seek to annually add volume to our program offerings but rather continually align them to priority objectives. For example, this year's programs have been designed to integrate math learning objectives. We also pruned from our offerings those programs having low impact, as revealed on our teacher surveys.

Our schedule of changing major exhibitions and programs becomes our compass for the year, setting the course for all primary actions and communications we pursue that year, from programming and gift shop merchandising to marketing, newsletter publishing, and fund-raising. Additional featured programs and potential guest demonstrators are identified and the visitor experience calendar is updated for the full year. Program data sheets are then prepared for each featured visitor experience, detailing the program dates, targeted audience, what's fun about the program, what's the science, key promotional messages, any special fees, and so on. The annual operating budget materializes from this process as each area determines what it will cost to implement its part of the action plan and what levels of attendance and fund-raising resources it will achieve as a result of these offerings.

BUILDING RELATIONSHIPS

Providing a fun learning environment that changes regularly is not enough to sustain our efforts. Ultimately, it is the public's participation through attendance, donations, and rallying of government support that will determine the center's success with its mission. Therefore, establishing and building relationships with our targeted audiences and constituents is fundamental to every aspect of our daily operation. Our livelihood relies on continued, positive, word-of-mouth communications in the community, and keeping those with the potential to give aware of the impact Sci-Port Discovery Center is making on meeting the region's priority needs.

We've structured our organization to clearly communicate Sci-Port Discovery Center's focus on the public whom we exist to serve. Rather than diagram staff positions and internal reporting relationships, our organizational chart identifies our mission, our major programs, and the functions supporting our execution of the mission. Surrounding the organization and

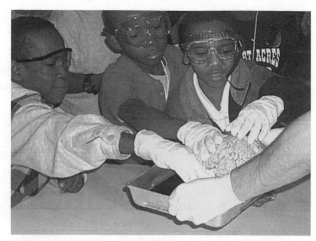

Figure 13.2. Boys touching brains at Sci-Port Discovery Center.

also identified in the chart are key audiences and constituent groups, such as schools, members, families, and contributors, who are primary to our mission.

Community Development

Internally, our Community Development staff leads our marketing, public relations, and fund-raising efforts, whereas building relationships is a function carried out by all staff members. Program and support staff is organized into Friendship Teams, a separate team for each major program, such as our birthday program. The team is made up of a program guru, responsible for every aspect of the program's design and delivery, and a sales leader, responsible for promoting the program to the potential participants and engaging their attendance. Support members on the team include a Visitor Services representative, responsible for fielding inquiries and processing the sale and an accounting representative, who provides timely collection of payments and reporting of results. A focus group meeting is held by the Friendship Team at the beginning of the year with Community Development staff and any interested staff to plan the program's attendance growth for the year and strategize how to achieve it. Program bookings and attendance are tracked and compared with the goals on a weekly basis.

Opening Events

Our opening events laid the initial groundwork with the community for our continued building of relationships. During the week prior to Sci-Port's opening, a series of evening events were held as appreciation and advance exploration for key interest groups. These preview nights included an Educators Night;

SCI-PORT DISCOVERY CENTER
ORGANIZATIONAL CHART
MISSION STATEMENT

SCI-PORT DISCOVERY CENTER PROVIDES AN EDUCATIONAL AND ENTERTAINING ENVIRONMENT
FOR PEOPLE OF ALL AGES TO EXPLORE AND ACTIVELY PARTICIPATE IN THE WORLD OF SCIENCE
AND TECHNOLOGY. THE CENTER DEVELOPS A COMPREHENSIVE VIEW OF THE WORLD THROUGH
THE INTERACTION OF SCIENCE AND HUMANITIES AND SERVES AS A CATALYST TO ENCOURAGE
LIFELONG LEARNING.

COMMUNITY

DONORS

BOARD OF DIRECTORS & CEO

Exhibit Fabrication and Support

Community Development

PROGRAMS AND VISITOR EXPERIENCES
Discovery Areas
IMAX Dome Theatre
Featured Programs and Events
Birthday Parties
Camp- In Program
Summer Camp Program
Teacher Ambassador Program
Teacher Workshops
School Workshops
Urban Development Program
Outreach Programs
Special Event Rentals
Discoveries Unlimited Gift Shop
Galaxy Cafe' and Catering Services
Member Events
Website
Visitor Services
Volunteer Program

MEMBERS

PROGRAM PARTNERS

Graphics and Publications

Member Relations

Building Operations

Marketing

Human Resources

Information Technology

Administration and Public Relations

Accounting

SCHOOLS

ELECTED OFFICIALS

ACCREDITATION
ORGANIZATIONS

Figure 13.3. Sci-Port Discovery Center Organizational Chart.

Sci-Port Members Night; Friends and Family Night, for key suppliers, city staff, architects, and employees; Media and Hospitality Night, targeting hotel/motel employees, cab drivers, and tourism representatives, and a Gala Celebration Evening for Sci-Port donors, board members, and elected officials.

MAINTAINING RELATIONSHIPS

Maintaining effective relationships and regularly interacting with our constituents are essential actions to increase our attendance and financial support of our mission. Methods we use to establish and continue building these relationships on a regular basis include the following:

- With schools—Back to school free admission weekend for teachers to sample the year's school programs; complimentary Sci-Port membership for all principals, including regular notice of upcoming programs and events; monthly lunch meetings with school system math and science administrators; school visits and hand delivery of our School

Programs catalog in the local area at the beginning of school year; host principal's faculty meetings at Sci-Port Discovery Center; CEO meet with area superintendents, presentation at regional meetings, principal and school coordinator meetings, and visit with staff at state department of education.

- With members—Free previews of major exhibitions; invitation to member-only family camp-ins; free member-only special programs such as Fourth of July celebration; advance notice of programs through special mailings and bimonthly newsletters; providing a red wristband to members when visiting Sci-Port, enabling easy identification and personal greetings from staff.

- With community—Club speaking engagements; CEO active member of downtown Rotary Club; membership manager active in local sales and marketing professional association; vice president of Community Development teaches marketing/public relations at the state university; public relations manager regularly visits media stations throughout a 200 mile area; marketing manager visits with key group leaders such as area church leaders, scout agencies, organizations and air force base public affairs personnel; membership manager calls or visits every small business owner listed in the chamber directory; program leaders contact individuals with expertise or experience in the topic area of our featured programs.

- With donors—Annual dinner with our board; meetings at their home or business at the beginning of year, providing our annual report and hearing about their current interests; special recognitions such as mixing up a batch of liquid nitrogen ice cream for their birthday; private lunch at the center with the CEO and a board member; CEO providing personal tours of Sci-Port to their special guests; invitation to special programs such as IMAX film screenings, exhibition openings, or member-only family camp-ins.
- With government—One-on-one meetings between CEO and government administrator or elected official; update at city council and parish commission meetings; providing portable exhibits, programs, and token gifts in state capitol's atrium when legislature is in session; board hosting reception with elected officials at Sci-Port Discovery Center.

Other Professional Contacts

Other relationships valuable to me as CEO are my contacts with our legal representatives and external auditors. As an employer of seventy-two staff members, a variety of complex personnel matters may arise at any time, such as employee health issues, requests for family leave, or work performance improvement needs. These situations typically demand immediate attention and careful application of current policies, employment laws, and fair business practices. Similarly, the high visibility of our organization in the community underscores the importance of our compliance with the laws affecting nonprofits. These laws, such as sales taxes on purchases, solicitation licensing requirements, liability for serving alcoholic beverages on premises, and board fiduciary responsibilities may vary considerably from state to state. My approach is to acquaint our lawyer and auditor with our operations and seek their advice with identifying and interpreting the various legal and tax matters that may affect us before they become an issue.

ANNUAL REPORTING

Among our annual reporting requirements, as a not-for-profit organization within section 501(c)(3) of the Internal Revenue Code, is the Form 990: Return of Organization Exempt from Income Tax. This informational return must be filed with the Internal Revenue Service (IRS) by the fifteenth day of the fifth month after the ending date of our fiscal year. We also must comply with the federal government's Public Disclosure Requirements for Exempt Organizations by providing a copy of our return to anyone requesting it. Rather than respond to individual requests, we comply by making our 990 widely available on a national information database for nonprofits: www.guidestar.org. The penalties for failure to file the 990 on time with the IRS or to make it available for public inspection can be stiff. Therefore, I keep the deadline date noted on my annual calendar and engage the tax return preparation services of our certified public accountant firm annually to meet this requirement.

TRACKING PERFORMANCE AND REPORTING SUCCESS

Limited resources require us to operate leanly but deliberately focused on achieving our objectives for the year, and it is hoped, surpassing them. An outstanding accounting and reporting process is important to our internal monitoring of progress, providing us with detailed financial and attendance results compared with budget figures for the month and year to date. These monthly financials are detailed by program, department, or project to enable us to monitor results by major activity area.

Daily Attendance Compared with Those in Prior Years

As our daily compass, we use computer spreadsheets to track the following information as indicators of where we're headed compared to where we want to be. Our managers, program leaders, and I have easy access to the information for a regular check of progress and timely action when needed.

Attendance is a critical driver of all our earned income sources, 55 percent of our total revenue. A drop in attendance produces an immediate drop in food, gift shop, membership, and other program sales. Sustained attendance declines can adversely affect all resources. Lower attendance can self-perpetuate from a reduced word-of-mouth promotion because of fewer visitors. Likewise, donor support may decrease from a weakened confidence in how much of the public's interests and priorities our center is serving.

The ease of viewing the prior year's attendance by day of the week also enables us to more effectively staff and program the center relative to attendance.

Daily Membership Sales Compared with the Month's Sales Goal

Our daily tracking of our Annual Pass membership sales, whether by mail, phone, or in person, tells us how well all staff members are interacting with the public and how much value the visitor has placed on our current and changing programs to attract their repeat visitation. We track the number and type of memberships sold each day and how many were renewals versus new memberships.

Weekly Tracking of School Group Bookings by Month and Year to Date Compared with Goals

Sales of our other major programs, such as camp-ins, birthday parties, outreach, after-hours rentals, day camps, and gift shop are also tracked weekly. Typically this information triggers us to do more targeted marketing for the program or change what we are offering to better match the targeted audience's interests

Personnel Costs

Staffing costs represent 55 percent of our total annual budget. Traditionally, we had difficulty pinpointing the cause when personnel costs were reported over budget on the monthly financials owing to the lack of supporting detail. To help us better monitor this sizeable cost in real time, our annual budget is prepared on the basis of the hours of staff coverage needed daily to provide the visitor experiences we've planned for the year. This method provides our program director with a template for scheduling staff throughout the year. Weekly, our program director receives a report from our time clock system that summarizes actual hours worked compared with budgeted hours. When over budget, the director knows exactly how many hours to catch up on future staffing schedules and pulls a detailed time clock report to identify the reason for the overage to prevent its reoccurrence.

Daily Purchases

Each major program and department has a spreadsheet for recording daily purchases and monitoring the status of their budget. Together with the program's sales tracking spreadsheet, this information helps us track how well the program's net results to date compare with the budgeted net return.

Key Benchmarks Tracked Periodically

We measure key benchmarks periodically to track our success. These indicators include the cost per visitor served; the number of full-time equivalent staff; ratio of earned versus contributed sources to total revenue and support; ratio of our attendance to our area's population; membership renewal rate; ratio of general versus school versus member versus other attendance to total on-site attendance; and the ratio of visitors from local area versus other parts of the state versus other states versus other countries to our total attendance. When compared with other science centers, this information helps us identify other effective models or programs to consider.

Other Monitoring Methods

Aside from our data tracking, we regularly review visitor comments and rotate all managers to work weekends as the manager on deck, responsible for the overall visitor experience that weekend. These additional methods help us stay focused on our mission, be alert to cues from our visitors and community about the impact of our activities, and identify major shifts in the public's priority needs, attitudes, or interests, which may reveal our future direction.

Visitor Comments

Fan mail from children and parents and comments from teacher and visitor surveys are posted in the break room for staff and volunteers to enjoy. Excerpts are reprinted in our newsletters and annual report as testimonials of our center's impact on the region.

MANAGER ON DECK PROGRAM

All managers, including the CEO, rotate weekends as manager on deck, responsible for the overall visitor experience that weekend. Duties include supporting staff in opening, closing, and programming the center that day, interacting with visitors throughout the day and reporting results, including comments from visitors. This program has been very effective in keeping our managers in touch with our public and knowledgeable about the programs we provide, and helps create new ideas and solutions from the diverse perspectives provided by our managers walking in the shoes of the visitor.

WEEKLY HOWIE MEETINGS

The "Howie" (how well are we) meeting, held every Thursday morning at 8:30 a.m., is designed as a quick, checkpoint of how well we are prepared for our public and school group visitors over the next week. A staff member from each area attends this meeting, and the weekend's Manager on Deck facilitates the meeting. Key visitor experiences planned for the week are reviewed, the attendance outlook is discussed, and the sharing of visitor comments or participation in this week's programs is encouraged. This weekly gathering has proved invaluable in keeping everyone on the same page concerning upcoming programs and staying focused on our visitors as plans and activities are discussed.

REPORTING OUR SUCCESS

Beyond setting the right course for advancing our mission each year, keeping our attention on the public we seek to serve, and tracking our progress, our most critical factor in sustaining our operations is communicating our success. I engage the board's participation in keeping our various constituents informed by infusing their daily conversations in the community with their personal excitement for our activities and future plans. This involvement from the board is achieved by providing a CEO's report monthly in advance of each board meeting and, at the meeting, having them participate in a fun but brief demonstration or hands-on activity from one of our current or upcoming programs. The CEO's one-page report provides bulleted updates of key activities, progress, and results in our four critical areas of fund-raising, community relations, programs/operations, and strategic initiatives.

ANNUAL REPORT

To quickly and succinctly acquaint grantors, donors, and the interested public with the progress their support makes possible, we take great care in producing and printing an annual report. Without this report, our constituents would be relying heavily on our Form 990 Information Tax Return filed annually with the Internal Revenue Service or fragmented pieces of information they obtain from the media. Neither source shares a full view of our results and the impact of our programs on the people we served during the year.

Our annual report presents the outcomes of the year more in pictures than words to involve the reader in a comprehensive and enjoyable visual journey of the impact our educational activities had on the lives of young people and families. Throughout our annual

report are reprints of comments received during the year from some of our visitors, teachers, donors, and children. Other components of the annual report include the following:

- Our mission statement, purpose, key features and description of our organization including 501(c)(3) status and structure of governance
- A letter from our board chairman and president/CEO providing the year's highlights, including recognition of major gifts received and listing of board members
- Program highlights and participation results from the variety of major programs and visitor experiences offered on site and off site during the year
- Educational highlights, particularly the results of our partnerships with schools
- School and overall attendance results mapped to show where visitors traveled from
- Recognition of all $250 and above donors for the year, highlights of major gifts and giving programs
- An overview of our future plans
- Our financial and operating results taken from our annual audit for the most recently completed fiscal year and pie charts showing the sources and uses of our total support
- Recognition of our volunteer support and the diversity, educational backgrounds, and experience of our program and management staff

Again, the information listed above is presented more visually than by written text in our annual report. Overall, we want the look and information in the annual report to reflect our identity, conveying the right message about who we are and the progress we're making with our mission. In other words, the report is designed to reflect the fun, ever-changing, people place that we strive to be.

The annual report has become my most valuable tool in building and sustaining relationships with supporters and prospects. Ultimately, communicating our success everyday, with the aid of the tangible results revealed through our annual report, keeps the attendance and resource pipeline flowing and the community's passion for our mission ignited.

CHAPTER 14

Over the Top: Building Resources to Open and Grow

MARILYN HOYT

EVERY SCIENCE CENTER STARTS WITH NEEDS TO BE met and resource opportunities with which to meet them.

Most often, a group of leaders and volunteers have been organizing around the idea of a science center for some time when a significant resource opportunity *suddenly* emerges. It's not perfect, but if it is not taken, the enthusiasm already pulled together will be lost. So the group takes it.

This launch decision then dominates the path to the science center's opening, and its sudden arrival ensures that the path is not articulated from the outset. That said, a rational marketing, public relations, and development process can be employed to maximize success.

MISSION

Missions are all unique in some ways, but there are two goals nearly every science center seeks to accomplish:

- Promote the public understanding of science through opportunities for real-time, personal discovery, and learning
- Enhance the quality of science education in the community

Although the needs science centers seek to meet are similar, leadership and resource opportunities vary. Leadership, staff capacity, and funding available at the very outset will determine the speed with which a new science center unfolds, its reputation and influence, and its size at opening and over at least the next decade.

SPACE

Donated space is the most common *sudden* launch point for a new science center. Success comes from working mindfully to maximize opportunities while spending as little as possible to neutralize challenges. Four key development and marketing perspectives need to be included when deciding whether or not to accept a donated space:

1. *Speedy first-phase opening:* Earned income makes up more than half of the annual resources for most science centers. It positively influences area donors—especially government and business donors—as an indication of public interest in the center's services. With this in mind, what will it cost to put the building into service quickly as a well-attended science center?

2. *Ongoing subsidy for operating costs:* Major donors, foundations, and businesses do not generally give significant, ongoing gifts and grants for operations. Earned income, special events, the annual fund, member dues, and investment proceeds must produce unrestricted monies for frontline operations, facilities, security, utilities, insurance, marketing, public relations, fund-raising, and the finance office. Unrestricted income is always in short supply.

3. *Securing the science center's investment:* A multi-decade $1 per year provides maximum security for the science center's investment. A five-year offer of space "until you get your building built" is a much more difficult challenge, as the new science center can never hope to recoup the cost of its investment in facility renovation.

4. *Destination building costs:* Is the site centrally located with other nearby organizations or stores that will help draw the target school and family constituencies? Or will the center have to open with a large and exciting program to build destination status all by itself? Adequate parking and accessibility are also important considerations.

An offer of donated space deserves a very hard look if it is located in a low-traffic area with an undesirable reputation. The extraordinary costs of ongoing marketing, traveling exhibitions, films, and special events needed to draw families and schools to a forgotten or badly located site will last for decades. A donated space with these kinds of handicaps should be accompanied by additional donated cash to provide operating support at the 50 percent level for the first three years and then continue at a level of some substance for at least a decade following. After all, space is not a definer of a science center's launch. A science center can operate for years as a 501(c)(3) incorporated entity without a building. It can recruit leadership, raise funds, and meet community needs with a successful program of traveling exhibitions, an outreach science assembly and classroom program for local schools, a science day-care and/or after-school program in a church basement, and a science summer camp contracted to a county parks department. Such a center will meet needs from year one *and* build earned income, leadership, and a fund-raising network at the same time. At the end of year five, it will dominate every programmatic public science-offering niche in its area. And, it will be well positioned as an institution to win a donated building in a good location, or substantial funding from government, business, foundations, and individuals to buy or build phase one on its own property.

LEADERSHIP

In order to open and grow a science center, it is critical to tap into the highest level of leadership possible. Every layer up in the power hierarchy of the community brings more credibility and more access to funding, technical assistance, and door opening with other leaders.

These leaders have many calls on their time and money. They will need to know that, when they call in a resource for the science center, they can make the donor of that resource feel that a good choice was made. It is critical to plan many tangible success points into the months and years before the science center opening. Finishing a survey or master plan and announcing key findings, prototyping exhibit modules or programs in a public space, or delivering programs in school or community settings are all opportunities to demonstrate success and tangible results. Before the launch campaign is announced or the first shovel is in the ground, the science center is already positioning itself as a community leader and rewarding those who are investing in it.

FUNDING

In most instances, newly launching science centers don't have access to all the giving sectors: government, individual, foundation, and corporate funding. So it is essential to identify strengths and build on them to produce early, tangible results.

A key to campaign success is to dedicate each staff and volunteer hour and related fund-raising dollars to the next largest gift or grant where a door opener is available. By definition this means that largest gifts and grants will be secured first, then medium, and finally smaller gifts. At the outset, if extra time is needed to identify a door opener, this is time well spent. Cold calls and cold proposals are a very inefficient way to raise money. And small gifts at the campaign outset can so damage confidence in the prospects of the science center's success that it fails.

Gaining Access to Funders

Both staff and volunteers can serve as door openers. The kinds of funding they can have access to are different because of the different motivations for giving.

1. *Staff:* Access and credibility are based on academic and professional credentials. These credentials are of particular interest to competitive grant makers in government, corporations, and foundations.
2. *Business and social volunteers:* Access and credibility are based on relationships and their related demands and loyalties. Exercising these relationships yields major gifts from individuals or family foundations, corporate funding from both the foundation and marketing sides of the company, government discretionary funding (sometimes called *pork*), and fund-raising event donations and ticket purchases.

Costs of Fund-Raising

Another key to campaign success is to raise the least expensive dollar in the shortest period of time so that the maximum amount of dollars raised goes to the project rather than fund-raising expense and cash-flow problems do not force early borrowing.

The benchmark for budgeting fund-raising expenses is 10¢ per dollar to be raised. This should be a category in the campaign budget.

It may be helpful to benchmark the typical cost per dollar raised for each kind of financial support. Although these typical costs per dollar raised may not be accurate for every campaign, the hierarchy of inexpensive to more expensive sources of support can be depended upon.

Major gifts and planned gifts from individuals who are asked by a volunteer they know can come almost instantly at a cost of less than 3¢ on the dollar. However, for most new science centers, the board itself is the

Figure 14.1.   The New York Hall of Science's red-aproned Explainers answer all the tough questions and are the human face of the Hall.

only group that has been cultivated. There has been no time to identify and cultivate a larger circle of wealthy individual donors. Their gifts—each giving at a level that others would respect as a major gift given their capacity—will be very important to the success of the campaign.

Cultivating major donors over the period of the campaign should lead to major gifts before campaign's end at a cost ranging up from 8¢ per dollar raised if there is a group of potential donors with whom to work. If there are only one or two major donor prospects, the cost of cultivation prior to receiving the first major gift will likely be higher.

Competitive grants from funders who have funded this effort before or where there is a strong door opener can be obtained in three months to a year depending upon how often the decision makers meet. The cost is 8¢ to 12¢ on the dollar raised. Procuring the first grant from these competitive funders can easily cost 25¢ on the dollar as it includes the time needed to identify door openers, and the likelihood that proposals may be prepared and submitted, but not funded upon the first approach.

Discretionary government monies raised through volunteer door openers can be 8¢ on the dollar or less with funding coming in the first budget year. However, if the elected official representing the location of the science center at the city, county, state, or federal level is not well positioned on a budget committee, it will likely be necessary to work with a lobbyist. Funding should come within the first or second budget year, but the cost will climb to 25¢ per dollar raised or more.

If volunteers donate to or underwrite fund-raising events, these dollars can come in at a very low cost. But

the benchmark for a fund-raising event where there are not major donations or underwriting is 30¢ for every dollar raised.

Direct mail fund-raising from donors who have already given should cost less than 25¢ per dollar raised. If mailing lists are traded with another nonprofit whose donors are a good match, costs typically climb to over 60¢ per dollar raised. And if lists are rented, costs will climb to $1.14 or more per dollar raised. Why would anyone spend more than $1 to raise a dollar? This is a way of investing in identifying new donors. Or the true purpose may be marketing.

To understand cost per dollar raised, it is important to keep in mind the following:

1. *Dollar cost:* The first dollar is always more expensive than the next dollar from the same donor. The high cost of identifying, cultivating, and involving the donor lies with this first gift. So early dollars raised must be big dollars or a center will never reach its goal in real time.
2. *Techniques for keeping cost/dollar raised low:*
   - Build the potential for volume by writing very few (one to three) base proposals once a year, and then tailoring them to the needs of specific donors each time an "ask" is prepared. Base proposals are written around key priorities of the science center plan where there is the largest number of potential funders. These may not be the highest priorities of the campaign such as land acquisition or the costs of architects and fees. These grants often fall in the areas of education, after-school or early childhood programming, and staff development.
   - Work from the top down and the inside out—always investing time to seek the next largest gift from the next closest donor prospect.
   - Drive fund-raising with relationship-based asking supported by quality desktop-published materials rather than investing large funds up front in glossy materials that may or may not be on target with what donors find most appealing.
   - Use those closest to the project to open doors. Circulate to board members, selected donors, and other friends of the center the names of foundations, businesses, individuals, and government leaders who should take an interest in the project but are not yet acquainted with it. When a connection is made, talk with the current supporter about how best to approach the new prospect. An invitation to lunch, a letter or call, or even just the supporter's name as a "cc:" on the proposal can make a substantial difference in initiating a first gift.

- Hang on to donors already won. Find inexpensive, high-volume ways to keep in touch with the donor base. Time for building personal, ongoing communication can only be spent with the top tier of donors. Include them and all the rest in mailed updates, postcard updates, hard-hat construction site walk-throughs, or family "test it" days as the exhibit components go into prototype. Use e-mail to share with them good things that leaders in the community say about the center's work or exciting moments from behind the scenes.

3. *Set up a base from which to work smart:* Raise $100,000 from those very close to the center in order to quietly begin the launch campaign without a constant scramble for money just to get started. And do not start out until it is known what the key donor circle is willing to commit in terms of leadership and money.

Look before You Leap

The tool for assessing the pool of leaders and donors willing to commit money and leadership to a campaign is the feasibility study. There are many excellent development consultants who produce these. Be sure to pick someone who has significant experience with first campaigns and with local cultural campaigns. Ask your consultant to include a campaign plan as part of the report.

PUBLIC RELATIONS AND MARKETING

1. *Marketing is an ongoing effort* to gain information on what the center's constituencies value and will pay for. The final component of this process is its most well-known targeted media placements informing those constituencies that the center has what they want.

    Some centers will need to commission professionals to help with this process. Others will win pro bono support from a local market-research company. But these tools, even if used, are adjunct to the useful and effective assessments the center performs for itself. Mine real-time data from existing local activities used by the center's constituencies. A computer camp or arts camp can tell a lot about the family audience and pricing appropriate for youth services. Likewise, a look at the local YMCA's after-school and weekend programs will tell a lot about appropriate scheduling and pricing for families in the area. The key is to look at how area families, schools, and tourists behave when they are looking for fun and learning. The content that they choose to participate in is not as important as the values of enrichment, learning, and fun that they seek.

Of course, best of all, is the center's own experience with prototype programs or traveling exhibitions brought in for a time at a donated public site. Here the center can formally and informally survey attendees and get a sense of how to find and attract them and what they are willing to pay.

With this information, the center can craft effective marketing materials using paid or pro bono advertisements and other media placements. It will also know with whom it wants to build marketing partnerships: A sports science quiz embedded in the local radio game broadcasts? Partnering with banks that target families for first mortgages? An organic grocery store? Or a science and technology company? These all include the adults who make decisions about what their children do and see. Which ones will be natural partners for the center's marketing efforts?

2. *Public relations involves meeting the needs of the media as well as those of the center.* These placements are successful to the extent that stories and photos meet the media's need to add readers, listeners, or viewers. With that in mind, know that not everything can be effectively placed through public relations efforts.

    Prior to opening, public relations placements are driven by the goal of institutional brand building. The center and its leadership are credible, smart, able, and have cachet. As a result, it can be assumed that what the institution produces will also be valuable. This institutional public relations effort continues beyond opening. But opening brings the first launch of the second public relations front—placements driven by the goal of achieving paid attendance by key family and school constituencies.

    Science center launches are driven by public relations activity. As the opening approaches, the center may not have the data or money to launch a substantial marketing campaign. The first task of building a public relations program is managing expectations—creating a balance between the excitement and awareness the center wants to build, the recognition owed to leadership and donors, and the need to maintain a low profile over the years of planning and fund-raising, so that the actual launch is exciting and new.

YOUR SCIENCE CENTER MEETS YOUR COMMUNITY'S NEEDS

The needs identified as the basis for a center's mission—quality education, a destination that brings families together, and perhaps tourism or workforce development—are real, quantifiable needs.

Figure 14.2. The New York Hall of Science's Science Playground is an outdoor laboratory packed with hands-on exhibits.

These needs are enormous. No science center can fulfill them for the whole community. So what component of them will be the center's focus? And how can impact be measured? By number of people trained or served? By changes in test results or assessable behavior? By funds raised and spent with area vendors? These will be important figures, representing a real impact in the community.

FOUNDING DEVELOPMENT LEADERSHIP

We all come to this from some other world, bringing what we know and who we are. As we get more involved in the process, each of us will likely feel what we have brought is not quite enough. But it is. The exceptional growth of new science centers around the world reminds us that our experience and tenacity, and the partners we draw in along the way, are enough to bridge the gap between our founding dream and the

hands waiting on the other side to receive what we have to offer.

During the launch years when everything needs to be done at the same time, here are some recommended resources. (Special thanks to Michael C. Savino, Headline Communications, for recommendations listed under Marketing, Research, and Public Relations.)

RESOURCES

OVERALL www.astc.org

DEVELOPMENT

Local press and newsletters related to the political, business, educational, and social activities in the area

Newsletters and event programs of every local charity that serves schools or paying families

Chronicle of Philanthropy: www.philanthropy.com

Foundation Center: www.fdncenter.org

MARKETING AND SPONSORSHIP

Museum Strategy and Marketing: Designing Missions, Building Audiences, Generating Revenue and Resources. Neil Kotler and Philip Kotler. San Francisco: Jossey-Bass, 1998.

Mission-Based Marketing: Positioning Your Not-for-Profit in an Increasingly Competitive World. Peter C. Brinckerhoff. Hoboken, N.J.: John Wiley & Sons, 2003.

Marketing the Arts: Every Vital Aspect of Museum Management. Eds. Simon Blackall and Jan Meek. Paris: International Council of Museums, 1992.

American Marketing Association: Joining is the easiest way to stay on the inside track of research and trends for marketing overall.

RESEARCH

Marketing Research That Won't Break the Bank. Alan R. Andreasen. San Francisco: Jossey-Bass, 2002.

Mass Media Research: An Introduction. Roger D. Wimmer and Joseph R. Dominick. 7th ed. Belmont, Calif.: Wadsworth, 2003.

Effective Public Relations. Scott M. Cutlip, Allen H. Center, and Glen M. Broom. 8th ed. Upper Saddle River, N.J.: Prentice-Hall, 2000.

Visitor Studies Association: www.visitorstudies.org

PUBLIC RELATIONS

Public Relations: The Profession and the Practice. Otis W. Baskin, Craig Aronoff, and Dan L. Lattimore, 4th ed. New York: McGraw-Hill, 1997.

This Is PR: The Realities of Public Relations. Doug Newsom, Alan Scott, and Judy Vanslyke Turk. 8th ed. Belmont, Calif.: Wadsworth, 2003.

CHAPTER 15

Marketing Basics: Applications for Small Science Centers

KIM L. CAVENDISH

HAVE YOU EVER HEARD THIS FROM ANOTHER MUSEUM professional? "Marketing a museum is totally different from marketing in the for-profit sector. We aren't selling a product; we are preserving and interpreting the culture. It isn't a business." From my perspective, if we begin to define our exhibits, programs, and experiences as mission-related products, then we can learn a great deal from the for-profit sector, and we can apply the accepted marketing basics to make our museums more successful. On a macro level, we can define our capital campaigns and annual fund-raising efforts as products and apply marketing basics to those products as well.

The purpose of marketing can be said to be *getting the targeted public to exhibit purchasing behaviors*. That means, getting the public to buy admission to your museum, enroll in your programs, or contribute to your campaign. It can also mean getting the community to buy into the value of your mission and therefore provide support for your museum. You can build your marketing effort by treating your museum as a product you must sell; and to sell it, you must consider all the basics that go into marketing any product. These basics can be defined as follows:

- Research
- Product development
- Pricing
- Packaging
- Segmenting the market and choosing targets
- Advertising
- Branding
- Distribution
- Devising the marketing recipe

Following is a discussion of each concept, as it might relate to small museums with limited resources.

RESEARCH: THE SYSTEMATIC COLLECTION OF INFORMATION FOR THE PURPOSE OF DECISION MAKING

Collecting extensive and customized data about your community, existing audience, or potential museum visitor can be a very expensive proposition. The cost of obtaining information must be weighed against the benefits you expect to derive from the information. Fortunately, much valuable information about the demographics of your city or region can be found on the web on a variety of sites. Your city's chamber of commerce, lead newspaper, or local tourism bureau will usually have valuable information available at your request.

There is no substitute for learning about the audience you already serve. The simplest techniques often provide a great deal of information useful not only for decision making but for advocacy. For example, collect a sample of the zip codes of your visitors over a period of time, and plot them on a map to learn if your audience is based tightly around your facility or if you are attracting some from a broader range. The Virginia Air and Space Center in Hampton, Va., separated from Norfolk/Virginia Beach by a very wide body of water and a congested bridge-tunnel system, once was battling the perception that it was just a local interest facility and therefore not worthy of support from donors or corporations based on the south side of the region. Zip code surveys demonstrated that 50 percent of visitors were tourists, and 20 percent were residents from Norfolk and Virginia Beach. Those statistics enabled the center to command advertising support from Hampton's convention and tourism bureau and to increase its corporate support from donors from the broader region. When the Discovery Center in Fort Lauderdale was planning a major expansion in the 1980s, the local Junior League provided a major needs assessment, under the tutelage of the newspaper's research department chief. That assessment helped to garner support for the capital campaign and helped to target new exhibit development to community interests.

Detailed address information on your membership base can be used for a "psychographic" analysis—which can help you segment your existing audience into categories related to their probable interests, available income, and lifestyles. This can help you determine whether your existing audience base will respond to a new product offering, such as "mommy and me" classes, health lectures for seniors, or science competitions for teens. Such analysis might be available to you through in-kind support from your newspaper or a university. Small museums seldom have the cash necessary to commission customized audience research, but innumerable opportunities exist to find community partners who will provide either staffing for surveys or support for analysis if you articulate why the research is critically needed.

PRODUCT DEVELOPMENT

In a typical commercial endeavor, developing a product would mean inventing or imagining a product or service and shaping it to the public taste. It would mean looking at the surrounding environment, checking out the competition, identifying emerging consumer trends, and then adjusting the product to the marketplace. And all the while it would mean taking the company's own resources, financial and human, into account and adjusting the plans accordingly.

What I have just described is exactly what a museum must do in defining its mission. It must be distinct to your institution and must be adjusted to the realities facing you in your community. Your mission must take into account what else is already being offered locally; for example, how well are the needs for informal science education being met? It must also consider your strengths and challenges. It must take into account a realistic picture of the financial resources you can reasonably expect to muster, the talent you can attract, and what your community will truly support. And once your mission—your most important product—is defined, everyone connected with your organization must be able to speak with one clear voice in describing it.

With your mission in place, your product development continues as you define and establish your exhibits and program offerings. Again, you should

be balancing your choice of mission-related products against what the marketplace wants and what you can effectively deliver. You are balancing the ideal of what you want to present against the available financial and human resources you can bring to bear on it. You must fit your programs and exhibits to your own community, available resources, internal abilities, and institutional capacity. And finally, you must take into account what your competition is doing.

For a specific example, consider museum birthday parties as a source of earned revenue. Many museums offer a party option for a package price that includes museum admission and use of a private gathering space. But competition is fierce, from other museums, from fast food restaurants, and from entrepreneurial entertainers. So how can a science center compete in this milieu? By offering a complete experience, not just group admission. Think about developing a product that will appeal to today's busy parents by offering a total package. At the Virginia Air and Space Center, a party includes a cake shaped like the space shuttle, a private room with a private science demonstration, an IMAX film, and a personalized exhibit tour. At the Orlando Science Center, the room used for birthday parties has been decorated with fanciful and colorful wall murals and the birthday child gets a special welcome in the theater before the show. At the Museum of Discovery and Science, new party themes are now implemented each year, and parents of the birthday child buy theme-related invitations and party favors from the museum store. The idea is to create one-stop shopping for the event by one call to your reservations number. We have found that, in fact, the deluxe options for parties, while more expensive, attract more buyers than those with fewer bells and whistles. When you have developed a quality product that is unique to your institution, it will sell.

PRICING

Pricing is an important consideration not only for admission fees but also for classes and outreach programs, sponsorships for exhibits and programs, capital campaign naming opportunities, museum store products, and traveling exhibit offerings. Your prices, whether for experiences, objects, or opportunities for recognition, have to reflect the standards of your own community. Research what other science museums charge for similar offerings, but be sure to weigh this information in light of the household income available to the audience and the size of the museum relative to others, as larger museums often command higher prices. Study how competing museums, attractions, stores, and theaters in your area are pricing products similar to yours. Underpricing to beat the competition is tempting, but it can be a big mistake. Once you underprice, or give away for free what you could justify charging for, it is hard to increase the price or begin to charge for the offering later. Even when a program is completely subsidized by a sponsor, it may be warranted to assess a small fee to users, to protect your ability to charge for the program once the sponsorship expires.

During periods of weakened attendance, such as many museums faced following the September 11, 2001, tragedy, a reduced admission price is often suggested as a way of increasing visits and ultimately increasing total revenues. However, a detailed examination of attendance statistics and prices for science centers in the United States indicates that the demand for museum services is inelastic. In other words, while a lower price may result in increased attendance, the increase in visitors will not be enough to offset the loss of revenue from the decreased price per visitor. A more effective strategy can be to target particular audience segments with "call to action" discounts or coupons with a limited shelf life.

PACKAGING

With a commercial product, packaging might mean deciding what kind of box to put the product in to sell it, and deciding exactly how the product should look—its color, shape, and style. Think about how many of the products that you buy are more package than substance: perfume and cosmetics are perfect examples of products that are defined more by their elaborate packages than by their ingredients. The distinct look of a certain brand might be the very thing that attracts you and keeps you coming back to it.

In a museum context, packaging concepts can be applied to everything from the architectural design of your building, which can add distinction to the museum's image and create a glow of credibility, or, less happily, can discourage participation by being cold and unwelcoming, to how you describe your program offerings in your advertising, press releases, or newsletters. At the Museum of Discovery and Science, we have experimented with repackaging our live science demonstrations on the museum floor to ratchet up excitement and participation among visitors. In an area unsuitable for permanent exhibits, hemmed in by escalators and aquariums, we have created a flexible Science Café.

Our Science Chef presides over a cart with a colorful overhead umbrella, and small round tables with tabletop displays or activities create a welcoming ambiance. Each day we post a menu of science snacks available at specific times, and our Science Chef serves up tasty science that's fulfilling and fun. By packaging these demonstrations with a theatrical flair, we have taken standard programs and made them newsworthy again. A formerly empty spot has become a place where people gather in expectation of the next course.

Another example, drawn from Virginia Air and Space Center's history, can illustrate how a basic consideration of pricing and packaging can work together. When I joined the museum in 1995, it was offering generic tours followed by classroom-based educational programs for school groups with an option to buy an IMAX film. Flyers were being mailed several times each school year describing available programs, but attendance was slumping. To create an immediate sense of change, we repackaged existing programs while we worked on creating new ones. We broke up the ninety-minute standard experience into its component parts, renamed the components, and provided options for one-hour classes, thirty-minute demonstrations, several film choices, and tours of varying lengths and themes. We developed a list of pricing options that ranged from a basic themed tour, up to a deluxe combination of themed tour, one-hour classroom program, half-hour exhibit floor demonstration, and film. Then we constructed a booklet to distribute to teachers in the region, outlining the programs by grade level and subject, describing all the films and traveling exhibits coming up during the school year, highlighting themes of permanent exhibits and the related classroom programs, and including detailed information for teachers and chaperones on how to plan and supervise the field trip. Teachers understood that they could customize their visits by choosing their own product mix and price point. And by coordinating program offerings with new exhibits or films, we made a compelling case to encourage teachers to sign up their classes for several experiences. As a result, income from our programs doubled that next school year. By tying up the programs in a more attractive package, we encouraged our customers to buy more from us.

MARKET SEGMENTATION

Which segments of the public are potential visitors or potential contributors, and what will be their motivation to visit or to contribute? Once you have identified your potential markets, you must prioritize them in terms of how, when, and how often you can deliver your messages to them. This will bring you back to the old balancing act. Assuming your resources are not unlimited, you must expend your money and energy first where they will have the most positive impact. If your first priority for impact is to increase attendance and associated revenues, a basic rule of thumb is, *play first to your strength*. If you know that families with children ages four to twelve are your most natural target market, saturate that group with your message before you go after a group that will be more difficult to attract. Get your market penetration high in your strongest target market group before turning your limited resources to a campaign to attract, for example, singles in their early twenties, or senior citizens.

The Fernbank Museum of Natural History in Atlanta had a great deal of success with choosing a particular target market for a specific offering. They targeted the yuppie crowd living in the surrounding affluent suburban neighborhood by offering live jazz and cocktails at their Friday night film showings. The presence of the jazz and drinks made it clear that they were not looking for a nighttime family audience, but signaled to young professionals that they could have a sophisticated evening at the museum. Presumably some of those young professionals would come back during the day with their children in tow. This proved to be a successful effort to connect the new museum with its neighborhood in an unexpected way.

In a capital campaign context, market segmentation could mean targeting your biggest donors first, going after nothing but gifts in excess of $100,000 until you are within $1 million or less of your final goal. It will take just as long to cultivate a $50,000 donor as it would a $500,000 donor, so it should be easy to choose what to do first. Segmentation also means using different techniques to attract different types of donors. One generic campaign brochure developed at the initiation of a new project will not necessarily be your best approach for all donors at all levels. Major donors may want to see a proposal made up just for them. But other segments of your donor base might resent it if your literature looks too rich; they want their smaller gifts to matter.

ADVERTISING

Advertising is a way to reach the public that you have targeted with a specific message about your product, exhibit, or program. Advertising presents challenges to small museums because few have enough dollars to purchase the quantity of advertising necessary to introduce museum offerings to the public in a memorable way. It can be a real catch-22: until you generate enough earned income from admissions to boost your budget up a notch, you hesitate to spend dollars on expensive advertising. But, if you can't advertise, how do you get your message out and attract paying customers? There are several solutions:

- Develop *in-kind sponsorships* with local media for free space. They will often design ads for you and run them when space is available. In return, offer them credit as a sponsor of whatever event or exhibit you are promoting. As part of the original capital campaign for the Museum of Discovery and Science, a newspaper agreed to donate $300,000 in cash and provide $200,000 worth of free advertising over a two-year period. For this commitment, they received recognition at the $500,000 level as the sponsor of the traveling exhibition hall in our new facility. We used many of those advertising credits to promote the exhibits we put in the traveling exhibition hall, and when visitors came to see those exhibits, they also saw the sponsorship recognition in the gallery. It was a win-win situation, and it gave us critically needed advertising visibility as the new museum got established in the public eye.
- *Piggyback* on the advertising of a corporate sponsor. When a housing developer sponsored our dinosaur exhibition, strips promoting our exhibition were placed in the corners of their already-existing billboards.
- Some corporate partners will put coupons for museum admission in their *customer or employee newsletters* or *internal group e-mails*. For example, when a bank sponsored the Museum's SHARKS! exhibition, coupons for admission were distributed in all their local customer statements.
- Some movie theaters will provide free *on-screen advertising* in return for discount coupons for their customers.
- Sell discount *tickets at off-site locations*, such as mall or store customer service offices or the concierge desks of local hotels. The sales locations will then display your posters.
- Work with a *billboard* company to put up your posters on a space available basis, rent free. You pay the cost of the printing, but not the monthly rental.
- Get *transportation* companies to display your posters, in return for offering discounts to their passengers.
- Similarly, restaurants, stores, and parking garages will display posters or flyers in return for *customer discounts or coupons*.

There are endless opportunities to get your advertisements out without paying the going rate. Never forget this, *a satisfied visitor is your best ad!* How you treat your visitors is a key component of an effective marketing strategy. In the museum world, advertising becomes two things: the formal ad messages you design and place in various media and the visitor services you sustain to keep your visitors happy and well cared for. Good people-handling inside your doors will result in good one-on-one message carrying outside.

BRANDING OR POSITIONING IN THE MARKETPLACE

Branding is establishing the personality or character of your institution in the public consciousness. What sets you apart? How do you differ from other attractions or institutions? What exhibits, collections, programs, or special partnerships does your museum have that can be found nowhere else in the community? Does your museum have a distinctive personality? (Is it casual or formal?) How is its level of quality perceived? (Is it bargain or premium?) What kinds of people are associated with the museum? (Are they inclusive or snobbish?)

You must be able to answer those questions if you are to have a clear enough picture of where you want to take your museum. You must decide how you want the public to perceive your museum and what attitudes they will carry around about it. The right brand can go a long way toward establishing the good will you need in the community in order for your museum to attract an audience or attract campaign dollars. You can construct the image you want, provided you plan carefully and make sure everyone involved in your museum is singing the same tune. Elements of the brand construction include logo design and use, consistency in publications and advertising, sound bites used in media communications and editorial coverage, photos used in your press releases and publications, and web page design. This concept ties back quite readily to establishing that one clear voice to describe your museum's mission and character.

DISTRIBUTION

With a typical commercial product going to market, management will have to determine what distribution channels should be used to get the product in front of potential buyers. What kind of store should carry it and in what locations? Upscale or discount house? Mail-order catalog or Internet outlet? For a museum whose main products are concentrated in one building, distribution becomes a tougher concept, but it still comes into play. To what extent are you willing or able to take the museum beyond its walls to serve those who cannot or will not come to it? In distributing your museum, you will have to make tough choices in balancing these customer interests against your available financial and human resources. You may have to sacrifice customer convenience if you decide you cannot make ends meet delivering outreach programs around the region. You may find that the increased variety and quality you can provide by restricting your activities to just one site may overcome the visitor's price resistance or the perceived difficulty of coming to your site. On the other hand, a good exhibition off-site in a nontraditional space may be just the thing to call attention to your main facility for a visit later.

The Virginia Air & Space Center has taken its ride simulator to off-site locations such as state fairs and shopping malls for public relations and revenue generation. Posters promoted the museum, and the ride attendant answered questions from potential visitors and presented a friendly impression. The excitement

Figure 15.1.   Museum of Discovery and Science, Fort Lauderdale, Fla.

of the ride implied that a visit to the museum would be exciting as well.

In the early stages of the expansion plan for the Museum of Discovery and Science, we discerned a need to reposition the museum in the marketplace, to shift the image from that of a small and cozy children's museum to that of a major institution that could command extensive community support and attract large gifts. So distribution became an important strategy in driving the success of the capital campaign. First, we arranged to use 10,000 square feet of raw space in a vacant downtown store for installation of major exhibits that could not be accommodated in our small museum. We showed the community exhibits on a scale they had never seen before. Second, we contracted with the Broward County Fair, about 20 miles south of the museum, to feature exhibits in the main pavilion during the annual fair for several years. Next, we opened a 3,000-square-foot branch for six months in a new outlet mall, about 20 miles west of the museum, where we featured an exhibit and museum store. These mini-museums considerably extended our reach to new audiences. Finally, we put two self-produced exhibits on national tour.

These efforts did succeed in repositioning the museum. Not only did it become more visible over a wider geographic area by opening more distribution channels, but also it attracted larger audiences, more grants and sponsorships for operations, and more earned income from admissions, memberships, store sales, and educational programs. With more outlets and thus more revenue, the museum demonstrated that (1) it could create first class exhibits; (2) it could handle a much bigger budget and bigger operation; and (3) staff could stretch to meet the challenge. We increased our confidence in ourselves, and our community's confidence in us.

THE MARKETING RECIPE

The variety of methods and channels you use to get your messages to those you seek to reach and how you weigh in your use of each one can be called your *marketing recipe*. In the start-up stages of a small museum, your cash resources may be quite limited. Your recipe may include very little paid advertising, lots of in-kind media spots, lots of grassroots social or media-oriented events to attract publicity, and lots of one-on-one targeting of community leaders so that opinion makers can help you deliver your message and garner support. As resources develop, you should begin to segment your market so that you can choose where to spend hard-won advertising dollars: how much goes to radio, newspaper, rack cards, magazines, direct mail, and promotions? Also, you must consider the timing of paid advertising exposure to support the launch of new products and programs.

Public relations efforts can be used to sustain interest when cash is not available to expend over the longer term. You must determine which programs and exhibits best serve the audience you can attract and decide which distribution channels you can use most effectively in delivering your mission to new audiences. You should clarify your museum's brand image at every opportunity, realizing that everything from what exhibits you present to what entertainment you bring in for a fund-raising event will become part of your museum's public persona. Focusing on your mission, while remaining alert to the competition, will help you understand how to keep your pricing competitive, your products relevant, and your distinctive message clear and compelling.

As your reputation builds, you will find it easier to attract institutional or corporate partners who will believe in your mission and help spread your message. You can piggyback on their publications and advertising, exchange coupons between constituents, and engage in joint ticketing or programs. Thus partnerships become an important ingredient in your marketing recipe.

The opportunities to find your museum's constituents and to motivate their purchasing behaviors are only limited by your imagination. Your mission is your message.

CHAPTER 16

The Science Shop at the Science Spectrum

CASSANDRA L. HENRY

IMPORTANT FROM THE BEGINNING

Even from the Science Spectrum museum's early days as a demonstration project in an old grocery store, the inclusion of a gift shop was a part of the plan. Probably in the beginning, the motivation was more to generate revenue than to extend the educational museum experience. Nevertheless, the addition of a science gift shop not only gave us a small revenue stream but also an opportunity for advertising: we reached others when museum visitors shared their new treasures with family and friends.

GETTING STARTED

Kay Dudley was the only one of the original volunteers/founders who had any retail experience, and that was as a sales clerk only (today's sales associate). She was charged with finding merchandise and setting up a small gift shop for our demonstration exhibition—the Science of Sports. The topic of sports was a good one for generating lots of merchandise ideas. We obtained a state sales tax permit and began to seek items. Using local vendors, we monogrammed hats with logos, had logos woven into sports socks at the local mill, and printed posters of rat basketball. We ordered science of sports books. Soon we were in business. Our initial costs were high, meaning less profit, but we felt that the shop's presence and our experience were more important in the long run, as long as our losses were low.

When the opportunity came for our project to expand and to exhibit the robotic dinosaurs alongside the sports exhibit, we increased the shop's gift line to include dinosaur items. Where to get dinosaur merchandise? We had no accounts with any such vendors and we were not yet even an official museum entity. We found a gift wholesaler in Dallas who would sell to us if we drove to Dallas to pick up the merchandise. We did, and our expanded gift shop enjoyed increasing

popularity. The store was so successful that we had a hard time keeping inventory! When the demonstration project concluded, we had even made a little money and gained a fortune in experience.

The Science Spectrum museum opened formally to the public a year and a half later. Though still operating on a shoestring budget, and not professionally designed, the gift shop was larger, better appointed, and better equipped. Kay and I obtained buyer credentials and attended the Dallas Gift Market looking for inventory to sell. We began to build a vendor base with a network of sales representatives. We joined the Museum Store Association and attended the trade shows and educational workshops. Meeting other museum gift shop staff, hearing professional speakers, and attending a market especially for museums were very valuable experiences. As the shop matured, sales representatives came to the museum or called frequently to show new merchandise and take orders. Though we continue to maintain our membership in the association, we now attend annual meetings less often. For the beginning, or new gift shop manager, membership in the Museum Store Association is a must. The help and networking opportunities in all areas of store management will jumpstart the venture and save time and costly mistakes.

THE SCIENCE SHOP TODAY

In the fall of 1993, the museum moved to its permanent home, and the gift shop became the Science Shop in 1,650 square feet of new atrium free space. It is a little more sophisticated now with an even more attractive environment, better looking store furniture, and an electronic networked credit card machine. Gross sales average around $200,000 per year. Store operations are not computerized. Records are manually kept from cash register tape. Retail consultants, to this point, have advised that the shop is marginal regarding

Figure 16.1. Science Shop at Science Spectrum, Lubbock, Tex.

the benefit of computerization. We have grappled with this decision. Teachers want the students in and out of the shop in record time. We can sell faster manually. Computerization requires operator training in sales as well as managerial mastery of software capability and troubleshooting. Software support contracts mean additional costs. With staff turnover, training and software management become difficult, time-consuming, and costly for the small museum. However, there are trade-offs. One must use other methods to track customer demographics, inventory, orders, and accounts payable.

The only overhead expenses charted to the Science Shop on the financials are inventory and supplies. We have not tried to profile the gift shop as an independent entity. It operates in shared space with shared staff and is regarded as an ongoing program fund-raiser. We buy print advertising at the onset of the holiday season. Otherwise promotion is by word of mouth and a function of the traffic in the building. Members have always received a 10 percent discount on purchases.

TALKING SHOP WITH KAY
Through the years Kay has received many compliments on the Science Shop for the caliber of merchandise she buys and the neat, well-ordered display of its inventory. A key to her success is her ability to observe and ask what the buyer wants. In other museums, particularly ones that have a volunteer committee handling purchasing, decisions are made more often on the

basis of what the committee members like than on what the visitor looks for in that particular shop. This can lead to costly buying mistakes and merchandise left on the shelf that may not even be purchased later from the sale table. Kay knows her patrons and their price point, and she buys accordingly. Another benefit of the gift shop is that it is the museum's "town square." Informal observation and conversation with visitors reveal much valuable information about the museum's image and its operation as a whole, as well as purchasing trends. Visitors feel free to talk in the gift shop, and little informal survey conversations should take place often.

Kay recognizes that the museum is in the business of education and the shop's products must reflect this. We are all conscious of the unrelated business tax rules. Following the mission statement with items related to museum exhibitions and programs is very important. But Kay looks for items that are fun as well as educational. She says that she looks at the play value and tries to accommodate individual needs. She often receives requests from teachers and people living out of town. Supplies for science fairs and insect collection projects are popular. She finds that parents feel good about buying educational items as gifts for family members, and many patrons like to support nonprofit organizations.

Quality and price point are key concerns. Kay looks for high-quality merchandise that is affordable but not found everywhere. She feels that many customers would rather spend money on better quality gifts with play value and developmental benefits. In buying she tries to stay away from too many mechanical and battery operated toys. These toys may not work when the shipment is received or after the customer gets home. Much time and money is spent on servicing their return. Child-friendly display units with samples increase sales. They also help to keep the merchandise from being played with, opened, and damaged. Kay tries to carry enough higher-end gift items to appeal to adults but ones that do not tie up too much money in inventory of slow moving items. Jewelry, kaleidoscopes, and men's specialty neckties have been consistent sellers.

Sales in the gift shop are a function of museum attendance, which is a function of generated traffic through museum programs and events. Gift shop business is also affected by events in town, especially large sports events on weekends. Football games bring large numbers of people to town; however, a daytime game is an all-day affair with little time for anything else. The frequency of school group attendance is also a factor. March, April, and May see the most attendance by school groups and the best sales for the gift shop. There are many new commercial field trip options that now include grocery stores, farms, and pet and aquarium shops that compete for the school group trip.

A FEW WORDS ABOUT UNRELATED BUSINESS INCOME TAX (UBIT)

The Internal Revenue Service (IRS) monitors the sales of museum gift shops through various means, including the Internet. It continues to rule more narrowly on the tax-exempt status of the business income realized from these sales. The items sold must be substantially related to the purposes for which the exempt status of the organization was given. It is, therefore, very important that the shop buyer evaluate carefully the relevance to the museum's mission of every potential inventory item. Some museum gift shops go a step further to ensure understanding and include with the article sold a card explaining the item's relationship to the museum's mission of exhibits and activities. Some vendors are even making it easier for shop buyers by providing these cards ready-made with merchandise. The Science Shop uses these prepared cards when available and is careful about items for sale carried in the shop. Any effort to make clear to the public that gift shop items are related to the museum's purpose and an extension of the education experience can only help to free the institution from potential IRS review.

LAST IMPRESSIONS

The story of our gift shop's beginning is not meant as a model so much as it is an illustration of how relatively easy it is to establish a small museum gift shop with little retail background and resources. With some experience and advice, a sales permit, a cash register, possibly a credit card machine, inventory, and display, a museum shop is in business.

The museum shop has been in business for nearly twelve years now, and we are discussing a face-lift, including new carpet and some new store fixtures to improve the shopping experience for our visitors. Comfortable and appealing surroundings with engaging store displays and quality merchandise should result in more shopping enjoyment and increased sales. Also, we will address more promotion and marketing opportunities. As other sources of revenue decline, the money generated from gift shop sales is becoming increasingly important to support our museum programs. Our investment in gift shop improvements, while increasing marketing and promotion, will return a higher earned income in the future. Gift choices grounded in good quality science will encourage the joy of life-long discovery.

Part IV

EXHIBITS

Decision Making on Purpose: Translating Organizational Identity into Effective Experiences

HANDS ON! INC.

HOW DO YOU GO FROM AN ABSTRACT IDEA TO A finished exhibition? Many new museums are working hard to develop missions, core values, and strategic goals for their institutions. It can be hard, however, to take those philosophical foundations and translate them into concrete visitor experiences.

NEED FOR A STRONG DECISION-MAKING FRAMEWORK

Effective exhibitions are part of making a museum sustainable. In our experience, the key to effective exhibition planning is the creation of a strong decision-making framework at the beginning of the process. Before we design anything, Hands On! works with a museum to develop a set of criteria that will guide all decisions in the exhibition development process. These criteria form a tangible framework that gives shape to the mission, values, and goals of the museum. They might build on work that the museum has already done, or we might start from scratch to create them together. We have found that with this framework complete, every member of the development team has a shared vision of the final exhibition through which all ideas can be filtered.

A strong decision-making framework created at the beginning of exhibition planning does the following:

- Allows the team to focus creativity to an effective end
- Supports sustainability by increasing the efficiency and astuteness of decisions
- Builds consensus and trust among team members that is reflected in the quality of the final exhibition and supporting elements
- Gives the museum team something that can be articulated with confidence when communicating with key stakeholders and potential donors
- Narrows the task at hand so that the team stays focused on priorities rather than side issues

- Gives outside consultants such as exhibit designers and architects clear guidelines
- Helps reduce the last-minute crisis management and remediation that sap the energy and budgets of many projects

So how does a museum—small, medium, or large; new, expanding, or rejuvenating—build a framework for good decision making? There are countless ways to do it, as we've discovered in our work with museums such as Great Explorations; the Children's Museum in Florida; the Telfair Museum of Art in Georgia; the Fort Worth Museum of Science and History in Texas; and on the traveling exhibition RISK! for the Science Museum Exhibit Collaborative (SMEC). One of the most comprehensive examples of the framework building process is the collaborative work done by Hands On! and a new interactive learning center in Belfast, Northern Ireland, whowhatwherewhenwhy:W5.

W5: AN EXAMPLE OF TRANSLATING IDENTITY INTO EXPERIENCES

W5 is a unique interactive learning center located in the disused docklands of Belfast. It is one of the most tangible symbols of Northern Ireland's transformation—a $35 million statement about the kind of city that Belfast wants to be. Funded in part by the U.K. Millennium Commission, a government agency funded by lottery money, W5 and the multipurpose entertainment venue it anchors represent the largest urban development project in Ireland, north or south, and the only major U.K. Millennium Fund project in Northern Ireland. Because of its size and its location in a region best known for civil conflict, expectations were high and scrutiny intense. What could W5 do for Northern Ireland? After an intensive international search, the project developers called upon Hands On! to answer that question.

To create W5, Hands On! worked with a small development team that included Sally Montgomery, PhD, the museum's director. A select group of collaborators from the museum field joined us from time to time to respond to ideas at critical stages. The building architect was also included, allowing us to work together to make sure that the building meshed well with operational and exhibition needs. Surprisingly, we found that the small size of our development team, supplemented by additional resources when needed, made working on this large project very efficient. Decisions could be made quickly to keep the many tasks on track. Although this medium-size museum has 36,000 square feet of exhibition space, the process shared here can be applied to projects of any size.

AT THE CORE

At the start of the project, W5 had a conceptual outline, used for fund-raising purposes. The development team's first key decision, however, was to set that outline aside and take the time to really consider why we were making this museum. What is its core ideology, its reason for being? The core ideology would define the museum's purpose and form the foundation of every decision. As Montgomery put it recently,

> How do you know what kind of exhibits to have or what kind of museum to form if you don't know what and who you want to be? Creating the core ideology gave us the confidence to find designers and staff that fit with that core ideology to create a cohesive museum. It also took away decisions dominated by personalities. 'I want to do it this way' discussions became 'Does this fit with our core ideology?'

This change meant that the museum was about serving visitors instead of our individual interests.

Unlike a conventional mission statement, which can be sentences, paragraphs, or pages long, the core ideology of W5 was limited to a few essential words. A short, clear statement would be memorable and have impact, increasing the likelihood that the team, the stakeholders, and future staff could take ownership of it.

The team worked with organizational coach Roy Shafer to develop a decision-making framework, beginning with the core ideology. To craft the core ideology, the team first considered two main aspects of science—its existence as a summative body of knowledge and its function as a human process of discovery on which that body of knowledge hinges. Discovery, questioning, and experimenting are all creative pro-

cesses, fundamental to the growth and change of scientific knowledge. But these processes are equally fundamental to other disciplines and, in fact, to the definition of what it is to be human. Because of its potential depth and transformational nature, the group gravitated toward the idea that engaging people in a process of discovery should be a key aspiration for W5. The museum would do something inspirational to and with visitors. W5 would "fire the spirit of discovery." We had half of our core.

Realizing this was still an abstract concept, the team knew it still needed to go a step further and define the other half. What would our core business be? How would we pursue our aspirations, our core ideology? How would W5 carry out this idea of firing the spirit of discovery? When the team explored the idea of discovery and discussed science as a process, we saw that people already have varied scientific skills, many of which are usually unrecognized as relating to science at all. If we were to fire the spirit of discovery in people successfully, the museum could help people discover that science relates to them because they already practice the processes of science—questioning, observing, experimenting, predicting—every day to understand and navigate the world. We would "unlock the scientist in everyone." W5 would fire the spirit of discovery by unlocking the scientist in everyone.

The team then spent time defining the values of the museum, or the essential principles that underlie the core and, therefore, all decision making. As a start-up, W5 could only have "strategic" values, principles defined by specific behaviors that everyone—from the design team to the floor staff—could understand and express. As the museum came to life, it would embed those values and behaviors into its core. For W5, the strategic values were determined to be "innovation, imagination, and integrity." W5 will fire the spirit of discovery by unlocking the scientist in everyone through innovation, imagination, and integrity. We had our core ideology and strategic values, the base of a decision-making framework on which to build the museum.

LOOKING AT OUR AUDIENCE

Like many museums, W5 would be an intergenerational social space, attracting people of all ages. Focus groups underscored the particular importance of this idea; from the youngest to the oldest, spending time with family and friends was cited as the most important thing to the people of Northern Ireland. The museum

Figure 17.1. Magnetic Pendulum, W5, Belfast, Northern Ireland. People work together to move this giant pendulum with nothing but the force of tiny magnets. Photographed by Oscar Williams.

needed to serve a broad section of this often-divided community in a comfortable and welcoming way.

The development team felt, however, that more specifically defining our target audience would help us determine the specific spin for the exhibition, which would manifest itself in the overall design and content of the experience. In the case of W5, a special exhibition area had already been set aside for children under eight; that audience would not be a primary focus outside that space. Focusing on the nine-to-eleven age group helped the team better envision the level of design sophistication, the types of activities, and the depth and breadth of content for the remaining exhibition areas. This more specific target audience became another layer of our decision-making framework, refining the way the team approached the exhibition plan.

PERSONALITY MATTERS

To support the core ideology and appeal to the target audience, the W5 development team had to consider the nature of the science center. If the museum were a person, what kind of personality would it need to convey to visitors to achieve its goals? We have all visited places and described them in personal terms—friendly, arrogant, fun, cold, cultured. The development team defined a character that would guide the nature of the exhibition spaces and experiences, the employment of staff, daily operations, the relationship of the museum with its visitors, and the museum's presentation of itself to the public.

To identify that personality, we looked at some faces in magazines, chose an evocative one, and described the personality traits that attracted us to that image. We focused on words we felt best expressed the core ideology of the museum: approachable, authentic, humble, experienced, thoughtful, gentle, respectful, open, kind, supportive, moving, beguiling, magnetic, unselfconscious, classic, and full of joie de vivre. Of these, "moving" rose to the top as the most essential. These traits became benchmarks against which we checked each aspect of the exhibition and environmental design as it developed.

Our team went further by defining three specific characteristics of an exhibition environment that would successfully support the core ideology:

- a physically, emotionally, and intellectually compelling space that is attractive, engaging, and transforming for visitors;
- a personal space where visitors feel welcome and comfortable trying new things; and
- an environment that has a sense of belonging and a regional appropriateness that helps distinguish it from other science centers in the world and helps visitors make connections between their museum experience and everyday life.

We also countered these three ideas with a list of qualities to avoid—ones that would ensure that the core ideology and business would not be achieved: boring, monotonous, predictable, mediocre, ugly, dogmatic, didactic.

AN APPROACH TO LEARNING

The core of W5 focused on discovery, using science process as an entry point to the excitement of finding things out for oneself. With that in mind, our team needed to find an approach to learning and content that supported that idea. This "educational philosophy" would be another layer of our decision-making framework, helping us develop exhibits, labels, and programs, and guide staffing throughout the museum.

At W5, the most important thing we could help people understand was that science relates to them because they already practice the processes of science every day. The skills used in the pursuit of scientific knowledge are the same life skills and thinking skills that help us understand the world and each other. W5 would help people recognize the life skills and thinking skills they already have and help them learn to use them to discover new things about the world and themselves. Through this, we could unlock the scientist in everyone and open people to the excitement of discovery.

Choosing the development of life skills and thinking skills as our educational philosophy also meant that we could broaden the subject matter of our exhibits beyond traditional science fields. Focus groups in Northern Ireland showed us just how important a factor this would be in attracting a broad visitorship. Historically, the Catholic community in Northern Ireland has felt more comfortable with the arts, while the Protestant community has felt more comfortable with the sciences. At W5, we would try to achieve a balance between those two foci, so everyone could feel comfortable and have an entry point to explore ideas.

In creating our educational philosophy, we also considered the museum's sustainability. Northern Ireland has a small population, and repeat visitation would be critical to the museum's long-term success. Open-ended experiences, in which the visitor's actions determine the outcome, would give the museum a good chance at maintaining public interest; people could come back repeatedly and have a different experience without significantly changing the exhibit base. The focus on life skills and thinking skills supported this kind of open-ended approach because each person would bring a unique set of skills and experiences to the museum.

MAKING DECISIONS

Our development team now had the foundations of decision making built—filters that took into account the museum's overriding mission, the people it would serve, the character it should express, and the way it would help people learn. This framework had been developed over the course of a few short months. Now it was time to pull the original outline of the museum off the shelf and compare it with the new concept.

W5 had initially been conceived as a science center organized into four thematic areas called Communication, Energy and Motion, Ourselves, and The Changing Earth. Because our decision making tools were clearly thought out and expressed, it was reasonably easy for the team to decide that this organizing strategy seemed too limiting in its subject matter orientation. It was perfectly possible for people to have great discoveries following this more traditional model, but we felt we might be fighting people's preconceived notions of science as a compartmentalized body of knowledge. The team made the tough, but exciting, decision to throw out the original plan and start fresh, using our decision making framework as the guideline.

What might fire the spirit of discovery and unlock the scientist in everyone? The development team brainstormed multiple organizing strategies that reflected the personality of the museum and focused on process and life skills. The final result was an organizing strategy of five major experience areas that supported the core ideology: WOW (a small introductory area in the lobby), START (the area for children eight and under), and GO, SEE, and DO (the main experience galleries). These simple action words implied that W5 was a place about doing something, focused not on facts but on a process of discovery. Within these areas, we could provide experiences for visitors that crossed all subjects, including but not limited to the original content themes, and underscored the interrelated nature of all areas of learning and discovery. Visitors would be encouraged and trusted to make their own connections and discoveries within this exhibition structure.

This organizing strategy was a turning point for the development team, opening up the project to many possibilities. We used our previously defined personality traits to give each area its own character, while staying within the sensibility of the museum as a whole. GO embodied "joie de vivre," SEE was "beguiling," and DO expressed "authenticity." WOW was "magnetic," while START was "approachable." Our decision making framework guided the planning of a highly open-ended exhibit base, helping us develop and edit our ideas. It also inspired the approach to design, choices in color and materials, labeling, communication styles, and the development of programs.

GO, SEE, DO

With our organizing strategy and key personality traits defined, Hands On! conducted a series of charettes (brain-storming sessions with the project team) to establish the design direction and environmental feel for the GO, SEE, and DO areas. The energetic give and take in these sessions allowed ideas to percolate from a wide range of sources, and our decision-making framework helped us filter out the best matches for our core. Inspiration came from artists, designers, actors, educators, journalists, and perhaps most important, the citizens of Northern Ireland. Through focus groups, for example, we learned of the strong political connotations of color in Northern Ireland. This led to a palette that avoided those associations but still belonged to the physical character of the country. One of our favorite inspirations was an exotic beetle whose intense hues brought to mind the colors of Belfast—the greens of the mosses and the reds of the bricks. Paints and stains at W5 were matched to the "beetle palette" to create a beautiful, slightly off-kilter space still in keeping with the city surrounding it. Other inspirations were equally odd—a handbag, a cocktail napkin, even a toothpaste tube collection. Language, film, theater, and an abundance of specially commissioned contemporary artworks found their way into the mix of exhibitory and environmental elements to create a space where visitors are emotionally engaged and challenged to fill in the blanks.

The first and largest of the three main galleries, GO was purposefully designed to leave open spaces where groups can gather and choose a destination. Large-scale kinetic centerpieces anchor the space visually, provide clear points for navigation and gathering, and add excitement and an expression of the joie de vivre that this area was meant to embody. As the visitor moves farther into the space, the environment becomes denser. Sightlines and custom walls create a visual and spatial progression toward the pointed glass "prow" of this ship-shaped space—a visual celebration of the city that overlooks new waterfront development and a revitalized city center.

SEE, with its emphasis on perception, was oriented more strictly on the column grid of the building. In contrast to the open informality of the GO area, visitors enter SEE through four monumental arches. Using this play on classical formality, people gain a multisensory understanding of perception by moving through these frames and seeing how they define and alter space. Secondary placement of objects, openings, and sightlines introduces a sense of unpredictability and surprise that contrasts with the order established by the archways.

DO, devoted to creativity and "making," is the most experientially dense of the three areas. Small-scale worktables and a material-rich environment inspire invention and invite visitors to collaborate and communicate. Central to this space is a structure that is used by the staff as a preparation and demonstration area, and by visitors as a critical gathering point. This structure embodies the sensibility of the DO area, possessing a "built" quality with all elements expressed as part of its overall visual order.

Together, GO, SEE, and DO create an experience in which the visitor is in charge of the outcome. The exhibition encourages the use of imagination, communication, and collaboration in a population eager to turn outwards and emerge onto the world stage. The result is a social space that is friendly yet sculpturally elegant, refined yet relaxing, international yet respectfully local.

FEELING THE EFFECTS

The decision making framework also had an impact beyond the design of the exhibition. The museum's

Figure 17.2. Lever Tug, W5, Belfast, Northern Ireland. There's an advantage to good leverage, as visitors discover in this experience. Photographed by Oscar Williams.

original title, The Science Centre at Odyssey, no longer seemed appropriate to the development team in light of our much clearer vision of what this place could be. Working with a public relations firm, The Attik, in London, Montgomery developed a completely new identity for The Science Centre at Odyssey that reflected the ideas of questioning, exploring, and discovering; whowhatwherewhenwhy:W5 was born. The framework was also applied to marketing, staff hiring, and the selection of traveling exhibits to support the permanent exhibition platform.

As the project developed, we discovered that, because we had taken the time to build a clear framework at the beginning of the project, many of our later decisions had already been made. We could focus our efforts in the right direction and spend our time and money purposefully. Whenever we felt a little lost, we could check back with the decision making framework and determine whether we were on track, before a crisis set in. The result was a cohesive museum that opened on time, on budget, and to great success as a social learning space where people feel welcome and supported in the discovery process.

Since opening in March 2001, W5 has spearheaded a trend to marry science and art in dynamic experience, leading the *New York Times* to declare the design "an approach that straddles the line between science exhibit and installation art." Learning is embedded, rather than being the experience itself. Because of the early development of the decision making framework, Montgomery feels the team was able to "create an environment for all families and communities in Northern Ireland to enjoy—one that is mysterious, highly motivational, visually exciting and, most importantly, enjoyable and educational." In 2002, the cohesiveness of the design that resulted from our decision-making framework was recognized with a Gold Award from the Industrial Designers Society of America, placing the modestly budgeted W5 in the same ranks as the U.S. Holocaust Museum and the Rose Center for Earth and Space at the American Museum of Natural History. The Institute of Designers in Ireland also honored W5's exhibition design in its annual design competition.

The public response to the project has been overwhelmingly positive, confirming W5's role as a unique cross-community meeting space. W5's opening year saw 245,000 visitors in a country with a population just shy of 1.6 million. Repeat visitation has remained strong, particularly in the critical family and teen markets. As a mark of this success, W5 was awarded the Northern Ireland Tourist Board's Visitor Attraction of 2002, and was runner-up for the Chartered Institute of Marketing's award for new business marketing. W5 has also experienced increasing tourism from Europe and North America and has hosted several European science conferences—groups that would never have considered Belfast a destination prior to the success of this flagship facility. As one of the most successful Millennium Projects in the United Kingdom, W5 should continue to move people to laugh, wonder, think, and put aside their differences.

CHAPTER 18

Making Effective Exhibits for Rewarding Visitor Experiences

Kathleen R. Krafft

WHY WE DO IT

"We don't make mistakes, we make prototypes!"

This motto, hanging in our workshop, summarizes a key approach to creating fun, interactive exhibits for a start-up science center. Choose an idea you are interested in, then dive in and build a quick prototype, which can often be made of duct tape, cardboard, or scrap materials. Try it out with visitors, and you'll watch them have fun with your prototype. You can then improve your prototype, move on to a new version, or decide that the idea wasn't really feasible.

And We have Fun Doing it!

I asked our staff and volunteers why they enjoy the exhibit process so much. Here are some of their responses:

"To inspire visitors."

"You are always exploring something different."

"There is humor in our mistakes and learning." (We make it fun by laughing at our mistakes, and sharing in our successes.)

"Triumphs."

"It is fun working together."

"You can always recharge your batteries by walking on the museum floor and watching kids." (They are always laughing, or squealing in surprise, or intensely engaged, and you'll hear "it's cool" a lot.)

We have a core group of volunteers and crazy staff, who having been doing this a long time and keep coming back for more. It is rewarding, with wonderful successes and surprises in store for anyone.

An Example of How an Exhibit is Built

One of our volunteers had seen a small demonstration where a spinning magnet was held over a metal disk and the disk spun without being touched. We created a prototype and made it work.

Then we built an exhibit for our museum floor. An upper plastic magnet disk has sixteen magnets of alternating polarity glued to its underside. Visitors turn the crank on top to spin the magnet disk to get either the copper or aluminum disks spinning. They can locate a plastic disk under the magnet disk, but it won't spin unless they pile up change on the plastic disk. They can also raise the magnets farther away from the disks with the handle on the left, and discover that the copper or aluminum disk turns more slowly.

GETTING STARTED

What Makes a Good Exhibit?

The Sciencenter developed a set of guidelines to keep in mind when planning and building exhibits—make interesting exhibits, with good underlying science or math content, where visitors can manipulate and affect the outcomes, and where they can be surprised.

Good links and summaries can be found at Informal Learning Review's website: www.informallearning.com.links.htm.

Sciencenter Exhibit Design Guidelines

CONCEPT

- Relatedness to everyday life: Visitors are more interested in something they can relate to, something that they might have experienced, something that matters to them.
- Capacity to promote group interaction: Visitors learn and explore more in group settings, conversing and exploring with parent or guardians or other kids; build exhibits that a group of people can gather around.
- Appeals to range of appropriate ages: Make exhibits that many ages can enjoy.

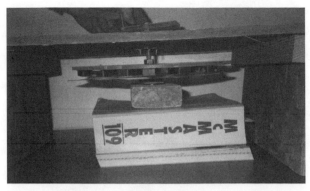

Figure 18.1. One of our volunteers had seen a small demonstration where a spinning magnet was held over a metal disk and the disk spun without being touched. We created a prototype and made it work.

EDUCATIONAL ASPECTS

- Science content: Exhibit should be based on real scientific (including mathematics and technology) principles, and must be accurate.
- Open-endedness: Provide lots of possibilities for exploration, not just a single correct answer.
- Interactiveness: Touch, manipulate, make something happen.
- Communication of science content: Science principles can be uncovered and come with sufficient signage and suggestions.
- Relation to process versus content of science: Let visitors come up with ideas to try out as they explore, generating and answering questions. Discovering, imagining, sensing, modeling and testing are parts of the process of science.

Figure 18.2. After creating a prototype, we built an exhibit for our museum floor. An upper plastic magnet disk has sixteen magnets of alternating polarity glued to its underside. Visitors turn the crank on top to spin the magnet disk to get either the copper or aluminum disks spinning. They can locate a plastic disk under the magnet disk, but it won't spin unless they pile up change on the plastic disk. They can also raise the magnets farther away from the disks with the handle on the left and discover that the copper or aluminum disk turns more slowly.

DESIGN AND CONSTRUCTION

- Buildability: Can it be built by the volunteer? Are there realistic plans for construction?
- Repairability: What happens if something wears out? Build in access panels, too.
- Maintainability: Can you keep it looking good and running without too much effort? Is it made with standard, easy-to-replace parts? Is it strong enough?
- Survivability: Is it durable? (Again, try out a prototype.)
- Affordability: Do a realistic estimate, and compare with funds available.
- Testability: Can a prototype be built to test the concept? Particularly when trying out a new idea.
- Networking: Has a similar exhibit been built before? See what you can learn from others.

OTHER CONSIDERATIONS

- Safety: It absolutely must be institutionalized, from idea conception to final fabrication. Anything that you think a visitor might do must be addressed— it can happen. Loose parts, choking hazards, parts that could fall, sharp corners are typical examples of safety concerns.
- Labels and signage: Plan for these from the beginning. Not too long, layer the text, and position them in a good location that has been tested.
- Handicap accessibility: Can a wheelchair approach it, or can the user reach it?
- Expendables: What will it cost to supply balls, paper, whatever?
- Special requirements: Power, water, drainage.
- Relatedness: Relating to other exhibits can enhance the visitor experience.

Where Do We Get Ideas?

- *Visit other museums.* Visit a number of museums, and take many photographs. I collect photos and put them in binders, in clear plastic sleeves, to refer to later—you won't remember as much as you'd think. Photos also provide a good resource for other projects later: exhibit design, construction, signage, accessibility, finish, and so on. Also observe how visitors use exhibits and their labels.
- *Copy good ideas. We say that "we don't steal ideas, we propagate them!"* The Association of Science-Technology Centers (ASTC) sponsored a couple of sessions on "Whose idea is it anyhow?" about who owns the exhibit ideas, at their annual conferences in 1996 and 1997. In summary, the consensus was that the science belonged to all of us, but if we saw an exhibit idea we liked, we would try and develop it further, and make it fit into our museum space. At the Sciencenter, we acknowledge the source for

exhibit ideas and volunteer builders on the bottom of the labels. Most museum folks are very willing to help you out if you call with questions.

- *Check museum websites.* They are good sources for ideas, such as the Exploratorium books (three Cookbooks, and Snackbook). ASTC has published Cheapbooks of easy exhibits to do. There are some wonderful Question & Answer books on neat science phenomena, many of which lend themselves to exhibits. Some museums and individuals have great websites. Some topics are particularly challenging, such as biology, which often requires living samples, or chemistry, which is generally done as supervised activity.
- *Don't forget animals.* Visitors love animals and are very interested in learning about them. Simple critters such as newts, or African giant millipedes, are easy to care for. But many animals, especially reef aquariums, are a lot of work, and need dedicated champions.

READY TO START BUILDING?

- *Find a champion.* Every exhibit needs a "champion" who is passionate and dedicated to it and who will keep driving it along until it works well for visitors.
- *Work in small groups.* One person can offer suggestions, note upcoming problems, and help problem-solve. Another may suggest a good location for a label, or know how to bend Plexiglas, or know of some good part you could use.
- *Start small.* Choose easy exhibits that your builders are excited about building. Expect that about half of them will become useable exhibits.
- *Recruit volunteers for specific tasks.* We know several artists who are thrilled to be asked to paint a Mars backdrop, or a tropical rain forest, or draw cartoon characters. Find a retired machinist who is willing to do welding and some lathe work.
- *Find science content or technical help if needed.* Check around your local community university, technology teachers.
- *Ask for help.* Most folks generously share their expertise, ideas, and experiences. I'm happy to share how we built something and what we would do differently now. The ASTC or webhead listservs can be valuable sources of information, and good places to post a query; I make a point of replying if I have any useful information to share.

Materials and Techniques for Fabrication

Finding materials may seem difficult when you are starting up. Visit a couple of other shops and see what kind of stuff, tools, and catalogs they use. Collect cata-

logs. Absolutely amazing and essential are McMaster-Carr and Grainger, which are both available online; you must, before finishing reading this article, look through their websites just to see what is available. The Internet is getting increasingly useful for material searches. A very comprehensive list of sources and materials, organized into categories, is included in the Resources section of this book. Please use it!

Acquiring Skills and Tools

Learn by doing and asking questions. None of us went to exhibit-building school. We have added to our bag of tricks as we have worked with other people, or saw something at another museum and inquired how it was done. Staff at your local lumberyard or hardware store are generally very helpful, especially if you mention you are building something for the local science center. A local cabinetmaker would cut lumber for us before we had a good table saw and had great suggestions about tools and techniques such as working with laminates (such as Formica). We learned about working with Plexiglas (acrylic plastic) from one of our volunteers. Acrylic plastic is a key material to work with. Learn how to glue it, get a blade designed to cut plastic for your table saw, Plexiglas-compatible drill bits (available from McMaster-Carr), and a set of hole saws. Tools are expensive—we made major purchases (such as a good table saw) with funds from grant projects. (Note: the Ann Arbor Hands-On Museum was able to purchase shop equipment with grant funds also— money well spent, for a small center.)

Prototyping

We strongly recommend making prototypes for all exhibits to avoid investing a lot of time making a beautiful exhibit that is a dud either for technical reasons or because it isn't of interest to visitors. Sometimes we build a cardboard mock-up just to get the layout right. We did this for the dividers for some five-foot diameter round tables for puzzles, for instance, after which the dividers worked really well for signage.

You can build really quick prototypes (foam balls hanging from an umbrella to mimic cells inside your body), and try out the idea with kids in a school setting, after-school program, and so on. We invite staff to comment before anything new goes on the floor, at any stage from early prototype to finished product.

For a math exhibition, we wanted to develop an exhibit where kids and adults counted how many of "their feet" long was something. Thus the cardboard mock-up of a dinosaur was created using an overhead

Figure 18.3. The cardboard mock-up of a dinosaur was created using an overhead transparency projected onto a piece of cardboard.

transparency projected onto a piece of cardboard. Visitors liked the activity once we added footprints on the floor to guide them in putting one foot touching the next.

Making Your Exhibits Durable, Repairable, and Portable

We build our permanent exhibits in-house, almost always with a laminated countertop and painted cabinet bases. We mostly use MDO (medium-density overlay) plywood as it has a smooth paper finish and looks good when painted carefully. We use auto-body filler for countersunk screw holes (two-part product such as Bondo), which holds up much better than wood filler. Include access panels and holes for electric cords.

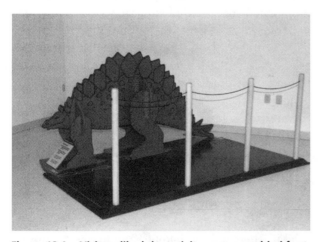

Figure 18.4. Visitors liked the activity once we added footprints on the floor to guide them in putting one foot touching the next. Photo 4 shows the final exhibit.

After a few months, you'll start to see what will make the next exhibits better. For instance, we found that plywood panels running all the way to the floor get easily damaged at the bottom when being moved; now we put a rounded skid of thick hardwood under such cabinets. Volunteers built some standardized tables with laminated countertops and painted legs. We later added some five-foot diameter round tables with removable metal legs for easy storage. These give us flexibility to move small exhibits around. Exhibit maintenance is ongoing, so anticipate it.

ONCE YOU ARE UP AND RUNNING

Once you get started, several themes will recur, no matter what your size or mission.

Need for Change

Visitors definitely appreciate changing exhibits—we get a lot of positive feedback in our comments book about this. When we were smaller, we would develop in-house mini-exhibitions of 800 square feet or so, using our volunteer exhibits committee. Puzzles, mirrors, and illusions, tabletop electricity and magnetism exhibits, and motion made great topics. These clusters kept up visitor and volunteer interest and provided us with a store of finished exhibits when we expanded our facility.

Maintenance

Visitors definitely notice and remember broken exhibits. We try to make repairs as quickly as possible, or we take the exhibit off the floor. We have cute "Exhibit Under Repair" signs that we'll put on exhibits that are out of order; otherwise visitors will be frustrated and wonder if they are doing something wrong when the exhibit doesn't work. We use our volunteer museum guides to fill water, add paper.

Safety Inspections

We take safety very seriously, as an integral part of our thinking from early brainstorming to finished exhibit. For example, we make sure to round corners and edges, and make exhibits stable. We don't allow pinch points, exposed wires, or dangerous protrusions. We add extra safety cables, clamps, or supports to ensure that nothing can fall down from up high, which can be particularly dangerous. We do not allow any serious choking hazards on the museum floor. Over time, we've become more knowledgeable about proper electrical wiring inside cabinets, and UL (Underwriters Laboratories) standards. Before a new exhibit goes on the floor, our floor staff coordinator, who is also our museum's official safety officer, inspects and approves the exhibit.

Our exhibits committee performs a semiannual safety inspection of each exhibit on the floor—noting any serious and minor safety concerns (worn rope, tippy sign, splinters, and so on) and maintenance needs. We address the most serious concerns immediately. We compile an annual running safety report to keep track of when the concerns were addressed, and what was done.

Invite a safety expert to look over your facilities. For instance, for our chemistry demonstration program, a safety expert from a local college came to our facility and made some important suggestions such as adding an eyewash, labeling every container, or storing some chemicals in a plastic tub in case they were ever to leak and corrode the metal shelving. Playground safety standards also provide some helpful guidelines; visit www.cpsc.gov/search.html and search for playground safety.

Accessibility

Accessibility is a very important consideration and becomes second nature to think about it as you design your exhibits. Smithsonian and other guidelines help you in choosing where to put label copy, how far a visitor in a wheelchair can reach, table heights, tripping hazards for visitors with low vision, and other similar issues. Check out the ASTC website for many good links.

Signage

Most of our labels are created in Microsoft Word—they don't need to be fancy for starters. Simple is good—maybe fifty words of instruction, then some more content, layered if necessary. Make a consistent location for the to-do part of the sign with a large font, then a smaller font for additional information. Clip art and color will jazz up labels. There are accessibility guidelines for fonts and sizes that are most readable. It is rare that the directions are very good on the first try, so plan to watch visitors, and be prepared to try several versions and change the label's location.

Design Exhibits that are Group Friendly

We try and design exhibits that encourage group interaction, such as rounded tables with access from several sides. Research on family learning at museums supports our experiences that visitors in groups talk to one another, encourage one another, brainstorm together, and so on.

Observe How Visitors Use Exhibits

A lot can be learned by standing back and watching five or so visitors use an exhibit. When they're done you can ask them for suggestions, and see where they had difficulty. Also, ask children to read labels. This will let you know if the text is age-appropriate. As you are watching visitors, keep in mind that some evaluators talk first about:

- The *context of invitation*: Is the exhibit attractive, inviting, intriguing?
- Then *navigation*: Can visitors manipulate and try out things successfully, or do they walk away to the next exhibit without being able to figure out what they might do?
- Next *engagement*: Are visitors engaged, conversing, and experimenting with the exhibit?
- Finally *learning*: Are visitors making connections to other experiences or to other exhibits?

OTHER CONSIDERATIONS

The following should be helpful rules of thumb for exhibit development and maintenance.

Exhibits Budget

We are very cheap—a few thousand dollars to develop new exhibits each year, and a few thousand dollars for maintenance and parts kept our smaller museum running; after our expansion, we now rely on grant-related projects to generate many new exhibits.

Staffing

When we had 4,000 square feet of exhibits, the equivalent of a half-time staff person was responsible for exhibit development and maintenance. This was done with the assistance of one full-time volunteer jack-of-all-trades who did much of the work for both exhibits and facilities maintenance, plus an assortment of other volunteers; additional staff was hired for the development of grant-funded traveling exhibitions that provided prototypes and finished exhibits for our visitors.

At 12,000 square feet, we now have one full-time staff member who is responsible for exhibit development and maintenance, plus the same amazing full-time volunteer, and a regular corps of volunteers. We have one director of exhibits (yours truly) who oversees the exhibits on the floor, budgeting and record keeping, maintenance and safety, and manages the development of traveling exhibitions from various grant-related projects with the help of additional staff and volunteers. Budgeting becomes increasingly important as your museum grows.

Shop and Materials Storage

We started small—maybe 300 square feet and a drill press—and we built most exhibits on boards spanning two trashcans. Storage space for parts and scraps, and exhibits off the floor, is wonderful. Parkinson's Law applies here: Junk expands to fill the available space. It really helps volunteers if you organize your shop so there are appropriately labeled drawers for tools, hardware, Velcro, hinges, long bolts, and other items. We have some shallow metal drawer units (donated) that are terrific for scrap plastic, plumbing parts, and the like. We keep spare parts and leftover materials that might be useful for other projects in labeled tubs and boxes, ranging from a conveyor belt, extra power supplies, and plastic bottles, to polarizing sheets. Exhibits staff makes sure there are adequate supplies of band saw blades, drill bits, hardware, sandpaper, and working tools. Manuals for all the tools are saved in a three-ring binder with dividers. As we have grown, we've made more custom racks to store lumber and sheet plastic, and other storage for potentially useful scraps.

Documentation and Record Keeping

You'll be forever grateful if you keep detailed, accurate records. At the Sciencenter, every exhibit or idea in progress is assigned a number. A simple Filemaker database is used to keep track of which exhibits are installed, in storage, in repair, and so on. This database is used to generate the lists for the safety inspections. The database can be used to keep track of exhibits that need work, dates, if the exhibit has a label, and so on. Each exhibit also has a file folder with name and number, where we keep wiring diagrams, manuals for parts, old labels, sketches, and photocopies of important receipts for parts and materials.

Digital photos are great for documentation. Prototypes are sometimes photographed to document the development process, including those in the "graveyard" that were tried and deemed unsuccessful. Critical repairs or components can be documented with a digital camera.

Separate files are also kept for safety reports and exhibits committee meeting minutes.

The maintenance log is a separate binder, in which exhibits are listed by number, with a separate sheet for each exhibit. Any non-routine maintenance is noted in the log. This is good documentation for safety purposes, too.

VOLUNTEERS

A volunteer exhibits committee has met biweekly for much of the past twenty years to brainstorm exhibit ideas and help each other. Although the majority of exhibits committee members have backgrounds in physics, chemistry, or engineering, to find folks interested in astronomy, biomechanics, and other areas adds a lot to the group. When someone new contacts us about an exhibit idea they would like to share with our community, we invite them to a meeting.

This group is very willing to generate ideas, and the rest of the group will happily brainstorm and ask lots of questions. No one is offended at the critiquing of their specific plans. This is an opportunity to bring up safety or maintenance concerns, or make sure correct science is presented in the signage. Ideas are always welcome, and it might be several years before someone gets around to a project.

We now have an exhibits listserv that allows the sharing of ideas and means that the exhibits committee members and interested volunteers, as well as several staff members, get notes so that they know what might be coming that is new.

Volunteers are interested in doing different things. Some are interested in adding new signage. Some keep records of our meetings. Some love to do software projects. Some will not attend meetings, but love specific projects such as welding. Some volunteers get online and do research. Some find unique ideas online and share them. Some will do minor repairs or paint jobs.

Many members of our committee have connections with Cornell University (five minutes away). They bring us their ideas, champion exhibits they are excited about, and also bring resources. National Science Foundation and universities are encouraging outreach, so we can often access Cornell University's machine shops or other resources. We have benefited from numerous class projects, and we get a lot of donated laboratory equipment.

SUMMING IT ALL UP

A True Story (January 2005) from One of Our Staff Members

I called one of our large national suppliers this morning to order some parts. The branch we call is located in Syracuse (an hour north). When I gave the woman who was helping me our account number, she began yelling "Oh my! You're from the Sciencenter in Ithaca! That is my son's favorite place in the whole world!" She went on to tell me how every time they go to visit her son's grandmother in Elmira (an hour south), she takes him to the Sciencenter. It is their "special place." She said they once even discussed the possibility of one day going to Disney World, and the grandson replied that they could go to the Sciencenter instead.

CHAPTER 19

Learning from My Mistakes

Paul Orselli

> My life is full of mistakes. They're like pebbles
> that make a good road.
>
> —Beatrice Wood

As I look back on the over twenty years that make up
my own road in the museum business, I am struck by
the similarities in the various situations I've encountered. Despite the different places I've worked in (Acton, Austin, and Ann Arbor, to name a few) and the
different types of museums I've developed exhibits for
(science centers, natural history museums, and children's museums), there are always similar "pebbles"
along the path that create the challenges in turning a
bright idea into a completed exhibit component. I've
gathered together a few rules of thumb that I continue
to find useful as I develop exhibit ideas.

MONEY MAKES THE (EXHIBIT) WORLD GO ROUND

Naturally, you need money to create exhibits, but, para-
doxically, having a tight exhibits budget often forces
more creative solutions to challenges that arise. Here
are a few things to consider:

Do Look a Gift Horse in the Mouth!

Be very careful when accepting donations of exhibits
materials from well-intentioned people. Your own institution should have a formal collections and acquisition policy, as well as a form that can be filled out with
pertinent details on donated objects. The most important part of any acceptance form is the inclusion of a
clause that gives your institution complete authority as
to whether to display, keep, or dispose of the donated
object(s). Outdated technology, including computers
and the like, can become a real nightmare to maintain.
At the end of the day, does the donation truly fit your
mission and approach, or is it merely a refugee from
someone's attic? A corollary to this is to be very careful
when sponsors offer to build exhibits for you. Just be-

cause a board member's company employs a phalanx of
engineers doesn't mean they know how to make things
that will hold up to the enthusiasm of museum visitors.
If a sponsor builds an exhibit, you should work closely
with them to have the final say on both content and
fabrication issues.

VOLUNTEERS, COLLEGES, AND COMMUNITY GROUPS

Retirees and volunteers with special skills and talents
are a valuable and often overlooked resource. These
folks can often complement your exhibits staff in important ways. Retired tradespeople often are willing to
complete projects or exhibit tasks for your institution.
Local colleges or universities are often willing to work
with a museum to create real-life projects for their students to complete as a class project. Engineering, education, and art students are all good candidates for
this approach. Are there community groups, such as
Rotary or Kiwanis clubs, that might join as a partner
with your museum on an exhibit project?

DEVELOP RESOURCE BASES

It is important to become familiar with exhibit resources (human, printed, and electronic) so that when
you are stuck trying to find the answer to an exhibit
development problem, you will have several avenues
to follow. Two types of printed resources are books
and catalogs. It is worth spending the money to create a bookshelf full of resource material for yourself.
Both the Association of Science-Technology Centers
(ASTC) and American Association of Museums sell
many excellent publications. Some of my favorites are
the Cookbooks and Snackbooks published by the Exploratorium, as well as The Cheapbooks published by
ASTC (and edited by myself).

If I could only pick one catalog to recommend to a
fellow exhibit developer it would be the McMaster-Carr
catalog. From the catalog, you can purchase anything

from a single washer to a wheel for a railroad boxcar. McMaster-Carr also has a website: www.mcmaster-carr .com.

The World Wide Web provides many resources for exhibit developers. You can find materials (don't overlook e-Bay as a source for unusual items) as well as local suppliers and fabricators. There are several museum-related internet newsgroups that allow registered participants to post questions as well as just following the conversations of others. Examples of such newsgroups include Museum-L, CHILDMUS, ASTC-ISEN, and webhead. Internet newsgroups provide an instant set of colleagues even for those people who are new to the field or are unable to attend conferences.

PEOPLE AT AN EXHIBITION
Even though I'm writing about exhibits, often the most important thing inside your museum is not the exhibit stuff, but how people (staff and visitors) interact with the stuff.

A museum visit starts before the front door. Was the museum easy to find, with good direction signs? Do your website and answering machine convey the information visitors need in a simple and professional manner? If a visitor becomes frustrated before they even get inside your museum, it doesn't matter how great the exhibits inside are—your museum is starting off at a disadvantage.

Little Details Add Up
Clean, attractive, well-maintained exhibits send a message of "we care, you should too!" The same goes for the entire museum environment. Grimy walls, ill-maintained bathrooms, worn-out exhibits with missing pieces—what messages do these things send to visitors? I have spoken with several colleagues who have corroborated that when their institution made a concerted effort to improve overall building maintenance, instances of exhibit vandalism decreased as well.

How Happy and Engaged Are Members of Your Staff?
Every staff member (not just exhibits people) should *want* to know what sorts of things happen inside the museum. Ideally, they should *want* to use the exhibits, too. There is nothing worse than asking a person at the admissions desk of a museum a question, and realizing they've never set foot inside! Make it a point to include *all* staff and volunteers on minitours of new exhibits and some training on the permanent exhibits. Encourage mechanisms for suggestions about exhibits

by visitors and staff to be recognized and responded to. A worthwhile use of wall space is to create a bulletin board where common exhibit suggestions or questions can be posted with answers from the exhibit staff.

MAKE IT A MARATHON, NOT A SPRINT
After the band stops playing, and the ribbon is cut on opening day, then what? What will your new exhibition (or new museum) look like five months from now? Five years from now? As Don Verger, the founder of the Discovery Museums in Acton, Mass., remarked to me after the opening of a new museum building, "Well, now we're open forever!" Make decisions about your exhibits program as though you'll be open for a long, long time. I've seen plenty of exhibits that were meant to be temporary are still being used and enjoyed by museum visitors years later.

BE THE BEST YOU CAN BE
Sometimes exhibit staff members from newer or smaller museums develop an edifice complex (especially after seeing the larger and older host institutions at museum conferences). I've heard colleagues complain that "we can't compete with the big museums or Disney World." There is room for a whole spectrum of museums, large and small. I would much rather visit a smaller museum that is well maintained and filled with thoughtful and creative touches, than a giant museum filled with gigantic exhibit galleries and many "Out of Order" signs. Large institutions often overwhelm their visitors. (How many gigantic museums have you left with a headache from the sheer noise factor?) Instead, as a small museum, look for ways to "whelm" your visitors by providing more authentic and intimate opportunities for discovery.

GROW SLOWLY (IF AT ALL)
Be careful as to how your museum grows. I firmly believe that there is an inherent "carrying capacity" or maximum effective size to an interactive museum. There *is* such a thing as a museum that has grown too large. Larger institutions are often forced into putting $5 ideas into $5,000 boxes simply because they are trying to fill space with things (they hope) that will never break because of inadequate staffing. If you don't have enough floor staff and exhibits staff to manage your interactive exhibits, you are creating a recipe for visitor and staff dissatisfaction.

Break tasks into chunks. Starting a museum, or an exhibition project, can be both an exciting prospect and a daunting one. Rather than looking at your project as an enormous, complex task, I have found it useful to break the project into smaller and smaller tasks, or chunks. By breaking your project into these smaller chunks, it is often easier to see clearly which parts of the project need more attention as well as making it simpler to delegate responsibility for specific aspects to others (if you are not a one-person department).

TRY, TRY AGAIN

Creating exhibits really is an iterative process. Every good exhibit developer has to have a little bit of a masochistic streak. Every day visitors are going to be pounding away at components that you have probably spent months (if not longer) developing. Save yourself some heartaches and headaches by following the tips below:

You Call That Prototyping?

Prototyping has become a word that means many different things to many different people. To me, prototyping should be a quick and dirty process. Scrap wood and duct tape all the way. Create a working "proof of concept" to show people in order to find out if they're even interested in your idea. You need to make mistakes in terms of label placements, table heights, reach distances, and so on *now* rather than after you've made welded frames or created custom-sized laminate tabletops. If you have already spent too much time and money on an idea, you may be unwilling, or unable, to give it up.

Beware of the Latest Gizmo

Be clear on what you want visitors to get from an exhibit component. If you don't know where you're trying to go, you won't clearly understand the best exhibit tools and techniques to get you there. Too often the "siren song" of technology becomes louder than the "voice" of the exhibition content itself. Make sure the technology with which you chose to create an exhibit becomes an elegant solution and not merely an answer in search of a problem.

Build It to Last because Everything *Breaks Eventually*

Choose materials and finishes carefully. If you don't have enough money for laminates, for example, avoid the trap of painting surfaces that visitors will come into contact with. Even high-quality paint fades and gets worn after thousands and thousands of hands rub and pound away at your exhibits. You will be much better off with choosing a sturdy material such as birch plywood carefully finished with a clear coating such

as polyurethane finish if you are on a tight budget. Then, even if a chunk gets gouged out of your exhibit furniture, a little judicious sanding and a reapplication of a clear coat makes it look almost as good as new. Also, don't forget to add a 4-inch or greater kick plate or recess around the bottom of exhibit furniture to prevent scuffs and dings from flying footwear.

Even if you build exhibit components that you consider bombproof, make sure you have left the people who may come after you with easy access to controls, parts lists, and repair instructions (including a copy inside the exhibit component itself). I'm reminded of a frustrating telephone call to the designer of a traveling exhibition when I couldn't figure out how to get inside a piece of exhibit furniture to replace a blown electrical component. "Well, you can't get in. The box is completely sealed," he said. I replied, "The component is rated for one million hours, so we never expected it to break down!" Needless to say, that exhibit was never repaired at our museum, or any other museum for the rest of the exhibition's run.

What Does ELVIS Have to Do with Prototyping Exhibits?

At the 2000 ASTC Annual Conference in Cleveland, I was fortunate to speak on a panel about exhibit prototyping chaired by Patrick Tevlin of the Ontario Science Centre. Part of my talk dealt with how to use the word ELVIS as a mnemonic (memory aid) for exhibit developers when they are developing prototypes and exhibit components. The meaning of each of the letters in ELVIS follows below:

E = Everyday Materials—Using everyday materials makes it easier to prototype and maintain exhibits.
L = Looseness—Providing open-ended opportunities during exhibit development and for the visitors.
V = Vermicious—The word means *wormy*. Good exhibits and prototypes should worm around on many different levels: old/young, art/science.
I = Interesting, two things: (1) Great exhibits/prototypes always seem interesting, no matter how many times you've seen them. (2) If it's not interesting to you, how can you expect your visitors to be interested?
S = Sharing—We should be sharing ideas with our visitors and one another as professionals.

A GREAT EXHIBIT IS NEVER REALLY FINISHED

After the blood, sweat, and tears that went into creating an exhibit component, you might feel like you never want to see the darn thing again. Later, after you catch up on showers, meals, and sleep again, watch the un-expected things that visitors do with "your" exhibit. It might take a while, but you will start to think of ways to make the exhibit even better. This iterative process of observation and improvement is an important part of exhibit development—enjoy it!

I'd like to finish with a quote from Dick Crane, an eminent physicist, who in retirement became a remarkable exhibit developer at the Ann Arbor Hands-On Museum:

A newly installed exhibit will break down in ways that could not have been foreseen. That is normal; it is the way we learn! But don't repair it back to the way it was before it broke. It will just break down in the same way again. Always think of a better, stronger, more fool-proof way to do it. If that is done, the frequency of breakdowns will approach zero exponentially (but never quite get there).

CHAPTER 20

The Use of Objects in a Small Science Center

CLAUDE FAUBERT

THERE CONTINUES TO BE CONSIDERABLE DISCUSSION about the place and role of objects in a science center. Some advocate that objects do not belong in a place dedicated to understanding the methods of science and the processes of technology through interactive exhibitions. Others on the other hand have said that objects do belong in science centers as products of technology, and have much to tell visitors about the very technologies that have produced them and about the people who have used them.

Many museums of science and technology, such as the Canada Science and Technology Museum (CSTM) in Ottawa, have in recent years developed approaches that allow the visitors to appreciate not only the beauty and the ingenuity of objects new and old but also the way in which they function through interactive exhibit components. This article presents the approaches used at the CSTM and shows how they can also be used in a small science center setting.

SOME DEFINITIONS: ARTIFACTS, DE-ACCESSIONED ARTIFACTS, AND PROPS

At the CSTM, we use the following definitions to differentiate between the three types of objects that are used in our exhibitions and programs:

(a) Artifacts are "accessioned original or illustrative items which include objects of (Canadian) invention, discovery, innovation, manufacture or applied foreign science and technology which have played an important economic or social role or which have contributed to the further development of science or technology, nationally or globally" (CSTMC Guideline on De-accessioning and Disposal of Artifacts, 1993). Artifacts include, for example, models, art, museum material, and type specimens without temporal limitations. Artifacts are what museums typically collect and preserve. Some cannot be easily displayed, requiring special climatic conditions and protection from the hands of the visitors. Fortunately, many technological artifacts can be displayed, having been designed and fabricated initially to operate in difficult conditions.

(b) De-accessioned artifacts are "objects that have been removed from the permanent collection because they are duplicate objects, the technologies that produced them are over-represented in the collection or because they are in poor physical condition" (CSTMC Policy on Collection Development and Management, 1998). De-accessioned artifacts can be used in programs or in interactive exhibit components.

(c) Props are objects that have been acquired for the clear purpose of being displayed, very much like props on a movie set, or of being touched or operated by visitors in a program or a public activity.

EXPERIENCE OF THE CSTM

The Canada Science and Technology Museum is Canada's largest science and technology museum. It opened in 1967 and welcomes about 400,000 visitors per year. It has a dual mandate: to maintain and develop a collection of scientific and technological objects, with special but not exclusive reference to Canada, and to demonstrate the products and processes of science and technology. The museum's structure and operations are geared to fulfilling this dual mandate. Much of its space is devoted to collection storage and many staff members work behind the scenes on researching the history of objects, cataloging the collection, and conserving its pieces.

From the very beginning, and particularly in the last ten years or so, the museum decided to include a strong element of interactivity in its exhibitions. Members of the staff felt that artifacts by themselves were not sufficient to tell the stories, and thought that interactive elements would enhance the experience of the visitors by providing different and more appealing ways of showing some of the basic scientific and technical principles that make the artifacts work. Today, we use objects, mostly props or de-accessioned artifacts, in many of our programs and as the focus of many of our interactive exhibits.

After years of using objects in exhibitions and programs, we have found that:

- Older objects often show the application of a technology in a clearer and more understandable way than a graphic or a simplified mechanism. Visitors often point to the steam locomotives on display at the museum as a good example of this.
- Objects have the power to affect people's emotions by evoking memories. This is particularly true when groups of visitors of various ages visit together. Often, an object will bring strong memories to a grandparent who will then share it with a grandchild.
- Objects provide a concrete and tangible link to the past. In today's world, where much of the emphasis is on virtual interactions, people are still very impressed by old technical objects, particularly the very large ones.
- Objects provide a link to the real word. Here again, people rarely get to see technology in action, be it a robot operating today in a factory setting or a printing press from the mid-1800s.
- Objects create reverence and respect for the ingenuity of past scientists and engineers. Many visitors have told us how impressed they had been by the intricate workmanship of many older products of technology.

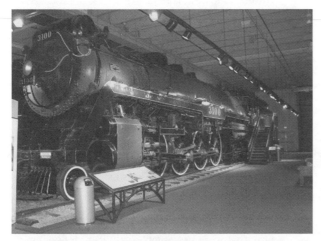

Figure 20.1. This steam locomotive was built by the Canadian Pacific Railroad in 1929. National Museum of Science and Technology, Ottawa, Canada.

HOW THE CSTM USES OBJECTS

At the CSTM, we use objects of all three types (artifact, de-accessioned artifact, and prop) in many different ways. Here are a few:

Artifacts on Display, Usually with Explanatory Text

- This is the traditional museum approach to displaying objects. Objects can be presented singly, in thematic groupings, or as part of a period setting. They are then used as the main element in telling stories about the past. We have found that visitors relate very well to certain types of objects: objects that are big or tiny, rare, beautiful, or expensive attract attention. As Stephen Weil wrote, "[w]hat museums have that is distinctive is objects, and what gives most museums their unique advantage is the awesome power of those objects to trigger an almost infinite diversity of profound experiences among their visitors"[1] (2002, 71). At the CSTM, we have a range of artifacts on display, from locomotives to royal touring cars to stylish domestic appliances.

Artifacts used in a demonstration

- We use artifacts in a few of our demonstrations. Such artifacts have been deemed by our conservation staff to be sturdy enough to withstand limited use in the controlled environment of a demonstration. For example, in a demonstration called Wash Day Blues, which shows how difficult it was to wash clothes at the beginning of the twentieth century, the demonstrator operates three washers from the 1920s to illustrate the early application of electricity to a demanding domestic chore.

Artifacts in operation

- A few artifacts on display in our exhibitions have been slightly modified to be operational in a semiautomatic mode. In the exhibition on domestic technologies, a Rider-Ericson hot air pump from the late 1800s has been fitted with an electric motor so that visitors can activate it by pushing a button. Such pumps were used near the end of the nineteenth century to supply homes with running water by pumping water to tanks placed on the upper floors.

De-accessioned artifacts or props used in an interactive display

- It is possible to do more with de-accessioned artifacts or purchased props. In a display on irons, we placed a small interactive that allows visitors to lift irons from different time periods. Simply replicating the weight of each iron would no doubt have given visitors an idea of the relative weight of each one, but visitors have told us that being able to lift a real old iron reinforces the message. In another example, visitors can sit in a modified Austin American, a very popular car in the United States in the 1930s.

De-accessioned artifacts or props used by visitors in a program

- Many such objects are used in our school programs, special events, day camps, and even sleep-overs. Evaluations of these programs have told us that participants greatly appreciate the opportunity to touch or use a piece of old or new technology.

SUGGESTIONS FOR USE OF OBJECTS IN A SCIENCE CENTER

This paper is not advocating that science centers, big or small, start collecting artifacts either for display or for storage. But it is suggesting that many science centers would benefit from including real objects of science and technology in their exhibitions and programs. The staff of the CSTM has found that it is a challenge to get the visitors to see the difference between objects that can be touched and used and others that cannot. But it is a challenge that is worthwhile. A carefully thought-out mixture of interactive elements and real objects in an exhibition creates for visitors a rich contrast between the experiential and the reverential. Far from detracting from the hands-on approach prevalent in science centers, the use of objects adds to it by showing visitors real examples of the tools and products of technology and science, right beside interactive elements that illustrate the processes and principles of the two.

TO COLLECT OR NOT TO COLLECT

Museums collect and, as a rule, science centers do not. The decision to collect, document, and preserve artifacts is not one that can be made lightly, as it has important consequences for the human and financial resources available to an institution. Artifacts require a lot of care, not only when they are acquired but for many years after. Collecting and caring for artifacts requires expert resources that are not usually available in a science center. In fact, many of the necessary resources are not available in many science and technology museums as well. The proper handling and storage of artifacts requires much dedicated space in an institution. That space must meet basic environmental standards that may not be easy or cheap to maintain over time. For example, in Canada, storage areas must be air-conditioned in the summer and heated in the winter. The expertise required to look after a collection is varied and cannot usually be combined with other duties. This expertise includes collection management, registration, conservation, proper handling for exhibition purposes, and basic research either on the artifacts themselves or on the subject areas of which they are a part.

Many institutions have set up what some call *teaching collections*. The term *working collection* would be more accurate, as the objects are not used solely for teaching purposes but are also displayed and used in public activities. Many of the requirements for establishing a collection of artifacts also apply to setting up a working collection.

Doing It Right

Once an institution has made the decision to establish a collection, be it a working collection or one of significant artifacts, it must put in place the following elements:

1. A clearly defined purpose for collecting, linked to the mission of the institution, its goals, and its resources. The statement of purpose should answer two questions:
   - Why are we collecting? Are we collecting to preserve a part of a region's technological heritage? Or are we collecting objects that will be used in programs and exhibitions?
   - What are we collecting? Are we collecting objects from a particular time period, or a specific region or related to one technology in particular?

     A clear purpose for collecting allows the institution to acquire only those objects that it really wants through donations, purchases, or exchanges. A large part of collecting is to have the ability to say no to objects offered to the institution and to be able to justify the use of funds for specific acquisitions. At CSTM, collection development starts with basic research in the form of historical assessments. These assessments then serve as the basis for defining the ideal collection and identifying the gaps and the redundancies.

2. A collection policy covering the acquisition of objects, their preservation, their use (internally in programs and exhibitions and externally through loans), and finally the de-accessioning and disposal of some of them. At CSTM, artifacts are only acquired if there is a reasonable expectation that they will receive adequate long-term care. CSTM, as a rule, will not acquire objects that have conditions attached to them. Examples of conditions that are not acceptable include the following:
   - Restrictions on access to artifacts for research, exhibition, or loan

- Requirement that the object(s) be on display within a prescribed period of time
- Requirement that a donor plaque be attached to the artifact; and
- Restrictions on disposal

In signing the gift agreement provided to all donors by the CSTM, each donor understands and agrees that the objects donated "may be retained, displayed, loaned, disposed of, or otherwise dealt with in such a manner as the [CSTM] may deem to be in its best interest." (CSTMC Gift Agreement) By making its position clear at the start, the institution removes much of the risk of misunderstandings later. In the case of a working collection, it is imperative that potential donors be told up at the onset what the possible future uses of the objects they are planning to donate will be. A donor may not be willing to make a donation if the object is to be used in programs until it is no longer of use.

3. A set of processes outlining the role of the board of directors, of management, and of staff in acquiring and disposing of objects. These processes should be written up and reviewed regularly. These must help avoid, among other things, conflicts of interest. The institution should set up a collections (or acquisitions) committee to oversee the acquisition or disposal of artifacts. Acquiring or disposing of objects should not be the responsibility of one individual in the organization. Also, there needs to be a process in place to properly appraise the fair market value of an object, not only for insurance purposes but also for cases in which the donor wants to receive a tax receipt for his or her donation. CSTM uses the average of two independent appraisers known for their expertise in the specific object or subject area. The disposal of an object deemed no longer relevant is in many ways more complicated than its acquisition. Some of the reasons for doing so include the following:
   - Lack of relevance of the object(s) to the subject area
   - Irreparable damage
   - Research reveals doubtful authenticity or ownership
   - Presence of object in the collection poses environmental problems (presence of chemicals, and so on)
   - Overrepresentation or duplication of a particular type of technology

   Methods of disposal may also require time and effort. At CSTM, three principal ways are used for disposing of an artifact. The preferred method is to offer the artifact to other institutions either as a straight transfer or as part of an exchange. An artifact may also be sold. If the artifact has an estimated fair market value that exceeds a predetermined minimum, it is sold through public auction. Finally, if all reasonable efforts have been made to dispose of an object through a transfer, exchange, or sale, and no transaction occurred, the object can be destroyed. Note that in general, no member of an institution's staff or board should ever acquire a de-accessioned object except in a public auction, after it has been offered openly to other museums and refused. Even this can expose the institution to suspicion and criticism.

4. Put in place the resources necessary to run a collection program. These resources include the following:
   - Expertise in collection management, from the acquisition of an object to an assessment of its condition to setting up a system that will accurately track its movements over time.
   - Expertise in conservation. The wide range of materials present in technical objects makes their conservation a real challenge. At a minimum, an object should be stabilized, that is, it should be treated to stop any further deterioration. Further work may also be required if the object is to be displayed or used in a program.
   - Expertise in research. Knowing more about the technology in question and the object itself (provenance, history of use and of ownership) is extremely useful in making the decision to acquire and keep a particular object.
   - Proper storage spaces, secure and safe both for the objects and for the staff who will work on them. Such spaces need climate control, shelving, swing spaces (a must when handling large objects), a space for photography, and so on. The cost of setting up and maintaining such spaces makes the decision about whether to acquire specific objects or not that much more important: they will be with the institution for many years to come.

In recent years, a few science centers have been debating whether to collect or not. The experience of science and technology museums such as the CSTM underlines the fact that the decision to collect should be made with full knowledge of the future implications for the institution.

OBJECTS AND INTERACTIVITY

It is still true that most science centers have been reluctant to display historical objects, arguing that such objects belong in museums and not in the active, interactive atmosphere of a typical science center. On the other hand, museums have been slow in adopting the interactive approach championed in science centers. Exhibitions and programs developed at the CSTM in the last few years have shown that objects and interactive components not only go well together but in fact support each other in creating an enjoyable learning environment. While it is not suggested that science centers start collecting artifacts, it is suggested that they seriously consider displaying more objects than they currently do. Real objects will add richness to the experience of all visitors.

NOTE

1. Weil, Stephen E. 2002. *Making Museums Matter.* Washington and London: Smithsonian Institution Press.

Part V

CREATING SPECIAL EXHIBITS SPACES

CHAPTER 21

Discovery Spaces: Small Museums within a Large Museum

Lucy Kirshner

In 1987, i came to the Boston Museum of Science from the Ann Arbor Hands-On Museum. I was overwhelmed by the big museum with its venerable history, huge budget, large staff, and millions of visitors. A spectacular view down the Charles River from the museum's lobby included several universities and city sights and impressed me with the power of science and technology to shape our culture. During my first days in this institution, I walked through the exhibits and attended stage demonstrations that further inspired and elevated my ambition to talk to the public about science. In a place like this, educators can build huge exhibits and design stage shows for three hundred people at a time with props that use the latest in astonishing technologies.

Soon it was time for me to start teaching, and I recognized that my personal style, gained through experience in a small museum, was not the amplified production of a stage performer. I am more comfortable teaching through conversation. I appreciate the props and exhibits that museums use, but I like to let everyone touch and manipulate as they investigate. Rather than stand in front of an audience, I like to stand with the audience exploring something together. Personally, I gravitate toward smaller props and smaller audiences.

Thus, soon after my arrival at the Museum of Science in Boston and still slightly homesick for the Ann Arbor Hands-On Museum, I could appreciate the differences between small and large museums and could recognize that there are trade-offs and even educational costs that came with growth. I had known that there'd be less flexibility and that decisions about programs and exhibits would take longer to make with so many colleagues involved, but I hadn't known, until my own teaching style was challenged, how large museums tend toward large performance and how my preference of educating through quiet conversation would need defense.

When I arrived at the Museum of Science my first job was to work with volunteer interpreters stationed throughout the halls, augmenting the exhibit experience through conversation and shared exploration. Happily, I found in this large museum an opportunity for all different styles of teaching and learning and this more personal approach suited me well. It brought the small museum style into the large museum and was the first step toward my interest in discovery spaces.

My tenure with the interpreters on the halls both in Ann Arbor and Boston filled me with admiration for volunteer teachers along with exasperation over the challenge of providing such an enthusiastic group with the training and materials they need in order to respond to visitor interest. In one day, people's curiosity can go in many, many directions, and while we train ourselves to confess to what we don't know, we are obligated to fuel visitor curiosity with the science skills of skepticism and inquiry. We needed resources to provide volunteers with teaching skills as well as additional information. This task was overwhelming in such a large museum with so many concepts and discoveries displayed. I looked with some envy at my colleague's job training volunteers in Boston's first Discovery Room.

Fifteen years ago, at the Museum of Science, when I first began, there was one Discovery Room, a small enclosed space where the youngest visitors could explore real objects representing exhibitions throughout the larger museum. The manager of this space trained volunteers in how young children learn as well as giving them information about the objects in the room. The great success of this Discovery Room was causing traffic problems. Every visitor, not just the young ones, wanted to gain admission and touch and explore the collection the small room held. The enclosed, staffed space of this "little museum within the big museum" gave it special value. Visitors within were protected and secure and responded to the privilege with enthusiasm by slowing down and taking care. Their explorations took on extra wonder.

In response to this discovery space's success I was given a new assignment. I was charged with making a discovery area for an adult and family audience. We felt that teaching about science using medical issues as a theme would benefit from slower exploration and would appeal to an adult audience. I was delighted with the challenge. It allowed me to continue to work with talented volunteers but gave a focus to our science education mission. My daily teams of volunteer interpreters were recruited for their interest in and understanding of human biology. They were delighted to have a place and an opportunity to share what they knew.

Even in our earliest, tentative stages the Medical Discovery Area, as it was first called, was a huge success. Volunteers were mobbed with people of all ages testing their own eyesight, exploring real skeletons, dissecting sheep hearts, and experimenting with taste and smell. Some interpreters were retired doctors; some were research scientists in biology taking an afternoon or morning off to have fun in the Museum of Science. Slowly over the first two years, our collection of exhibits and activities grew, largely thanks to the wisdom and enthusiasm of the volunteers. All of us learned much more about human biology than we had understood in years of college or graduate school.

The experimental stage was a success and so the Museum decided to invest in a larger permanent Human Body Discovery Space: 2,000 square feet of exhibits built to the standards and specifications of the museum's exhibit designers. The designers enjoyed a new freedom with this assignment. Some displays that would have been too fragile or prone to misuse in the rest of the museum were possible in a staffed area. A sink with water, open storage cabinets, a density of displays, many loose objects, well-lit reading areas, lots of chairs for sitting and studying—all of these possibilities made the discovery area look and work differently for visitors and for us as teachers.

As educators, we moved beyond the simple catalog of body parts and systems. We developed a new relationship with our visitors and a great respect for their curiosity. We felt a need to respond to their questions and concerns about more complicated issues: immunology, HIV/AIDS, genetics, nutrition, and organ transplant. We also started to use the busy environment in order to observe and study how the public learned. The Human Body Discovery Space provided an excellent opportunity for prototyping displays and researching how people learn. One volunteer designed a system for observing within museum exhibitions and was able to use her study as a PhD thesis.

While the Human Body Discovery Space went through this expansion, another special space within the Museum of Science was added to the discovery room list: the Cahners ComputerPlace. Here visitors had the time and supportive staff that allowed them to explore new computer software and hardware. Meanwhile, the original Discovery Room still enjoyed huge popularity and so was moved and expanded. All three spaces drew loyal audiences as well as being advertised specially for first time visitors.

At about this time some of the challenges that discovery spaces create became apparent. The first concern was a need for new staff. Enthusiastic teams of volunteers in all three areas expanded in number and ambition. With such a large, volunteer crew there was a need for several additional part-time employees to maintain a daily, active presence. The biggest misunderstanding in staffing discovery spaces with volunteers is the thought that volunteers substitute for staff. In fact, they need daily support and communication from the professional museum staff in order to maximize their contribution and in order to protect the mission of the institution from reinterpretation by eager but uninformed enthusiasm.

Secondly, because of the growing popularity of discovery spaces, exhibits within each were suffering wear and tear and the need for a replacement/refurbishment budget grew. Staff and volunteers in each space were inspired to create wonderful, inventive demonstrations and displays, which an eager audience used until they were worn out as well. Maintenance crews that

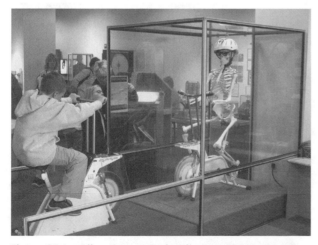

Figure 21.1. Bike, Human Body Discovery Space, Museum of Science, Boston. Photographed by Lucy Kirshner.

circulated throughout the entire museum found extra work in each discovery space where the density of exhibits was high and the hands-on nature of exploration caused frequent breakage.

The popularity of discovery spaces had more subtle consequences. Museum exhibit designers delighted by expanded possibilities that came along with staffed areas began to create components for unstaffed exhibits on the basis of what was successful inside the discovery spaces. Marking pens and tanks of water and small pieces and parts became part of the regular exhibit menu. The very presence of staff within a discovery space prevents a certain level of abuse. Some casual misuse, such as leaving a cap off a pen, can be corrected easily during the course of the day. This level of maintenance is not easy in an unstaffed exhibit.

Another subtle consequence of discovery spaces was that the public grew to expect staff attention and a level of exploration that had not been possible previously. Learners of all ages began to demand attention from staff throughout the entire museum and exercised their right to touch and manipulate every exhibit display the way they were allowed to in staffed areas. In response, exhibit planners, designers, and maintenance crews asked for help in expanding staffing to other areas of the museum. Everyone could make a case for making a new Discovery Space or for altering the assignment of the current ones. Existing spaces were thriving, but with our hands full, we guarded resources and redefined our own goals, becoming more and more independent and isolated from the rest of the programming and exhibits.

There was further tension for the small museums within the larger museum as discovery spaces were asked to conform to the expanding hours of the larger institution. The Human Body Discovery Space and the other spaces as well had existed with doors that could close—and did close at 4 p.m. every day. This gave the staff time to clean and restore displays, but it presented a problem for visitors who arrived late in the afternoon and still wanted to participate in the activities that the popular spaces offered. Discovery spaces were asked to stretch their hours and keep their doors open. This created an additional strain for the staff and display materials.

It was at this time that the Human Body Discovery Space made its next move to a lovely, larger exhibit area with a view down the Charles River. This was the same view that had inspired me on my first days in the museum and was now a daily inspiration, encouraging expanded educational goals.

I could see that the successes and the challenges of discovery spaces in the Museum of Science demanded new growth and change. Some changes were straightforward. The new space has permanently open doors and displays with built-in hidden cabinets. The Reading Room that can be locked is the only area that does not remain open after the volunteers leave in the evening. Certain expensive, delicate items are gathered each evening and kept safely in the Reading Room.

Conceptually there were changes as well. The discovery space was renamed The Human Body Connection, a name that represented commitment to a connection with the rest of the Museum of Science. The room features exhibits related to human biology but is charged with the goal of relating these exhibits to the rest of the museum displays and, more significantly, to the museum's mission of promoting the skills of science and their importance to individuals and society.

With this step in the room's evolution, the staff took on additional responsibilities within the museum, contributing to programming outside the room as well as preparing packets of curricular material. Volunteers were required to be familiar with and to teach in other exhibit areas within the museum as well as the discovery space itself. The Human Body Connection welcomed educators from all other parts of the museum, encouraging them to use the exhibits and stages within the space for their conversations with visitors and allowing some of the displays and learning materials to be taken from the room for classes elsewhere.

Our history of prototyping exhibits and gathering data on visitor interactions has expanded as well. We've come to realize that the visitors themselves are interested in interpreting data, especially when they have contributed a personal data point. Thanks to handheld computer technology, our volunteers can collect, graph, and display information teaching the science skills of evidence interpretation to an enthusiastic audience.

I am proud of the growth that I've been part of in discovery spaces at the Museum of Science. These areas have been instrumental in stretching the educational ambitions of the whole institution. Now, we are ready for the next step in our evolution. At the museum today there is a new initiative: to continue our success in informal science education while growing in our ability to teach about engineering and technology to schoolchildren, families, and especially, to an adult audience. There is an additional commitment to doing active research in how people learn in informal education centers. This initiative will bring dramatic changes to all three discovery spaces.

As part of the Boston museum's technology initiative, staff have been reorganized and my own assignment has changed. I have been made the manager of all three discovery spaces faced with the challenge of helping the three programs grow while protecting their intimate teaching style. Cahners ComputerPlace will undergo major change as part of the technology initiative, dedicating its mission to teaching the fundamental concepts of computer science and putting those concepts into the context of exploration and engineering. The Discovery Center has already begun to gather research on how very young children learn the fundamental design and invention skills that go into engineering. They will turn that research into new museum experiences. The Human Body Connection is the first discovery space slated for large, physical growth as we integrate many displays on current technologies and reframe our current exhibits with a new focus on the impact technology has always had on our biology.

There is an important feature of our reorganization that has made my new job possible. The three spaces are working very closely together. Together discovery spaces in this big museum will be able to articulate the strengths and benefits of small, intimate learning situations where visitors have the opportunity to touch, experiment, and converse with the staff. We are learning how to share our strength among ourselves and with staff in the rest of the museum and will be able to continue as educational experiments and a resource for learning about informal science and technology education. I believe this parallels the strength of small museums working together, sharing challenges and successes.

The skyline of Boston, one of the oldest, established cities in the country, is changing outside my window. The Museum of Science is leading the change, pushing an educational agenda that should equip our community for the rapid changes in our future. These are dramatic declarations full of excitement and ambition. I am reassured to know that quiet conversations in discovery spaces will be at the heart of that change.

DISCOVERY SPACES AT THE MUSEUM OF SCIENCE, BOSTON

Approximately two hundred volunteers work among the three museum discovery spaces and their effort is part of what makes these exhibits rank at the top in visitor popularity. Each has a unique content focus and each has a distinct style but all three feature creative exhibit experiences and intimate learning experiences. In a sense, they are all small museums within the large museum.

THE HUMAN BODY CONNECTION

The Human Body Connection is approximately 2,500 square feet of staffed exhibit space devoted to human biology. A large, enclosed central kiosk affords volunteer interpreters the opportunity to share wet or fragile demonstrations with visitors of all ages. Routinely, sheep heart, lung, and eyeball dissections are presented. DNA extraction, taste testing, owl pellet exploration and yeast experiments are special activities always available on request.

Volunteers circulate among more than a hundred exhibits assisting and augmenting the visitor experience. Their charge is to use the public's interest in human biology to stimulate an interest in science and technology.

CAHNERS COMPUTERPLACE

The mission of Cahners ComputerPlace is to engage learners in the fundamental concepts of computer science and to help our public understand how and why computing power has been applied to many facets of our world. The staff, volunteers, and exhibits of Cahners ComputerPlace provide the opportunity to learn about topics such as robotics, artificial intelligence, and programming and to explore computers as a tool for helping us understand and change our world. The exhibit features an informal presentation area and a variety of themed computer activity clusters.

THE DISCOVERY CENTER

The Discovery Center provides a unique venue for examining how young learners and their advocates learn together through inquiry and play. The mission of the Discovery Center is to provide concrete experiences with real objects and interactions with volunteer interpreters, through which visitors are encouraged to practice science process skills—to observe, classify, hypothesize, and model—and engage in activities that encourage technological literacy—to ask, imagine, plan, create, and test.

CHAPTER 22

Creating Developmentally Appropriate Early Childhood Spaces in Science Centers

KIM WHALEY

WHEN PEOPLE THINK OF SCIENCE CENTERS, THEY typically think of places that families take their children once they hit school age. It is often assumed that up until they start school, children are not really interested in science and that families with younger children are better served by more typical children's museums. As it turns out, science museums are the perfect place to locate exhibits geared specifically toward the early childhood age-group. After all, young children are the original scientists of the world. They spend their days asking questions, trying out new ideas, and forming theories about how the world works. This scientific process occurs when children have opportunities to play and directly interact with the world around them. These interactions are the basis for all science learning that will follow during the formal schooling years, and giving infants, toddlers, and preschoolers the space to do this is the essence of the science museum mission.

Research continues to emphasize the importance of the first five years of development. Large amounts of brain development and growth happen between birth and five years of age, and it appears that these early years provide optimal times for learning a variety of information (Shonkoff and Phillips 2000). This research has also led to a call for communities to provide a variety of appropriate learning opportunities for young children and their families.

Science center exhibits by necessity are designed to meet the needs of the general visitor population. This often makes the exhibits too large or too complicated for the youngest children to experience in ways that are meaningful. Creating a space dedicated to the developmental needs of young children opens the museum for new populations to experience. Including a space dedicated to young children can have positive economic benefits as well; statistics show that 20 percent of Center of Science & Industry (COSI) general public guests were five years of age or under in 2003. After children start formal schooling, many activities, ranging from sports to school events, compete for family time. Before school years, however, families are often searching for appropriate activities to do with their children. Creating a dedicated early-childhood space in the center can fill a community need while increasing revenue potential.

EARLY CHILDHOOD SPACES AT COSI—A SHORT HISTORY

COSI Columbus opened its doors in the spring of 1964 with exhibits designed for the general public for all ages. It was not until 1984 that COSI introduced the original KIDSPACE® in response to the expressed need by parents for a space where their young children could explore safely without competing with older children. KIDSPACE® was physically separated by low walls from the rest of the exhibits and served children from birth through first grade in a mixed-age setting specially designed for this age-group. The determining factor for entrance to the space was later changed from grade level to height—a more concrete and measurable factor than a stated grade level. While KIDSPACE® was successful, there were issues that arose that demanded responses. The exhibits in KIDSPACE® were appealing, perhaps too appealing, to children older than the height restriction would allow in, and families were often separated as one child could use the space while the other was excluded. In order to help alleviate this issue, the museum developed FamilySpace, a space comprised of replicate KIDSPACE® exhibits with some different pieces intermingled where entire families could explore together. Finally, BabySpace, a separated area for the very youngest children, was incorporated into KIDSPACE®.

In 1999, COSI moved to our new building and little kidspace® was born. Returning to our roots,

Figure 22.1. In order to help alleviate this issue, the museum developed FamilySpace, a space comprised of replicated KIDSPACE® exhibits with some different pieces intermingled where entire families could explore together. Finally, BabySpace, a separated area for the very youngest children was incorporated into KIDSPACE®.

little kidspace® opened, serving a mixed-age population of children from birth up to the start of kindergarten. Designed as a full-service, single-destination site for families with young children, little kidspace® is dedicated to the understanding of how humans learn and grow. The 11,000-plus-square-foot space supports the emerging sensory-motor, social, emotional, language, and cognitive capacities of infants, toddlers, and preschoolers. Drawing from early childhood research, we purposely did not replicate the age-separated areas of the old COSI KIDSPACE® but instead designed exhibits within the space that encourage children to self-select their participation level on the basis of their current developmental interests (Whaley and Kantor 1992). For example, in the Park area of the exhibit, ba-

bies learning to crawl are drawn to the low ramp area while new walkers are interested in the bridge on the other side of the same exhibit. The only separated area within little kidspace® space is the Hang-Out room, the area where children older than the age cutoff can spend time interacting with exhibits more appropriate to their age while their younger siblings experience little kidspace®.

Little kidspace® also responded to the need for appropriate programming by creating a space with the developmental need for play at its heart. In this space, children can engage in rich interactions with adults and other children, interactions that optimize development (Vygotsky 1978). Children and the accompanying adults are able to explore and experience a variety of developmentally appropriate yet challenging exhibits in little kidspace®, thus offering many opportunities for learning and development.

Little kidspace® was designed to be a unique site that would support the development of children across these earliest years of growth. Responses from guests to little kidspace® were mixed during our first year of operation. These responses were primarily positive with numerous compliments about the space being one where parents felt safe putting their babies on the floor to explore without fearing that they would be run over by older children. In addition, they liked the idea that children could interact safely with more competent peers around exhibits designed specifically for their learning level. In fact, compliments about the admission policy to little kidspace® constituted 32 percent of the feedback we received during this first year, making this the most commonly cited compliment for the space.

On the other hand, family demographics are changing and families more commonly have children with an age range of 5 years or more. For these families, the fact that the majority of the space was designated for their younger child while the older child was in the Hang-Out room was a source of disappointment. To this end, 21 percent of the complaints about the space were related to the entrance policy. In an effort to be responsive to guest feedback and our own observations, while still remaining true to our belief that the youngest children need a protected space in which to interact on their own terms, we changed the upper age limit for entrance to little kidspace® from entering kindergarten to entering first grade in an attempt to provide more enriching experiences for the older children while maintaining the integrity of the space for the youngest children. These changes have allowed us to design exhibits and experiences that are more appropriate for older children in the Hang-Out room while keeping those most appropriate for the youngest children available only to them. There was an almost immediate drop in complaints from parents following this age change.

The current flurry of research on children's learning has brought with it a call for more information on child development and learning to be made available for parents as well. Science centers have the opportunity to respond to this need daily through the interactions between the parents and the team members in the space and graphics placed in the exhibit areas. Little kidspace® includes a parent resource area called "info.kid," a place where parents can get the latest books and magazines on parenting and child development, view videos on development and parenting, and can access an extensive website created to help parents get to reliable, research-based information on child development. In addition, signage is located around the entire exhibit area and outlines specifically what infants, toddlers, and preschoolers are learning in each set of exhibits.

DON'T REINVENT THE WHEEL: LESSONS LEARNED THROUGH THE YEARS

Since the inception of the original KIDSPACE®, COSI has learned many lessons about developing exhibit areas and programs for very young children. The remainder of this chapter outlines the lessons learned and suggests ideas for creating your own early childhood space.

Lesson 1: Young Children Need a Space of their Own

While any museum naturally wants to be a place where families can explore together, a space designed specifically for young children that is separated from the rest of the museum is important. Having this separate space allows you to build on the research about early development and create experiences exclusively designed to meet their developmental needs. This, in turn, frees you to design more toward the older children and adults in other parts of the building. This creates a good balance of activity and experiences across the museum.

We have also learned that it is important to physically divide the early childhood space from the rest of the building if at all possible. If not possible to separate the area physically, the space should be separated visually. This separation gives children a sense of place

and parents a sense of security. Little kidspace® is divided from the rest of the building with wall-size panes of glass, thus allowing other guests to observe children playing while still keeping the space separate and secure. Children in the space have a sense of boundary without feeling closed off from the rest of the museum.

It is important, however, that while the early childhood space must be engaging and fun, it is not the most exciting and interesting space in the museum. If the space for young children has the coolest looking exhibits in the place, older children will want to go in, thus causing conflict within the museum. The trick is in designing exhibits that are engaging for young children but send the message to older children that this is clearly for children younger than themselves. Other suggestions for designing this separate space include the following:

- Be sure the space is large enough to accommodate and support the kinds of motor activity young children need to engage in—running, jumping, climbing, crawling, and walking.
- Fill the space with exhibits that support the development of process skills—skills such as predicting, observing, classifying, measuring, hypothesizing, and experimenting. These are all vital skills to scientific thinking and are what young children do naturally. Using process skills along with developmental milestones to design rather than designing around a science concept will create exhibits that are appropriate for infants, toddlers, and preschoolers.

Lesson 2: Parents Need a Break

We went into designing an early childhood space wanting it to be a place that encouraged, even tried to force, parent involvement with their children. To this end, the seating in the space is not permanent but is accomplished using portable soft seating that can be easily moved so parents can sit near their children to interact. In point of fact, parents have most typically moved these seats to sit next to each other and chat, read, or simply relax while their children play with their peers.

While we originally pushed for parent-child interactions, our stance on parent involvement has evolved over the years. Families today are more often separated from their extended family and more parents are working, leaving parents without vital support systems and downtime. We have come to believe that while we still want to encourage and support parent-child interactions, it is important that we provide parents with a place they can feel safe about their children while

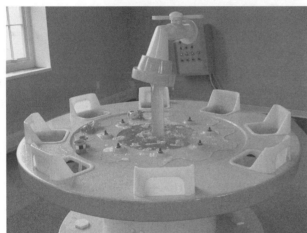

Figure 22.2.   In 1999, COSI moved to our new building and little kidspace® was born. Returning to our roots, little kidspace® opened serving a mixed-age population of children from birth up to the start of kindergarten. Designed as a full-service, single-destination site for families with young children, little kidspace® is dedicated to the understanding of how humans learn and grow. The 11,000-plus-square-foot-space supports the emerging sensory-motor, social, emotional, language, and cognitive capacities of infants, toddlers, and preschoolers.

THE PARK

EXPLORING IDEAS THROUGH PLAY

Infants, toddlers, and preschoolers are the world's original scientists. The methods of science – observing, experimenting, and testing – are mirrored in the way young children explore and build knowledge. Curiosity and the desire to explore are vital character traits of all researchers, and are also qualities which young children exhibit naturally. The park gives children the opportunity to test their ideas about how balls roll or where a tunnel leads. Just as important, it provides them an outlet to practice and build upon their emerging motor skills as they crawl, jump, and reach.

Young babies who are not yet mobile will enjoy holding the balls in this space, watching the various flowers as you spin them, and exploring the many different textures.

Older babies are beginning to test new motor skills such as sitting, crawling, and perhaps even walking. By exploring the park they learn about hard and soft, cool and warm, and the feel of different textures.

Young toddlers can use this space to develop ideas about how things work. Through their active, experimental play they learn how to make the same things happen over and over. Young toddlers love throwing balls and retrieving them or running to get the balls you throw for them.

Older toddlers are developing concepts about how things fit together. They enjoy matching and sorting objects - collecting all the red balls, for instance. Toddlers try things over and over and over and over . . . While this may be boring for caregivers, it is a vital activity that strengthens connections in young, growing brains.

Preschool children have well-developed motor skills but still need opportunities to test these skills in new ways. Parts of the park, such as the rope bridge, will let them do just this. Learning to walk on a surface that moves requires problem-solving and new ways of balancing. Preschoolers will also use this space to experiment with new ideas. They may race balls down the ramp, or investigate how the balls go up the tube, or create pretend scenarios from their own imaginations.

Figure 22.3. Signage is located around the entire exhibit area and outlines specifically what infants, toddlers, and preschoolers are learning in each set of exhibits.

having the opportunity to connect with other adults or simply relax and take a deep breath while their children are happily occupied. It is not unusual for us to see the same groups of parents arriving every Monday morning to spend their regular morning together chatting while their children play. In fact, parents of multiples and stay-at-home dads have become two groups of regular users of our early childhood area. To facilitate parents getting a break follow these steps:

- Provide seating that is easily moved around to allow parents to group themselves for conversation or interaction with their child.

- If space allows, provide a resource area for parent information that parents can pick up and look through while their children play.
- Encourage staff to introduce themselves to parents and interact with children (this not only allows parents to relax but can provide some valuable modeling for parents on appropriate interactions around learning).
- Support parents' current needs when they are visiting your museum. Help parents who want to be involved with their children during their visit to co-explore with their children, and respect those who need a break by spending time with their children and allowing the parents to have time to breathe and relax.

Lesson 3: People Make All the Difference

While the design of the exhibits in the space is certainly important, we have discovered that it is the people in the space that make all the difference. The primary job of the COSI staff is to build relationships—relationships with the children and with the parents. The more parents feel that they and their children are special and cared for, the more likely they are to return to your center. It is not always possible or even necessary to have a person specifically trained in early childhood on the floor, but it is vital that there at least be a person in the institution who has a strong background and understanding of early childhood and will advocate for the needs of young children. In addition, it is vital that the people who are hired to work in the space be people who enjoy playing with children, are comfortable interacting with parents, and are not afraid or embarrassed to get down on the floor, pretend to be another person, or sing at the top of their lungs.

Providing some training in child development and in interacting appropriately with young children for staff is vital for a successful space for young children. This training must include information on developmental stages of young children, early learning, developmentally appropriate practice with young children, and interactions with parents. To staff with great people do the following:

- Hire first for attitude—you can train them in the rest. You want people who interact naturally and easily with young children and who can model good interactions.
- Don't assume that just because a person has worked in an early childhood program in the community, he or she will be a great employee for this setting. There are many programs out there for young children that

do not use practices that are appropriate for young children and many that do not embrace the inquiry learning that is at the heart of informal settings.

- Develop a set of values about interacting with young children that everyone in your museum is well versed in and committed to carrying out.

Lesson 4: Keep It Simple

It is tempting to try to design exhibits for these youngest children that are complicated in an attempt to be educational. Young children have a typical and universal set of developmental needs, behaviors, and even interests. If exhibits are too complicated or not open-ended enough, children won't respond. We have discovered that the most popular exhibits are those that are designed simply around one or two overarching developmental skills. Remember the cardboard box lesson—children often drop the toy and go for the box because its simple design allows them to use it in a wide variety of ways. To keep your design simple do the following:

- Design around a skill rather than a concept. For example, a Bernoulli blower is a fun exhibit for young children but the goal is not to teach this concept. Instead, the goal is to help young children experiment with cause and effect. When you think of it this way, you might include a variety of different balls for them to test the air streams with or you might make the air streams moveable for experimenting.
- Keep the lines clean and the colors limited. Too much color and complexity only overstimulates young children.

Lesson 5: Worry less about education, more about development

It is much more important that the content of an exhibit for young children match developmental skills than teach science content. Early science is all about process, learning to learn, exploring, and trying new ideas.

The major change for us when we moved from our old building to our new was to go back to integrating all ages from birth to first grade into one open space. This was done deliberately on the basis of theories of learning and development and research on mixed-age interactions (Vygotsky 1978; Whaley and Kantor 1992). The space was designed with each exhibit having pieces that reflect the various developmental needs of children from birth through preschool. Young children self-select activities that are appropriate for them at the moment, moving to activities that support their next developmental level as they are individually ready. Older children serve as models for younger children; it is not unusual to see a young toddler watching an older child walking around the space intently before they try to stand and do the same thing.

As you design your space, divide your thinking into three developmental areas—infants, toddlers, and preschoolers. Think about what the developmental needs are for each of these ages and incorporate activities that match needs in each of those age groups. A grid such as the one shown in figure 22.4 can be helpful when designing or evaluating a design.

Lesson 6: Location is Everything

In our old building KIDSPACE® was located in the back of the third floor of the building. Thus, many families with young children did not find the space until they had already been in the museum for several hours. In addition, at the start we issued timed tickets for entrance to the space in an effort to control the number of people in the area. By the time many new visitors found the space they were feeling ready to leave and their children were often feeling tired. In our new building, little kidspace® is located right off the elevator on the second floor making it easy to find and get

Exhibit	Younger Infant Birth–6 Months	Older Infant 6–18 Months	Younger Toddler 18–24 Months	Older Toddler 2–3 Years	Preschooler 3–5 Years
Flowers	X	X	X	X	X
Slides			X	X	X
Ball Pit		X	X	X	X
Ant Farm			X	X	
Bridge			X	X	
Float Balls	X	X			
Nest	X	X			

Figure 22.4.   A grid such as this can be helpful when designing or evaluating a design.

to. Many families with young children will want to start their day in the science centers in the early childhood space so it needs to be visible or in a location that is both easy to find and accessible without having to go through the rest of the building first. We often direct a family with young children to little kidspace® first as they go through the box office for check-in to the museum.

Lesson 7: Attention to Safety is Vital

One thing that emerged from focus groups with parents in the conceptual phase of designing little kidspace® was the desire for a place that had considered security issues. Parents want to know that their child is safe in the space, that they can't get out, that they can't be taken. Little kidspace® is a "gated community" with a set of doors that one must be buzzed in or out of. Adults must have a young child with them to enter. Upon arrival, parents are matched with their children, and upon leaving, the person at the gate checks to be sure the children are leaving with the adult they entered the space with. Fire doors in the space are alarmed so a child cannot get out without there being a very loud noise. The attention to security has allowed for a space where adults can relax and let their children explore. While this level of security is certainly not necessary to a quality early childhood space, over and over we hear from parents that this attention to security allows them to interact with their children in ways that are different from spaces where they have to constantly be watching for escapes.

There are other, perhaps more obvious, safety concerns as well. Selection of materials and design of space must take into account that young children fall as they learn motor skills, are often in a hurry, put their mouths on things, and touch everything. Rounded corners, soft sides, and things to hold on to are all vital to the safety of children.

Lesson 8: Some Amenities are Really Necessities

Last, but definitely not least, is the lesson that those things we might consider to be amenities in our centers are in fact necessities to parents. Stroller parking, changing spaces in bathrooms, family bathrooms, nursing rooms, spaces to eat with snacks appropriate for young children are all vital to young children and their parents. Parents and children are better able to concentrate on the activities and learning when their basic needs are met.

Another amenity that we have found to actually be a necessity in our very large building is something that allows families to get back together when they separate. We often have a family who arrives with one child who can play in little kidspace® and one child too old to be in the space. While some of the older children spend time in the Hang-Out room, we have found that these families frequently either divide up or let their older children explore on their own. To this end we provide family members with pagers so they can contact one another when they are once again ready to meet. This amenity is clearly not a necessity for many museums, especially smaller ones, but in a museum of the size of COSI, families have come to depend on this and see it as a necessity.

Putting a specific early childhood space in your science center has many benefits, not only for the families in your community but for your center as well. While creating a space that is exclusively for your youngest guests is not without controversy, the benefits and the pleasure of the families who use the space far outweigh the concerns. Infants, toddlers, and preschoolers learn by interacting with people and things in their world, and science centers offer the perfect place for this active learning. Perhaps most importantly, early childhood spaces allow you to build relationships with families early—relationships that can last a lifetime.

REFERENCES

Shonkoff, J. P., and D. A. Phillips. 2000. *From neurons to neighborhoods: The science of early childhood development.* Committee on Integrating the Science of Early Childhood Development. Washington, D.C.: National Academy Press.

Vygotsky, L. S. 1978. *Mind in society: The development of higher mental processes.* Eds. and trans. M. Cole, V. John-Steiner, S. Scribner, and E. Souberman. Cambridge, Mass.: Harvard University Press.

Whaley, K., and R. Kantor. 1992. Mixed-age grouping in infant-toddler child care: Enhancing developmental processes. *Child and Youth Care Forum* 21 (6): 369–84.

CHAPTER 23

Outdoor Science Parks: Going Beyond the Walls

RONEN MIR

OUTDOOR SCIENCE PARKS ARE INCREASINGLY POPULAR at science centers worldwide. The outdoor setting provides both plenty of space for exhibits and flexibility in their arrangement. It facilitates the educational use of natural resources, such as sun, wind, and water. It complements and expands existing indoor activities with visually attractive (and cost-effective) outdoor correlates.

For visitors, science parks mean open air, fun, and a relaxed learning environment. Visitors enjoy the opportunity to observe, feel, and experience the many scientific phenomena that surround them. They appreciate that science can be learned in different locations, not just in classrooms and indoor laboratories.

Using the outdoor science park at SciTech Hands-On Museum, in Aurora, Ill., this article offers suggestions for outdoor exhibits to museums in the process of designing and implementing a science park.

Aurora is Illinois's second largest city with a population of 165,000. Nearby are two prominent government research facilities, Fermi National Accelerator Laboratory (Fermilab) and Argonne National Laboratory. Scientists from these laboratories established SciTech in 1988. Along with other community volunteers, they developed over 250 hands-on indoor exhibits, including a unique solar telescope, which projects a ninety-centimeter image of the sun into an interior viewing area. Over time, the museum has become a fixture in the community, attracting more than 50,000 visitors a year. SciTech serves an additional 35,000 people with outreach activities, such as our popular Museum in a School program.

From 1999 SciTech embarked on a program of indoor and outdoor expansion and renovation. With SciTech's location at the southern tip of an island where two branches of the Fox River meet, SciTech has a striking natural setting. A pretty outdoor area facing the water was being used as a museum staff parking lot.

Why not turn it into an Outdoor Science Park? I was already familiar with the power of outdoor exhibits to attract and inspire visitors, having served as Scientific Director of the Clore Garden of Science, in Israel.

SciTech's Outdoor Science Park is designed as a place where groups and families spend quality time, especially on sunny days, when they would not want to be inside a building.

MATERIALS AND SAFETY

Because science parks can be expensive and complicated projects, we decided to start with a pilot program and implement our plan in stages. The Outdoor Science Park was built over two years, with a few new exhibits introduced each summer.

Our first step was to do an extensive evaluation of the site, considering possible modifications to the topography to ensure proper water drainage. We calculated the path the sun would follow throughout the year—important for solar exhibits—and evaluated our water, electricity, and communication needs. We thought about visitor comfort—would our guests have adequate drinking water, shade, and benches for resting?

Our design called for exhibits to be arranged on a 5,000-square-foot brick terrace complete with planters, picnic tables, and sun shades. Those who plan to host special events in their science park may include food-service facilities and outdoor lighting.

Durability was a primary concern as we built our Park. Investment in quality materials is worth a lot after the fact. Noncorrosive metals—aluminum, brass, or stainless steel—are excellent, although these materials can be expensive to work with. Wood can be problematic, but high-quality oak weathers well over time, maintaining the look of the exhibit and its functionality. We went with aluminum and stainless steel, although finding fabricators who know how to work with these materials can be a challenge.

The extra you spend on materials may save you a lot in maintenance. That's not insignificant, because outdoor exhibits exposed to the elements need at least as much attention as indoor exhibits. Daily visual inspections and periodic tune-ups are critical for ensuring the smooth operation of a science park.

Safety and accessibility are equally important. Naturally, we wanted all visitors to be able to enjoy the Outdoor Science Park, and we wanted them to do so safely. According to the National Program for Playground Safety, each year more than 200,000 children are injured on U.S. playgrounds, with more than 75 percent of these injuries caused by falls. Since there are no U.S. or state codes regulating safety and accessibility for outdoor interactive exhibits, the codes for playground access, equipment, and surfaces are generally applied. To ensure the security of exhibits and visitors, the Outdoor Science Park is fenced, with controlled access. When visitors are using the exhibits, trained staff members are on hand. To ensure safety, a minimum age requirement for users is posted.

CHOOSING OUTDOOR EXHIBITS

Choosing interactive outdoor exhibits can be daunting—there are so many options available. Possible themes include botany, ecology, the environment, motion, music, optical illusions, solar energy, sound, space exploration, water, waves, wind, and more. For each of these, there might be several exhibits.

The Outdoor Science Park at SciTech currently includes five extra large exhibits and five smaller ones. Our decision was to include exhibits on motion, waves, music, solar, and water energies providing full body experiences for the visitors. We plan to add two exhibits every summer. The large exhibits are anchored to sturdy concrete pads covered by safety-tested cushioning surfaces. The small exhibits are moved into the museum for winter usage from December to March.

Our signature, and most visible, exhibit is the forty-five-foot-tall Weather Wave, which allows visitors to create their own vertical standing waves. Two decorative wind vanes on top demonstrate the wind's strength and direction. The Weather Wave is visible for a long distance. When it was first constructed, drivers stopped their cars on a nearby bridge to get a better look, resulting in extensive media publicity.

The Giant Lever consists of a ski-lift-type bench hanging at the end of a horizontal 33-foot pole. The pole rests on a fulcrum point with a ratio of 1:3. Visitors

Figure 23.1. Girl using SciTech's outdoor Wave exhibit at SciTech Hands-On Outdoor Science Park.

sit on the bench, and their friends pull on chains attached to the pole with a ratio of 1:1, 1:2, and 1:3 to lift them, thereby demonstrating the lever effect.

Everyone who rides our Bicycle on a Tightrope gets a firsthand—or, rather, first-foot—experience of the center of gravity, as the suspended weight keeps rider and bike stable. Illinois state senator Chris Lauzen rode the Bicycle on a Tightrope at the Outdoor Science Park's dedication; he was so happy that he decided to help fund the next stage of the park.

One of our most popular exhibits is the YouYo, an oversized inverted yo-yo that remains fixed while the user goes up and down. The flywheel representing the yo-yo is mounted at the top of the exhibit. The user pulls the rope, propelling himself or herself higher with every turn of the wheel. It takes practice and timing, but once you do it well, you can go as high as the limiting bar will allow—in this case 13 feet off the ground.

Our fifth large exhibit is the Coupled Swing. A single swing is coupled to a swinging 150-kilogram weight to demonstrate the energy transfer between coupled pendulums. A visitor starts to swing, but gradually her motion slows, as the weight begins to swing. The weight then slows down, "pushing" the visitor and starting her swinging again.

Among our smaller exhibits, the Lithophone is a favorite of mine. This marble percussion instrument, which resembles a xylophone, has a range of one full octave. The acoustic properties of the marble produce a clear resonating sound. Visitors play melodies on it all day long.

THE EXPERIENCE OF USING OUTDOOR SCIENCE PARKS

On one level, a science park is a marvelous playground, and children certainly enjoy it that way. To enhance their learning, we have developed some specific programs centering on science themes—ecology, environment, energy, and more. Story lines that incorporate treasure hunts appeal to most audiences.

Visitor responses to SciTech's Outdoor Science Park have been favorable. School groups and families interact enthusiastically with the exhibits and report positive experiences. The media has been attracted to the Outdoor Science Park's pleasant environment, and a number of articles have given the museum positive publicity.

My own experience is that science parks appeal to diverse audiences. All visitors seem to feel more relaxed about experimenting outdoors; perhaps it's because they are not exposed to the cultural messages that a building may project. As more science museums implement science parks, we are starting to notice that the outdoor setting invites a diversity of audiences to use the facility. These include visitors who shy away from the museum, such as middle school neighborhood kids who may be most at risk.

There is a need for more formal evaluation of outdoor science exhibit areas. A visitor-centered process, with front-end research to assess what the public knows and is interested in knowing, would assist in selection of themes. Summative evaluation would help to determine changes in visitors' attitudes or understanding as a result of their outdoor experiences.

The time is ripe for more science centers to provide their visitors with outdoor experiences. Science exhibits that engage children's bodies and minds whet young appetites for more. Outdoor exhibit areas add an extra dimension to an indoor science center and provide a challenging experience that complements the indoor visit.

CHAPTER 24

Traveling Exhibitions: Rationales and Strategies for the Small Museum

ROBERT "MAC" WEST AND CHRISTEN E. RUNGE

DEFINITION OF A TRAVELING EXHIBITION

Traveling exhibitions are here defined as those coherent exhibition experiences that are brought to a museum or other venue for a finite amount of time (usually one to three months) from a vendor or producer organization. The exhibitions may have the style and format of a typical exhibition, for example, objects and artifacts, interactive stations, art or photography, or any combination of the above, organized around a theme with an educational objective.

Traveling exhibitions are distinct from temporary shows produced by the institution's own staff or contractors, and they are developed with a single presentation in mind. However, successful in-house, one-time productions frequently are retooled and traveled to other interested institutions.

Traveling exhibitions also are different from short-term special events, even if an event is centered on objects, activities, or shows that move from place to place. Events in this category can include hobbyist weekends such as fossil shows, doll shows, car shows, or art gallery shows.

However, museum-sponsored traveling exhibitions need not be presented at the museum's sites. Many museums have developed collaborations with shopping centers, convention centers, or unused commercial spaces, to allow a traveling exhibition that is too large for the museum's spaces to be brought to a community. For example, over the last twenty years, many museums have featured robotic dinosaur exhibitions at off-site locations, much to the benefit of both the museum and the venue.

RATIONALES FOR PRESENTING TRAVELING EXHIBITIONS

It is well understood that repeat visitation is vital for the economic and educational success of a museum. The more exposure visitors have to the experiences and programs of a museum, the more they can benefit from and absorb the educational benefits of the museum and the more revenue they will generate for the museum. An important way to stimulate repeat or frequent visits is for there to be noticeable change at the museum. And, if the change is temporary, necessitating that a visit has to be made within a relatively short-time period, frequent visits are more likely. A menu of changing exhibitions thus is an audience stimulant, along with special events, celebrity visits, seasonal and cultural celebrations, and a variety of programs targeted at specific audiences.

Most museums cannot generate several fully developed new permanent exhibitions per year. Only some institutions have the staff and collection resources to produce several of their own temporary exhibitions per year. Thus, there is high reliance on externally produced traveling exhibitions to ensure media attention, repeat visitation, and community excitement about the museum.

An effectively marketed traveling exhibition ensures media attention to the museum—coverage of the opening event, interviews with staff, curators and subject-matter specialists, exhibition-related special events, and so on. It also is an opportunity to change the focus of the museum's advertising and promotional materials.

Traveling exhibitions can be an effective way to engage current and prospective donors and supporters. Many museums actively seek sponsorship of traveling exhibitions and, in fact, base their ability to afford to bring in the exhibitions on sponsorship income. Corporations, foundations, and individuals seek the positive association with high-quality community activities that are represented by traveling exhibitions and value a partnership with the museum around the exhibition.

Traveling exhibitions may make it possible for the museum to feature materials from its own collections as an elaboration of a traveling exhibition. On occasion a museum may have a few items related to a specific topic, but they are not adequate to mount a full exhibition or tell a complete story. However, they are helpful in the context of a traveling exhibition, or in giving that exhibition a local flavor. Similarly, local collectors may have objects or other materials that can be borrowed to expand upon a traveling exhibition.

Another aspect of presenting a traveling exhibition can be institutional capacity-building. Research must be done in order to provide quality educational programming around the exhibition, thus adding or updating the information bank. Some of the units in the exhibition may require learning some new IT tricks, conservation techniques, or program presentation skills. Likewise, volunteers will have to bone up on the new material. On occasion, there may be a physical leave-behind from a traveling exhibition—a curriculum, laboratory apparatus, completed experiments, and so on.

THE TRAVELING EXHIBITIONS INDUSTRY

As museums have become more dependent on a steady diet of traveling exhibitions, a system of exhibition development, production, and distribution has evolved. This has developed into a rather complex interlocking system of producers, distributors, consumers, funders, and marketers of traveling exhibitions.

The majority of traveling exhibitions are created by museums and shared with other museums. A number of children's museums and science and natural history museums have developed traveling exhibitions departments that prepare exhibitions dedicated to the travel market. In recent years, for-profit firms have increasingly entered the traveling exhibitions business, usually with large, high-priced offerings. This began with the developers of robotic dinosaur exhibitions such as Dinamation and Kokoro Dinosaurs. More recent robotic entrants such as WonderWorks, DinoMae, and Adventure Edutainment have created robotic versions of numerous extinct and living creatures. Corporate exhibit producers also include Clear Channel Exhibitions (formerly BBH), Advanced Exhibits, and K'Nex Industries. Simultaneously, several small businesses such as Clifford Wagner Science Interactives, Northwest Invention Center (Ed Sobey), Minotaur Mazes (Kelly Fernandi), Dimensional Imaging Consultants (Doug Tyler), and Holophile (Paul Barefoot), as well as numerous independent photographers, are producing smaller and less expensive exhibitions. The best way to see the fullest selection of traveling exhibitions producers is to use the ILE database that can be searched by producer (see website listings below).

Figure 24.1 shows the distribution of sources of 173 smaller exhibitions (those less than 1,500 square feet or 150 running feet) currently listed in the ILE database.

Quality traveling exhibitions have production qualities similar to those of permanent or in-house temporary exhibitions and, because of the need to be

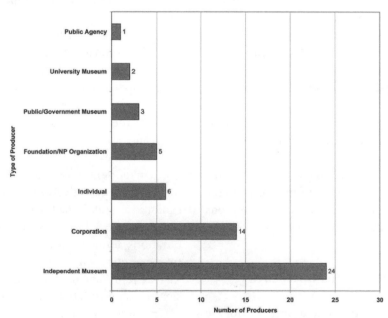

Figure 24.1. Organizations that produce small traveling exhibitions. Source: Informal Learning Experiences Traveling Exhibitions Database.

transported and set up in varying spaces, usually cost as much to design and build. External federal funding (from, for example, National Science Foundation [NSF], National Endowment for the Humanities [NEH], and National Endowment for the Arts [NEA]) generally is accessible only to large exhibitions, which will travel to larger institutions, thus generating substantial attendance. Clear Channel Exhibitions has developed an exhibition development model that relies upon national corporate sponsors. For example, their current exhibits Genome and Brain are sponsored by Pfizer, Chicano Now is sponsored by Target Stores and Hewlett-Packard, and Earth Quest is sponsored by Ford Motor Co., Hertz, and IBM. As an apparent complement to this, the NEH has a late 2003 initiative to support the resizing of large NEH-supported exhibitions so they can be shown at smaller venues.

More modest exhibitions generally are funded out of institutional operating budgets or by local sponsors as part of the temporary exhibition program or by external funders at the state and local levels. State Arts and Humanities councils are important supporters of many exhibitions, as are corporations and foundations with close relationships to specific individual museums. The budget plans for many exhibitions depend upon significant revenue from participation (rental) fees paid by museums that will receive the museum as it makes its tour.

Most producers also manage the tours of their exhibitions, using a variety of outlets for their marketing. The Association of Science-Technology Centers (ASTC) manages a stable of about fifteen exhibitions produced by member organizations, and Smith Kramer, Inc., ExhibitsUSA, Blair-Murrah, Curatorial Assistance Traveling Exhibitions, and Landau Traveling Exhibitions also manage exhibition development and touring for a wide array of mostly small producers. The Smithsonian Institution Traveling Exhibits Service (SITES) circulates over fifty exhibitions from the various Smithsonian museums.

ASTC: www.astc.org/exhibitions/index.htm
Blair-Murrah: www.blair-murrah.org
Curatorial Assistance: www.curatorial.org
ExhibitsUSA: www.maaa.org/exhi_usa
Landau Traveling Exhibits: www.a-r-t.com/lte
Smith Kramer Inc.: www.smithkramer.com
Smithsonian Institution Traveling Exhibitions Service (SITES): www.si.edu/exhibitions/traveling.htm

Information on most traveling exhibitions is available from the websites of the producers and distributors. In addition, several searchable databases are available: Informal Learning Experiences (ILE) has about 650 exhibitions on its database at www.informallearning.com, and the Centre for Exhibition Exchange lists about 300 largely Canadian exhibitions at www.cee.ca/english/exhibit.htm. Exhibition halls at national and regional meetings of the American Association of Museums (AAM), Association of Children's Museums, and ASTC feature booths staffed by traveling exhibition producers and managers, and the Humanities Exchange publishes frequent updates of its *Guide to Organizers of Traveling Exhibitions*. Informal Learning Experiences facilitates informational roundtables on traveling exhibitions at each AAM and ASTC annual conference.

A large but unknown number of traveling exhibitions currently are available. In the absence of a single comprehensive listing of currently traveling exhibitions, we have analyzed the rosters of exhibitions on several major databases, which admittedly do not include many productions from smaller players in the industry or most international producers. We estimate that there are somewhere between 1,000 and 1,200 traveling exhibitions in the areas of ILE's concern (which does not include the fine and decorative arts), and perhaps as many as 3,000 in total. (These are very rough estimates of highly dispersed data; reality may be significantly greater or somewhat less.)

The content and style of traveling exhibitions is as diverse as the museum industry. In Figures 24.2, 24.3, and 24.4 we summarize the breadth of exhibitions in the ILE database (which is biased by not including exhibitions dedicated to the fine and decorative arts, but which does have strong history and anthropology/culture listings).

Figure 24.2 gives an indication of the distribution of exhibitions among the general categories in the current ILE database as selected by the producers/distributors. Figure 24.3 is the twelve most common keywords selected from a list of about eighty descriptive words or phrases. Figure 24.4 reflects particular attributes of the same set of exhibitions. It is helpful to note the emphasis that is put on interactivity; this reflects the needs of museum audiences, but also requires more costly development, maintenance, and transportation than for static or flat, two-dimensional exhibitions.

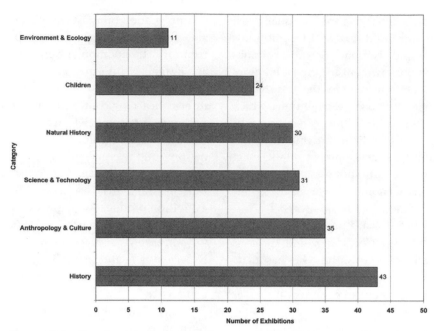

Figure 24.2. Broad categories of smaller traveling exhibitions. Source: Informal Learning Experiences Traveling Exhibitions Database.

TRAVELING EXHIBITIONS AND SMALLER AND NEW MUSEUMS

Small museums, at least as much as their larger colleagues, require continuous novelty both to generate essential visitor numbers and revenue and to provide services to their community. Thus, despite chronically tight budgets, limited space, and small staffs, chang-ing and traveling exhibits are a desirable part of the operations of most museums. Given these challenges, how can smaller museums obtain and present a regular schedule of traveling exhibitions?

Fortunately, the small museum market is not being completely overlooked by museum producers and smaller independents (although, because that market

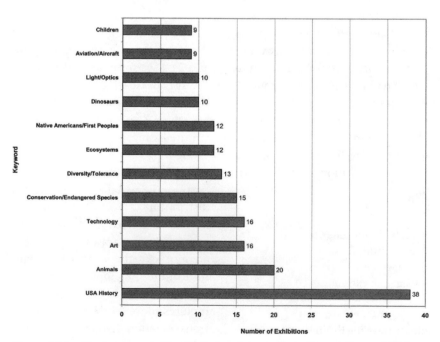

Figure 24.3. Most common keyword descriptors of small traveling exhibitions. Source: Informal Learning Experiences Traveling Exhibitions Database.

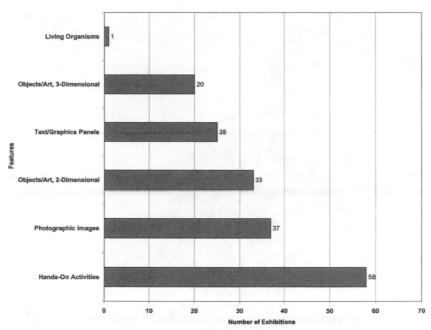

Figure 24.4. Attributes and presentation styles of smaller traveling exhibitions. Source: Informal Learning Experiences Traveling Exhibition Database.

does not generate a significant profit, the corporate producers generally ignore it). Several large museums often produce exhibitions in two sizes: large and 2,000 to 2,500 square feet; Clear Channel Exhibitions has done so for several of its exhibitions. But, there are only a modest number of traveling exhibitions that are smaller than 1,500 square feet or 150 running feet. The ILE database is distinctly skewed toward exhibitions for larger museums—68 percent of those listed are larger than 1,500 square feet or 150 running feet.

We have looked at the costs (rental only, as different museums make vastly different investments in marketing, development of programming, and enhancement with local resources) of smaller exhibits. Figures 24.5 and 24.6 show the range of costs and sizes for exhibits up to 1,500 square feet and up to 150 running feet. All the rental costs have been normalized to a three-month venue; many exhibits are priced by the month, so venue lengths are somewhat flexible. However, the field has generally settled on a three-month venue as long enough to attract a significant audience while being short enough to encourage frequent repeat visitation and media attention.

Partnerships with local institutions are an excellent way to develop and fund an exhibition that can travel within a limited area or to a set of related agencies such as libraries or community centers. Museums and community-based organizations can collaborate to develop and place exhibitions; the Oregon Museum of

Science and Industry (Portland) received grant support for development of exhibitions that traveled to libraries across the state. The Institute of Museum and Library Services promotes such partnerships, as do state humanities and arts councils.

Collaboratives of similar size museums with similar foci are another way to generate exhibitions for all participants. An excellent current example of this is TEAMS (Traveling Exhibitions at Museums of Science), a group of small science centers. Now completing its second million-dollar-plus NSF grant, TEAMS has produced nine exhibitions that have made or are making their way around to all of the participating museums. A 1996 NSF grant of $1.2 million (plus matches from each museum) enabled the production of five 1,500-square-foot traveling exhibits, which went to each of the five charter institutions (Ann Arbor Hands-On Museum, Ann Arbor, Mich.; Catawba Science Center, Hickory, N.C; Discovery Center Museum, Rockford, Ill.; Montshire Museum of Science, Norwich, Vt.; and Sciencenter, Ithaca, N.Y.). A second NSF award in 2000 generated similarly sized traveling exhibitions from four of the charter museums, each of which is mentoring a partner. As of late 2004, three of those first five still are traveling as part of ASTC's roster of exhibitions; the newer set is moving around within the collaborative.

A final way for small museums to obtain exhibits is to be alert to distribution of excessed or decommissioned

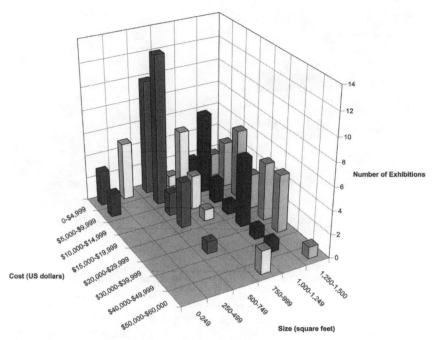

Figure 24.5. Cost for smaller traveling exhibitions measured by square foot size.
Source: Informal Learning Experiences Traveling Exhibitions Database.

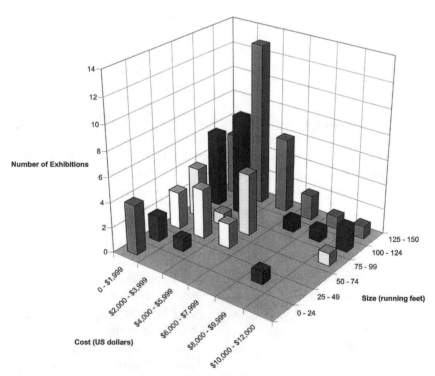

Figure 24.6. Cost for smaller traveling exhibitions measured by running foot size.
Source: Informal Learning Experiences Traveling Exhibitions Database.

exhibitions from larger museums. There often are opportunities to obtain—at very low cost—elements of large exhibits, which with an infusion of creativity can become stand-alone exhibition experiences in smaller spaces.

REFERENCES

Buck, Rebecca A., and Jean Allman Gilmore. 2003. *On the road again: Developing and managing traveling exhibitions.* Washington, D.C.: American Association of Museums.

Howarth, Shirley Reiff. 2003. *Guide to traveling exhibition organizers.* Largo, Fla.: Humanities Exchange.

Regier, Charles. 1994. Rethinking traveling exhibitions for small museums: Observations on the exhibit production process. museumstudies.si.edu/Regier.htm.

Sixsmith, Mike, ed. 1995. *Touring exhibitions: The touring exhibitions group's manual of good practice.* Oxford: Butterworth-Heinemann.

West, Robert M. 1995. Profile of Traveling Exhibitions. *Informal Science Review,* no. 14 (September–October): 14.

West, Robert M. 1995. Impact and importance of traveling exhibitions. *Informal Science Review,* no. 14 (September–October): 14–15.

West, Robert M., and Christen E. Runge. 2002. Traveling exhibitions: A virtually untapped resource for leisure entertainment facilities. Presentation at Trends in Leisure Entertainment Conference, Berlin, Germany.

West, Robert M., and Christen E. Runge, 2003. *Traveling exhibitions database.* Washington, D.C.: Informal Learning Experiences. www.informallearning.com.

Robert M. West, president, and Christen E. Runge, (former) associate, Informal Learning Experiences, Inc., PO Box 42328, Washington, DC 20015. Phone 202.362.5823; fax 202.362.3596. E-mail ile@informallearning.com; web www.informallearning.com.

Part VI

EDUCATION PROGRAMS

CHAPTER 25

Developing an Educational Plan

Laura Martin

SMALLER MUSEUMS USUALLY HAVE ONLY A FEW professional educators on staff who, by necessity, are jacks-of-all-trades: running workshops and camps, guiding school groups, conducting outreach, writing curricula, buying materials, caring for animals and plants, calling teachers, training volunteers, sometimes even cleaning up and repairing equipment. They don't have much time to reflect on their programs.

An educational plan that reflects the museum's approach to learning, however, is something that museum educators report they feel is important to develop. It helps provide consistency in their work and justifies what they are doing; unfortunately, they may feel that they don't have the time to organize and write a plan. If an educator is working alone, it is still valuable for him or her to have such a plan; working with input from staff in other departments, volunteers, scientists, and teachers, is also a good way to stimulate one's thinking and review a plan's appropriateness. What does an educational plan for a science museum look like and how can it help in your work? Why should you take the time to write one?

A plan can be simple or can include a number of components, depending on its scope. However, it usually includes the following:

1. The institution's general mission and its specific educational mission
2. Its educational approach or philosophy
3. The nature of experiences to be provided
4. Lessons from research that support the program
5. Specific activities to be provided with rationale and time line
6. How you will gauge success

The plan may have different versions, depending on the intended audience, such as members or stakeholders. It may be one person's vision or can be constructed by a group, over time. In contrast to a catalog or calendar of programs, a meaningful educational plan reflects the approach to learning adopted by the museum. It can list goals for various age-groups and how the museum experience will achieve these in age-appropriate ways. Moreover, any good plan should be both a record and a tool.

As a *document of record*, the plan articulates the approach to educational programming that staff adopts. It lists the goals for the institution's programming and justifies the choice of priorities, for instance, preschool classes, member offerings, or teacher workshops. It may spell out the basis for why exhibits are presented (for instance, tying into state standards).

As a *tool*, an educational plan helps accomplish the institution's mission, laying out the timetable for developing various educational capabilities. It also can be used to inform fund-raising meetings, community presentations, volunteer recruiting, and staff development. If a museum chooses to join as a partner with schools, the plan can help school administrators understand the museum's approach to teaching and learning.

ADVANTAGES OF HAVING AN EDUCATIONAL PLAN

One advantage of creating an educational plan relates to your audience. A professional plan demonstrates to the people who will participate in your programs that there is a method to the madness (actually, to fun, informal learning experiences). A clear statement of how you view education, the ways in which you propose to design learning experiences, and examples of programs indicate that this museum is a reliable resource.

Another advantage to having an educational plan is that program underwriters and board members need to feel they are supporting effective work in the community, whether those efforts are inviting exhibitions or effective educational outreach. A written plan helps them see the structure underlying the menu of programs and events you're asking them to support.

The mission of many science centers includes helping local schools teach science differently and better. If you are marketing school programs to districts and principals, an educational plan tailored for this group can reassure schools that you are providing meaningful assistance.

An educational plan can also help when you are conducting an evaluation of your programs or of the museum's impact. It specifies what you hope to achieve and is a benchmark that can help you revise as you go, in case you need to change course. For example, as you develop relationships with various community groups or as funding sources become available you may decide to focus on a particular sector (for instance, after-school groups) for which you hadn't originally planned.

Most important, a plan can help staff members focus and continually improve their programs. Museum educators often adopt similar practices in their programs, for instance, workshops a museum has traditionally offered or classes adapted from other museums' repertoires. These programs often continue without being examined for their effectiveness, particularly if program participants leave happy (if they're smiling, we've succeeded). Currently, however, there is more demand from funders, schools, and even museum visitors for accountability and the evidence of lasting impact from programs. Thus it is more important that educational staff be aware of what effective practices are and why these techniques are effective. If you understand the dynamics of a successful, memorable museum experience, you can more easily create new ones. Preparing an educational plan can help spur this reflection and review process. It can also help the staff decide what is important to concentrate on, since the tendency is to do everything for everybody.

Although we are busy and short-handed, it is important that museum professionals learn about the science of learning, about what program evaluation is, and about differences in interactive styles between different audience segments. With this kind of framework, exhibits and programs can be reviewed and evaluated in a more meaningful way than merely on their appeal. Some good ways to keep updated in this area are to attend teachers' conferences, read publications recommended by the Association of Science-Technology Centers (ASTC) or the American Association of Museums, or talk with professors of education or psychology in your community. You can consult the reading list at the end of this chapter for some basic readings in informal learning theory.

DEVELOPING THE EDUCATIONAL PLAN

In developing a plan, start with a skeletal form, or notes, and fill it in as you learn more and think things through. In addition to the content of your programs, think about how you will define the role of the expert (museum staff, parent, teacher) in your programs; the nature of the materials and opportunities you will provide; whether the learners will be in family groups, peer groups, or solo; and what the time frame will be for different experiences. These elements have all been shown to make a difference in learning. For instance, exhibits that allow several members of a visitor group to interact with one another leads to richer conversation about the exhibit. Short interactions with material, too, can lead to memorable impressions, especially when signage helps the visitor clearly understand the purpose of the exhibit. In guided classes and workshops, chances to get hands on materials and to ask open-ended questions about them also seem to lead groups of many ages to gain new insights. How much of each kind of experience will you provide? How will this fit within your mission and budget?

You would be wise to write out your ideas in both paragraphs and bulleted form since some people need to read the details and others want only the big ideas. Be prepared for either, when presenting to the board, a district, or to staff; the occasions differ but the message should be consistent.

COMPONENTS OF AN EDUCATIONAL PLAN

As suggested earlier, here are sections of an educational plan that you might think about including:

1. *The Institution's General Mission and its Educational Mission*

    Mission statements are succinct and set the stage for the details of an educational plan. For example, the Phoenix Zoo's mission statement is: We inspire people to live in ways that promote the well-being of the natural world. However, our educational mission is not only to inspire but to inform and engage people in nature-based activities of many kinds.

2. *Your Educational Approach or Philosophy*

    Your philosophy might be a stance that everyone can actively participate in science, or, that your institution's role is to introduce the most cutting edge science experiences to visitors. Below, as an illustration, I have summarized the educational approach of the Arizona Science Center, an institution I worked at previously. I attempted to place

Figure 25.1. Alex and Tortoise at the Phoenix Zoo of Arizona.

responsibility for the learning environment on the staff; took a stance with regard to inclusiveness and to balance between experiences; and defined the scope of our subject matter. I based it on what I know about learning theory and from the body of research that has been conducted on learning in museums. You can get informed by looking at various websites (for instance, informalscience.org); checking the ASTC publication listings; and looking at journals such as *Curator and Science Education or the Journal of Research on Science Teaching.*

3. *The Nature of the Experience you Hope to Provide, Including the Exhibits, Events, Programs, and Classes*

You might decide that you want experiences to comprise three levels: those that create a sense of wonder (exhibits, presentations on the trail, and

---

A SAMPLE OF AN EDUCATIONAL APPROACH IN SUMMARY

Adapted from the Arizona Science Center Educational Mission Statement, 1994

- Learning to like and take part in science activities is possible for everyone. The informal setting of the science center allows visitors to have fun, play with ideas, experience the wonder of science, and encounter new experiences and people connected with science.
- Learning how to learn about science is a sociocultural practice, fostered by specific types of interactions between novices and more experienced people, mediated by materials and activities.
- We will introduce participants to both discoveries and knowledge that our culture has accumulated and upon which our science progresses, as well as to the ever-changing frontiers of technology, exploration, and self-reflection that move us forward.
- We will convey the idea that science as a practice is something you can do; that the language of science belongs to you; that scientists are people like you; and that science is part of your world. This can be accomplished, in part, by creating personal connections for visitors. For example, we can call on role models with whom people identify, highlight familiar uses of science and technology in the community, and celebrate people in our community who are skilled and informed.
- Exhibits, demonstrations and activities, and programs should provide links between topics and concepts and point the visitor to further explorations.
- If visitors are given opportunities to reflect on their experiences, it will help them make sense of and recall their visit.
- We will offer a variety of experiences for people of different ages and backgrounds and provide multiple entry points for visitor through open-ended explorations, guided pathways, and casual or concentrated activities. Exhibit and programs will emphasize hands-on inquiries, demonstrations, puzzles, and explorations. A variety of technologies can provide interactive graphics representations of data and systems.
- We are committed to exploring ways to make the subject matter inviting to all sectors of the community. For instance, we want to reach the positively predisposed public as well as those for whom science and technology have not been friendly areas of experience. These groups include girls, minorities, and middle school students who are at the point of making choices that will affect their opportunities later in life.
- Every kind of experience we provide will be based, as much as possible, on proven methods about how people learn.
- We will evaluate our performance, effectiveness, and reach at each step of our development.

EXAMPLES OF RATIONALES FOR DIFFERENT PROGRAMS

a. Potential Activities and Materials for the General Public

A large sector of our audience is made up of adults in the community who are interested in though possibly reticent about science, want to explore science, and may be in a position to share their excitement with others. For them, we will offer a wide range of classes, events, and materials, some designed for the knowledgeable public and others inviting people into science by building bridges through the arts, current events, projects, trips, and stories.

By one year: speaker series, adult-only nights, astronomy classes

By three years: trips, elder hostel class, public forums, seminars

b. Children in the Middle Years

The middle years of childhood (which span fourth through eighth grade) are a critical time. This is when children make significant choices that determine their futures: about interests, course work, recreational pursuits, and habits of mind. We can meet their developmental needs by enhancing their feelings of self-esteem, respecting their considerable mental sophistication, introducing them to exciting individuals who can be role models, and providing social experiences that help make science cool. We will offer after-school and weekend classes, adventure challenges, and outreach programs.

By one year: revised summer camp program, mechanics for girls, weekend park outreach

By three years: science by mail, web activities, weekend science quest challenges

c. Preschoolers and Their Caregivers

It is never too early to encourage young children's curiosity and explorations—part of developing scientific thinking. We will have exhibits and programs geared for young children and offer classes in science with little ones for their caregivers.

By one year: preschool science class, staff training

By three years: preschool teacher workshops, targeted gallery activities

d. The Business Community

We are in a position to offer support to members of the business community who want more information about improving educational opportunities in the area. We can help them gather and disseminate information, learn about new developments in technical education, and link them to projects in the community by networking with schools, universities, and community resources. We can involve them in projects, hold meetings and events, and participate in community planning that supports educational initiatives.

By one year: participate in committees and public discussion, do presentations to chambers of commerce, and so on.

By three years: distribute reports on critical community issues in science education

e. Teachers

Teachers are critical mediators of children's experience with science. We can play a special role for teachers as a non-school learning environment. For example, we can help them learn about inquiry-based learning through workshops, how to get the most out of field trips by offering advance materials, activities to do during a visit, and follow-up materials such as scrapbooks. We can help future and current teachers learn to have more fun with science by offering pre-service placements and in-service internships at our facility. We can serve as a site for performance-based assessments in which students can demonstrate their knowledge of science in action at our facility.

By one year: inquiry workshop, intro to the museum

By three years: summer institute, resource room

demonstrations perhaps), those that create awareness (for example, signage, keeper talks, themed events), and those that involve people in more in-depth experiences (for example, camps, classes, conservation projects). Since we would like to be as effective as possible in our mission, we would like to offer experiences that support the visitor or client as an active participant. We will therefore focus on the visitor observing, handling, investigating, and actively thinking about the things we offer. We will make sure teachers have pre- and post-visit support for their field trips and that we extend the animal viewing in as many ways as we can.

4. *Lessons from Research about Learning and about Museum Learning*

The domains that have been explored in museum learning research are varied. Ideally you would address these research findings as you plan your programs. Here are some examples:

THE STRUCTURE OF A VISIT: Studies have looked at how a trip to the museum can be organized and how that affects what visitors get out of it. Research has shown, for instance, that signs can organize what the visitor gets out of a visit by identifying a theme that otherwise may not be clear.

THE ROLE OF SOCIAL INTERACTION: People in museums have a lot of strategies that encourage engagement and understanding. One finding here is that exhibits that allow families to work together show more exploration and explanation; families engage in a lot of educational exchanges in a museum.

MEDIATING EXPERIENCES AND DEVICES: The role of interpreters and activities has been studied. A notable finding is that teachers who prepare and follow through on class visits get more benefit for their students from field trips. Students who have a workshop-type experience also derive a lot of benefit.

5. *Activities you Provide with Rationale and Time Line*

In thinking about activities for different audience segments you want to think about audience needs, the appropriateness of the activity, and the weight you would give it in terms of resources.

6. *How you will Gauge Success*

How will you decide whether you have accomplished the goals of your educational plan and programs? Will you use evaluation, the appeal to visitors, the number of customers that attend, and so on? These days, program evaluation is a critical component of planning. For your plan, think about how much you will do, how it will be conducted, and under what circumstances (before, during, or after a program, for instance). You would do well to consult with a professional evaluator or museum colleagues if you don't have the expertise on staff

---

EXAMPLE OF STATEMENT ON THE ROLE OF RESEARCH AND EVALUATION

There is a rich literature on informal learning environments and what works for them. We will develop and improve our programs with awareness of evaluation and research findings and contribute to the body of knowledge by studying what we do.

Several kinds of evaluation and research are possible to conduct and will be included as needed:

- Background Evaluation/Research—what people know about the content of our work, what they are interested in, have done in zoos, want to know, and other questions help us develop exhibits, experiences, and programs.
- Formative Evaluation/Research—how people respond to prototypes that test ideas we'd like to develop for signage, programs, educational materials, and so on. Improvements are then made in the final phases of development.
- Program Evaluation—looking at appeal, effectiveness, and ease of use of a program, exhibit, event, or materials that are being used.
- Summative Evaluation—assessing the ultimate impact of a program or experience.
- Basic Research—fundamental studies of how people use the zoo and how experiences there contribute to family and individual identity.
- Market Research—how people perceive us, what they like about our services, where they hear about us, and so on, to help in our marketing and customer service efforts.

because valid evaluation is not as easy to undertake as you might think.

Conducting research at an even more basic level than tallying visitor response is a wonderful thing to consider if you are interested. Research projects or questions can lead you to collaborations with local university folk and can contribute to the knowledge base of the field.

Borrow ideas and the lessons of other science centers and other educators. You do not have to recreate the wheel to create your educational plan. In fact, it is good to use precedent, to avoid unsuccessful approaches and mere intuition. You know your own community best, however, so you know what they will need to hear and what you need to provide in the way of detail. An educational plan is a terrific tool for developing connections to your audience and for doing exemplary work for the public.

The plan can help you prioritize, too. As you articulate the reasons for working with each sector and identify objectives for your work, you can think about what is feasible now or in a year or in three years. You can think about resources you may need to add or ideas you want to check, either by asking around, conducting market research, or setting up pilot programs. If you're thinking of "importing" ideas, writing a plan can help you decide which might work in your situation. The exercise can also help you garner resources in the community—such as people willing to participate in a study—since you can point to specific ideas in the plan that you want to verify. Finally, educators in smaller museums can combat isolation by making the development of an educational plan a shared activity.

SUGGESTED READING

Falk, John H., and Lynn D. Dierking. 2002. *Learning from museums: Visitor experience and the making of meaning.* Walnut Creek, Calif.: AltaMira Press.

Paris, Scott G., ed. 2002. *Perspectives on object-centered learning in museums.* Mahwah, N.J.: Lawrence Erlbaum Associates.

Schauble, Leona, and Karol Bartlett. 1997. Constructing a science gallery for children and families: The role of research in an innovative design process. *Science Education* 81 (2): 781–93.

Schauble, Leona, Gaea Leinhardt, and Laura Martin. 1997. A framework for organizing a cumulative research agenda in informal learning contexts. *Journal of Museum Education* 22 (2–3): 3–8.

CHAPTER 26

Building Capacity to Work with Schools

COLLEEN BLAIR

THE MOST NOTABLE FEATURE OF MY COMMUNITY IS ITS complexity. The 500,000 people living here represent a range of backgrounds, experiences, and circumstances. However, we all share a commitment to our children and their schools because we know our future is with them. As you work to create your institutional vision, spend extra time thinking about the relationship you want to have with your K-16 formal education system: schools, students, teachers, parents, and the universities preparing your community's future teachers.

Ask the following critical questions:

1. Why do we want to be involved with schools and why does this matter to us?
2. What kinds of relationships do we want with them?
3. How do we responsibly build these relationships?
4. How do we bring our unique assets to the larger community through these relationships?
5. What are our financial expectations for these programs?
6. What costs are we willing to assume to build relationships with schools?
7. Does the board of trustees agree that work with schools and their support systems is a strategic priority?

Does the board of trustees understand the basic economics, demographics, and politics involved in working with schools? Do they know the competition, trends, and debates inherent in working with schools? Do they have a sense of industry norms?

In *Schools That Learn*, Peter Senge writes about understanding the "patterns of influence" that flow among elements of communities. "Some have direct influence on schools; with others, the influence is less direct but there is always interaction...an effectively operating community is one where people recognize the webs of invisible influence, seek to strengthen them, and feel responsible to everyone connected to them.

When that web breaks down, children fall through the cracks and are lost." Many of the Fort Worth Museum of Science and History's successes with schools stem from understanding the role the museum can play in its patterns of influence and interactions. Two ways of thinking have been highly productive for the museum: a learning communication strategy with educators and Family Science Night, a program that affects the entire school community. These programs are notable for their low implementation cost, sustainability, relevance, utility, and continued success.

The Fort Worth Museum of Science and History's institutional culture has shaped our behavior and work with schools. The museum, a sixty-three-year old, collections-based institution, brings a variety of assets and resources to North Central Texas including 74,000 square feet of interior public space, 30,000 square feet of exhibition space, a 390-seat Omnimax Theater and an 80-seat planetarium. Eighty-five percent of its annual $6 million operating budget is generated through ticket and program sales; annual program attendance is about 700,000 and the museum has about 4,000 members. It employs sixty-five full-time staff, two-thirds of whom are dedicated to retail, finance, marketing, development, theater, and building operations. Five percent of total full-time staff and 3 percent of the total annual operating budget are dedicated to school services and work that supports the formal education system.

CREATING THE CAPACITY TO WORK WITH SCHOOLS

Our capacity to work with schools began in 1997 when the museum hired an organizational coach. Two fundamental changes occurred: the museum's leadership and staff refined its sense of institutional identity, and they began learning about learning organizations. This focus enabled us to understand our unique value and build capacity to deliver our value. After a year of work with our coach, staff came to the decision that our core identity was "learning to change the world," our core purpose was "to provide extraordinary learning environments" and our core audience was five- to ten-year-old children and the people who cared for them.

We articulated five strategic objectives to help us create these extraordinary learning environments:

- Define our audience and their expectations
- Communicate effectively to our audience
- Become a learning organization
- Create new, extraordinary learning environments
- Provide safer, user-friendlier environments.

Our core values were respect and integrity; our strategic values were being warm, friendly, and accessible to children and families.

This institutional identity work set our gold standard for our relationship behavior. It forced us to think about the kinds of learning relationships we wanted with visitors, including students and educators on field trips. It led us to learning organization literature and systems thinking. It provided the clarity and strategic vision we needed to move ahead with the formal education system. We realized that in our twenty-first-century economy, our museum would probably be best served by the staff who understood how to provide educational value and service in a very competitive educational market. We'd heard about an experience economy but what about a learning economy? The Dallas-Fort Worth Metroplex is filled with an array of museums, science and nature centers, and zoos and botanic gardens. In addition, for-profit leisure venues such as Six Flags, the Palace of Wax, and even professional sports teams offer educational field trips for students and programs for teachers. How could our museum with its limited resources compete?

BUILDING RELATIONSHIPS AND CREATING NETWORKS

Years of practice have strengthened our belief that if you create and invest in building a welcoming and respectful institutional environment that openly values educators as lifelong learners, your money is well spent. Educators can become your very best marketers, your strongest advocates and allies, and the invisible motivating force responsible for hundreds of annual visits to your center. Keep in mind that educators just don't do field trips. They are also the influential voice continually advising parents that if they really want to spend quality time with their families, they will take them to the science center.

How is it that an institution learns to communicate and work with the K-16 system? We learned by participating in several sustained university partnerships and professional development initiatives that resulted in multiple network opportunities. These relationships forged our initial understanding of the state's political and complex education landscape and existing patterns of influences. Our biggest challenges were learning how to use our very limited resources in productive ways, aligning with the state standards and most important, retaining our unique informal learning identity. We had to understand how to add value to the existing leadership and reform efforts, how to use emerging research in informal learning and function as a learning organization.

Our strategic outcome was the creation of professional development program portfolios for leadership administrators, educators, and pre-service students. We wanted educators to begin relationships with us on their terms. We offered free preview events, workshops, and full-day in-service and weeklong inquiry institutes. These programs have taken almost a decade to develop, implement, and become sustainable. We started small, moved slowly, created strong, consistent feedback loops, and watched closely for the sweet spots, the places where staff's and educators' passions intersect. We carefully defined and nurtured our assets that were absolutely unique experiences in the professional development arena.

COMMUNICATING AND WEBS OF INFLUENCE

Our ability to sustain and grow a robust school audience reflects the museum's willingness to work with educators in new ways. Over time, we had forgotten how to listen to educators. In our marketing march to fill theater seats, we lost touch with our most unique asset: our ability to create learning relationships. We said we were about extraordinary learning environments, but how did we add value to the teachers' world? How were we helping teachers connect their museum experience to their classroom reality of high-stakes testing? How were we making the museum and its learning environments accessible? And just how user-friendly and safe were we? We surveyed educators and conducted interviews for an entire year and radically changed the way we were communicating. We realized that educators and their advocacy of our museum were priceless. In many ways, they were informational gatekeepers to webs of influence in our community.

Figure 26.1. Fun with Kitchen Chemistry Teacher Workshops at Fort Worth Museum of Science and History.

We identified our message: "We are in the business of creating extraordinary learning environments for students. We want to build a lifelong learning relationship with you, their teacher. We want you to connect with us and maximize both your and your students' learning." A few strategies that we found highly productive are the following:

- We've gone to local and statewide science conferences and take portable exhibits.
- We give away as many freebies at conferences as we can afford and offer free museum admission to all attending teachers; our personal contact along with a taste of the museum experience sends the message that our institution cares about science for all kids and not just those living in our hometown.
- We advertise our field trips in our Metroplex newspapers, placing ads in early fall and early winter, and regularly "tag" other regular ads with the reminder to "book your school field trip now."
- We've hosted state and local science meetings for free: there is nothing as effective as an educator's first-hand experience with our staff and our exhibits; goodwill with our statewide science education leadership is a very good thing.

- We've aligned our theater, planetarium, and exhibit programs with the state standards, and we mail teachers pre- and post-visit program planning guides that are also available on our website.
- School services staff offers telephone planning support to all teachers trying to align their visit and special "TLC" sessions at the museum for first-time teachers; they are encouraged to bring their family or school colleagues to these planning sessions. With teachers constantly changing schools and leaving the system, we found that yearly, one-third of our teachers field tripping with us were first-timers.
- We give all teachers who book field trips two free Educator VIP return admission vouchers. These vouchers enable educators to return on their own terms and focus on their learning experience. Complimentary vouchers send the clear message that we are interested in building relationships and getting to know them better.

Field trip evaluation forms match our core values and vision; returned evaluations are posted in our main administrative offices where all staff can read the feedback. Our continual feedback loops and focus groups provide our evidence that teachers believe that our value is in our unique learning environments and our staff's attitude about learning.

What about our challenges with school field trip programs? Our major challenge in this high-stakes testing state is making the case that their students' museum experience will improve their test scores. We don't have answers to this issue as we have not tried to assess this. However, we do closely follow the growing body of informal science learning research and are actively involved in the Center for Informal Learning in Schools. Our biggest economic challenge with schools? We oversell our exhibits resulting in the diminished quality of the students' learning experience. Perhaps we could have avoided this had we spent more time discussing financial expectations for school field trip audience.

FAMILY SCIENCE NIGHT: BUILDING CAPACITY TO WORK WITH FAMILIES AND SCHOOL COMMUNITIES

Our school services group has been building relationships with schools for nine years. Over time, we've increased our capacity to do the work. We added a second, full-time school services staff in 1998 and a third in 2000. We've built a small, core staff with shared values about teaching, learning, and the museum's role in the Texas educational infrastructure. Along the way, we've taken a lot of risks and tried to create pockets for program innovation. Perhaps our most successful risk innovation with schools has been Family Science Night, a simple program in concept, design, and implementation. Elementary schools have the opportunity to buy out the museum's exhibits and planetarium, weekdays between 6:00 p.m. and 8:00 p.m., for $600. Children must be accompanied by an adult and come in a family unit. Campuses register on a first-come, first-served basis and must provide bus transportation for families. The museum requires each campus designate a Family Science Night team to attend an orientation. At orientation, campuses get planning information and promotional materials.

During Family Science Night, the campus faculty Family Science Night team manages the family check-ins, orientation, and wayfinding. As families check in, they each receive an activity agenda and Come Back Soon! voucher for four free admissions to Omni, exhibits, and planetarium. All materials are bilingual (English and Spanish).

A Family Science Night's evening attendance has ranged from 40 to 1,500; an average attendance is 500. Since its inception, 70,000 Title One children and their families have participated in the program. Neither school district administration, the original planning team, nor museum staff ever envisioned the program's success, impact, and longevity. In its pilot year, the museum offered two Family Science Nights. By 2004, Family Science Nights had been scheduled for forty-two evenings.

The program's initial planning and implementation costs were minimal; the school district's Eisenhower funds paid stipends to the teacher development team. The museum contributed one full-time staff for seventy hours of development and implementation time. Consumable expenses for the original two evenings were $75. The museum now dedicates one-third of one school services full-time staff to manage and implement the program. Four additional part-time staffers facilitate the families' exhibit and planetarium programs. One security staffer is also on-site for the two-hour program.

Why has Family Science Night been so successful for the museum? It adds value to community and simply is the right thing to do. It has provided a low cost mechanism for families to enjoy science in a safe and joyful environment. Initially, the school district approached the museum with a problem: how could they support campus-based family involvement? The museum stepped up and offered facilities and staff and the school districts provided bus transportation and

program fees. Schools repeatedly tell us that Family Science is "the best program our school does all year long." In retrospect, museum staff would argue that it is the museum that has most benefited. Family Science Night is a living laboratory for our responsible community growth. When we began nine years ago, we knew little about cultural diversity and had no bilingual signage or staff. We were somewhat clueless about the barriers to museum participation: many families didn't have transportation, couldn't afford our admission fee, and did not see us as a welcoming place for them. Slowly, over time, our bilingual signage has improved and our staff is more diverse. The program has generated faithful donors who understand and appreciate the program's mission and value to community.

As you envision your work and think about your responsibility and involvement with schools and ways to bring your assets to your community, you might consider a simple, sustainable concept such as Family Science Night. It could be a potential program that helps keep your institution relevant and builds your capacity to do generative work with your community.

If you've decided working with schools is a priority, consider these few final suggestions. It is profoundly important that your team become familiar with the emerging body of research and literature about the nature of informal learning and schools. It's essential to build an understanding of why learning in informal environment differs from the more formal environment of schools. Seek out online program evaluation reports by firms such as Inverness Research Associates (www.invernessresearch.org) and research from the Center for Informal Learning in Schools (www.exploratorium.edu/cils). Get your staff Sheila Grinell's *Starting a Science Center and Keeping It Running*. Give them opportunities to participate in high-quality professional development. Urge them to visit other museum's programs and to grow a network of colleagues also engaged in work with schools. Convince your board of trustees that your relevancy and value to schools and community relationships will grow because of their participation in these networks. Recommend to your staff who work with schools that they study business and organizational dynamics literature such as Peter Senge's *Fifth Discipline* and *Schools That Learn*, and Margaret Wheatley's *Leadership and the New Science*. Working with schools is an economic and a political enterprise; staff needs to know how to articulate programmatic need to potential funders.

In 1999, Mark St. John wrote a paper titled Critical Supports for Elementary Science Reform: The Top Ten

Action Items for Superintendents. His work has motivated our museum leadership and may offer insight for your work with schools and community:

> Action Item #1: Have Courage . . . *I am convinced, as I think are many of you, that a rich hands-on inquiry based elementary program is a very good thing for children. But even though we may know that, I think it is going to take tremendous courage to create an elementary science program in today's world because ultimately, there is probably not enough external political incentive to justify doing it. But ultimately you invest in the people and the resources I described earlier because hands-on science is a wonderful contribution to the children. So it is going to take real courage to keep doing the steady quiet work that is necessary.*

As you and your board of trustees envision your emerging institution, spend more time than any of you have thinking about the relationship you want to have with your K-12 formal education system. There may or may not be the external political incentive to do it, but if you ask staff in Fort Worth, focusing on K-12 in your new center is a very good thing for children and the right thing to do.

ENDNOTE

With gratitude and affection, I thank the staff of the Fort Worth Museum of Science and History, Monta Bates and Anne Herndon, my colleagues in School Services, and Charlie Walter, my colleague and an incredibly passionate advocate for relationships, kids, schools, teachers, and communities.

REFERENCES

Grinell, Sheila. 2003. *Starting a science center and keeping it running.* Washington, D.C.: Association of Science-Technology Centers.

Senge, Peter. 1990. *Fifth discipline: The art and practice of the learning organization.* New York: Doubleday/Currency.

———. 2000. *Schools that learn.* New York: Random House.

Wheatley, Margaret. 1999. *Leadership and the New Science.* San Francisco: Berrett-Koehler.

CHAPTER 27

The Why and How of Doing Outreach Programming: Fulfilling My Fantasy

DENNIS SCHATZ

I HAVE A FANTASY

I have a fantasy—that someday science will be as pervasive as sports in our society. Just think what it would mean to have intramural science, after-school science, and even that pickup science activity at the local park. And how will we know when this fantasy has become a reality? When you go to the local department store and find hanging next to the sweatshirt that proclaims, "stolen from the athletic department at the university of (insert your favorite school here)," sweatshirts that state, "I made billions and billions with Carl Sagan" or "I did time with Steven Hawking"—and people would get the humor. The ultimate test for knowing when science is as pervasive as sports will be when everyone has to rush home to see Monday Night Science.

FANTASY VERSUS REALITY TODAY

Today's reality is far from realizing this fantasy. When I have asked children what they know about science, they typically tell me what they studied in school. Science is not seen as something we do outside of school. Many people think that science centers and other informal science education experiences are changing this. But the statistics say otherwise. Surveys of most science center visitors show that we primarily serve the following:

1. The highly educated
2. Those with high incomes
3. Those already interested in science

This is not a phenomenon that is special to science centers. For art museums, just change the word "science" to "art."

BRINGING REALITY CLOSER TO FANTASY

As long as our business is one of having people come to us and our "cathedrals of science" today's reality is not likely to change greatly. If we can't bring all of the people that we want to the science center then we need to take our science to the people. To broaden our reach and to strive to make my fantasy a reality, we need to take our message to places where people who are not necessarily interested in science are found:

1. Community centers
2. Community festivals and celebrations
3. Shopping malls
4. County fairs
5. Schools without easy access to science centers

An aggressive science outreach program should be a key element of all science centers if we want to attain our basic mission.

One challenge associated with an effective science outreach program is that the experiences need to be presented in a wide variety of circumstances and at a variety of distances from the home institution. But there are also a number of benefits to effective outreach programming:

1. Lower Capital Investment—Exhibits designed to travel are usually smaller, lighter versions of exhibits that already exist.
2. Lower Fixed Operating Expenses—There are no ongoing expenses for a physical plant and staying open during low-attendance times.
3. High Flexibility—Experiences are usually designed to be offered in a wide variety of locations, time periods, and audiences.

The rest of this article offers a look at five typical models that science centers offer as science outreach programs, followed by a list of Key Learnings that should help you as you develop your science outreach programming.

TYPICAL SCIENCE OUTREACH PROGRAMMING STRATEGIES

Model 1. Science Center in a Van—
The fifteen-passenger van pulls up to the only school within fifty miles of a small rural town in (you fill in the

name of the state). The van's colorful graphics, Blood and Guts, provide a hint to the subject matter in which students will immerse themselves for the rest of the day. Several students and parents rush to the van to offer assistance unloading the precious cargo. One group of students carries large cardboard boxes to the cafetorium, where eight six-foot-long tables stand ready to hold twenty tabletop exhibits. In less than thirty minutes, the suitcase-like wood boxes are opened to reveal the activity in the bottom and the exhibit copy in the lid. They are ready for the six groups of children that will come through on thirty-minute visits throughout the day. Organella is laid out in an open space on the floor, ready for students to explore the human body's system. The room is quiet as science center staff explain to volunteer parents their role to help students learn during their exploration of the exhibits. It will not be quiet again until the end of the day when students have gone home to tell family members about their experiences. Staff again come through to close the exhibits, stuff Organella and her body parts back into a bag—ready to go on the van and off to the next school the following day.

The exhibits are only one part of the day's events. Another group of students helps unload the props for an assembly that kicks off the day in the gymnasium, which often doubles as the auditorium. For thirty minutes, two science center staff members lead the assembly program in which students and teachers cheer, applaud, volunteer their services, and learn about the activities of the day that will transform the school into a miniscience center.

After the assembly, science center staff moves from room to room leading forty-five-minute inquiry-based lessons chosen by each teacher to best connect with the classroom's ongoing curriculum and the state's science standards. It is a special day. Science center staff often feels like royalty. Children come up to ask them questions at lunch. Many students stay after school to help

Figure 27.1. Engineering Outreach van, Pacific Science Center, Seattle, Wash.

load the van so that it can be on its way to the next school fifty miles away.

Not all days are this satisfying, but many are or even more so, reaching children who have little opportunity to experience science in the individualized, enriching, and inquiry-based way.

Not all science center van programs include all three of these elements. Many just do science assemblies; others set up activities centers where students move from table to table to explore science concepts. No new program starts as large as Pacific Science Center's is today (see Science on Wheels information in box). Science on Wheels started thirty years ago with one used van provided by AT&T. We packed it with demonstrations and classroom lessons we typically did at the science center. The first years, we did about twenty days of programming, during which we spent one week going more than fifty miles away. Start small and start with programs you already have. These are two of the lessons learned from our Science on Wheels experience. See the key learnings at the end of the article for a summary of these lessons learned.

Not all outreach programs deliver exclusively to schools as Science On Wheels does. That would not be sufficient to make science as pervasive as sports. The experiences described in this section are well suited for the order and structure at a school, but the basic programs are easily adapted for use in a wide variety of settings as you will see in Model 2, Have Programs, Will Travel.

Model 2. Fee-Based Community Education—

In the early 1990s, the main outreach program of Pacific Science Center was the Model 1 Science on Wheels program, but it was fully booked by schools the spring before the delivery year and it was optimally designed to serve K-8 schools.

Every two weeks at our education managers meeting, the head of registration would bring in her stash of pink phone messages.

> Is anyone willing to volunteer to take one of our van programs to the Renton River days? They saw the van at their local school and think this is just the kind of science they need at their River Day events.
>
> Here is another one from the Boys and Girls Club. Can anyone do a combustion demo for a Saturday afternoon event?

We had to turn down most of these requests since most of them were asking the van to come at no cost to the organization and Pacific Science Center could not underwrite the program. Given these circumstances, it didn't seem appropriate to ask staff to volunteer their time to present at these events.

The challenge of serving these events led to Pacific Science Center's fee-based community education program, commonly known as: Have Program, Will Travel. It started by taking existing presentations from the science center's floor and our Science on Wheels program and packaging them to travel. Little modification was necessary—Combustion demo and Supercold demo from our floor programs, the Starlab inflatable planetarium from the Space Odyssey Van, and a spare set of tabletop exhibits from the Rock and Roll van.

Once we decided to offer this type of program, it wasn't sufficient to just serve the requests from the

various groups that knew our Science on Wheels program. Again we started small, by making contact with groups we already knew. The King County Library system provides regular programs during the summer for its younger patrons and asked us to provide a number of these.

Once your community education programs are running effectively and you have tapped into the "easy" markets, it allows for time to consider new strategies and new collaborations. This opportunity occurred several years ago when talking with representatives of the Washington Mutual Foundation. Washington Mutual's emphasis at that time was on improving literacy, so a proposal to them strictly focused on science was not an option. Washington Mutual representatives were well aware that improving literacy, especially related to nonfiction reading, required reading and writing within a real-world context. Thus was born the idea for our Storybook Science program that coupled reading and science in libraries across the state. In brief, it had children reading science-based books and doing science activities related to the concepts in the books. Families received follow-up suggestions for readings, and the libraries received a set of science-based books.

We price our community education programs by the program, with travel expenses added on top. There are probably as many ways to price programs as there are institutions that do programming. As you develop a business plan for your outreach program as suggested under Key Learnings at the end of this chapter, you may want to check the websites of a number of institutions to see how they price their programs. Putting out an inquiry on the Informal Science Education Network and Association of Science-Technology Centers (ISEN-ASTC) listserv will also give you more responses. Go to astc.org/profdev/listserv.htm to see how to subscribe to the ISEN-ASTC listserv.

As you begin thinking about starting your own community education program, here is a list of suggested venues for you to bring science experiences to people who wouldn't normally attend a science museum:

1. County fairs—great place for exhibits and demonstrations
2. Community fairs, celebrations, and festivals (for example, Cinco de Mayo, Renton River days)
3. Libraries, Boys and Girls Clubs, YMCAs, city community centers—activity centers and lessons can be added to the lessons and demos
4. Gatherings of Boy Scouts, Girl Scouts, and Campfires

Model 2. Community Education

Example: Have Program, Will Travel

1. Program offerings include:
   a. Science shows (40 minutes) — $425 for first show; $200 each additional show on same day
   b. Hands-on exhibits — $550 for up to 6 hours
   c. Starlab planetarium — $300 for up to three 45-minute shows; $100 each additional show on same day
   d. Storybook science workshops — $300 for up to two hour-long workshops; $150 each additional workshop on same day
2. Until recently, staff used their own cars for travel (reimbursed for mileage). We now have a fifteen-passenger van and two minivans for general use.
3. Community Education provided over 250 separate programs, serving approximately 70,000 people in 2003–2004.
4. Three full-time staff, plus five part-time staff, including time to coordinate, develop, market and schedule the programs
5. Annual budget = ~$140,000 in revenues; ~$127,000 expenses
6. Cost to participants (see information above)

As you contemplate the challenge of serving locations far from home, keep in mind that you don't have the challenge faced by the Imaginarium in Anchorage Alaska that serves locations that have below-freezing weather, have roads that go nowhere, and are only accessible by small planes with skis instead of wheels.

Model 3. Education Kit and Equipment Rental—

The expensive part of most outreach programs is the staff cost to develop, deliver, and maintain a program. Some science centers minimize the delivery portion of outreach programming by training members of the community to give the program. Examples of a few programs include the following:

1. New York Hall of Science loans Starlab Inflatable Planetariums to local teachers for use in their schools. Teachers must go through a training program at the hall before being allowed to rent the equipment.
2. Science Museum of Minnesota loans Discovery Kits on various subjects related to themes in the museum. Kits include items from the museum's collections.
3. Museum of Science, Boston, loans curricular kits that cover important concepts in science and technology. Teachers use these to supplement the textbooks used in their programs.

Each of these was developed with teachers as the primary target, but can easily be adapted to serve non-school markets. These programs work best when involving teachers or leaders who are highly motivated to use the content, so that they can effectively use the equipment and kit materials.

Model 4. Virtual Outreach Programming—

The use of electronic-based outreach has increased as the ability to quickly send images and video footage has advanced. Access to appropriate equipment in schools and other organizations outside science centers has also grown. The results are new outreach services based on the web or teleconferencing technologies. Examples of a few programs include the following:

1. The Exploratorium regularly provides webcasting experiences (times when people can join an event happening in distant and various locations through the web). Examples include events held during solar eclipses, landings of spacecraft on Mars, or a visit to Antarctica.
2. Some centers, such as the Exploratorium, have extensive resources on their institutions' websites that can be accessed by anyone with a computer and web browser. TryScience (www.tryscience.org) is a joint project of IBM, New York Hall of Science, and ASTC that allows science centers with limited web capability to participate in a collaborative website that provides web-based science experiences and also spreads information regarding what science centers have to offer in their home institutions.
3. Some centers have set up teleconferencing-based learning experiences, through which students in a classroom can experience a program or event at the science center or at another location. Probably the most dramatic of these programs is Liberty Science Center's open-heart surgery, which is televised

live from the operating room to schools in the area. At the New York Hall of Science, the video and science experience equipment are delivered to the school where students get to perform experiments based on a presentation being made through teleconferencing back at the hall.

Most of these programs have depended on significant outside funding because of the high cost of the capital investment. The video conferencing-based activities seem easier to envision as fee-for-service activities that could ultimately cover their costs. Many people believe the web-based activities will be harder to do without continued external funding.

WOULD YOU LIKE ME TO SUPER SIZE THAT? Success with any program usually leads to three questions:

1. Can we do it in more places?
2. Can we do it bigger?
3. Can we do it better?

And if we can do it bigger and better and in more places, all the better. Thus, Model 5 was conceived.

Model 5. Traveling Exhibitions—

As our Science on Wheels program continued to grow during the 1980s and 1990s (going from a few weeks on the road to more than 650 delivery days in the mid-1990s), we developed ideas to tour major exhibits to towns across Washington State. Our Dinosaurs: A Journey Through Time exhibit (1987–1989) was so successful in Seattle, we decided to set up temporary showings of the 8,000-square-foot exhibit in county fairground buildings in other Washington State cities—Spokane and Tri-Cities. In Spokane, the building held the Flower Show in the spring and hogs during the summer. For the winter it was five robotic dinosaurs accompanied by numerous interactive exhibits exploring these extinct creatures. In six weeks, more than 100,000 people from northeast Washington, northern Idaho, and western Montana braved the snow and freezing weather to see Dinosaurs in Spokane.

The first two sites were so successful we did a second four-week run in Spokane and took the exhibit to two smaller locations in the state. People from these smaller communities were equally appreciative as other locations, but neither was a financial success. On the basis of these experiences, we showed our next robotics-based exhibit (Whales: Giants of the Deep) in Spokane and Tri-Cities. These were both a programmatic and financial success.

From our Dinosaurs and Whales exhibitions, we learned large traveling exhibitions are only financially viable in larger population centers. This led us to contemplate a smaller, more "nimble" science exhibit that could travel to small-population locations in the northwest. The result was a 5,000-square-foot exhibition (Science Carnival) that set up in two to three days in tents. The development and trial testing of Science Carnival was underwritten by the National Science Foundation (NSF) and a variety of sources in locations visited by the exhibition. It could visit city parks, shopping center parking lots, or almost any open space accessible to the public. The approach was based on the successful Discovery Dome project of Science Projects Ltd. in England. The devices were based on tried-and-true exhibits that existed in many science centers across the country—many of these were adaptations of the exhibits provided in the Exploratorium's science exhibit Cookbooks. It visited forty-six locations over six years before being retired. The goal for Science Carnival was to develop a traveling science center where the entrance fee would cover the cost of traveling the exhibit. Although Science Carnival was a programmatic success—bringing a science center exhibition to small-population areas—it always required significant local underwriting in order to make it financially viable. (For more on Science Carnival see Lynn Dierking, Foreword; Stephen Pizzey, chapter 10; Adela Elwell, chapter 4; Karen Johnson, chapter 6; and Cassandra Henry, chapter 8.)

In 2002 we decided to show our latest robotic exhibit (Aliens: Worlds of Possibilities) in Spokane, Washington. Unlike our previous experiences, the exhibit drew only 28,696 people in eight weeks. Why did it draw

Figure 27.2. Outreach staff conducting liquid nitrogen demonstration, Pacific Science Center.

so poorly? We think it was because the economy was in a slump and education reform was limiting school field trip opportunities. Plus robotic aliens don't have the same appeal as dinosaurs and whales. It did highlight that the bigger the project, the greater the risk and possible financial drain. Clearly, bigger and better is *not* always better. This result shows how important developing a business plan is (see Key Learnings at the end of this chapter) for any outreach project—but also that a plan does not guarantee success.

Our latest effort to serve underserved populations across the state—and ultimately the country—is the Space Spot. The development was underwritten by the NSF. The goal with the Space Spot is to have an even more nimble exhibition than Science Carnival that could be set up in an hour in 2,000 square feet at a shopping center. With only a couple of staff members and spending only four days (Thursday through Sunday) at the shopping center, we hope shopping center operators will see the fee as a great way to market that shopping center as a place to visit for more than shopping. So far, interest is high from across the state, especially from those shopping centers with Kids Clubs. Our trial test of the Space Spot at three shopping centers shows great enthusiasm from the shopping center operators and store managers, in addition to the visitors to the exhibit.

Most important to me is that 30 percent of the visitors say that they do not usually visit museums and 30 percent say they are not interested in science. These are the kind of results that are needed if we are to really make science more pervasive in our society. Maybe there really is the possibility that someday people will want to rush home to see Monday Night Science. For now, the big challenge is how to make these programmatically successful experiences financially viable.

LESSONS LEARNED—KEY LEARNINGS

The variety of ways to initiate, organize, market, and implement outreach activities is almost endless and usually highly dependent on the contextual situation at a given science center. So one size fits all is not possible, but there are some key learnings from the informal science education industry that are useful to keep in mind when planning your outreach experiences. May they be helpful to you as you work to make science pervasive in your community.

1. Start small, but keep in mind how your small effort will scale up.
2. Build on what you already do and offer.
3. Check what other science centers are doing—and how much they charge.
4. Begin by offering your programs to people, places, and organizations you already know.
5. Meet the demand already "knocking at your door" before expanding your marketing to serve new areas and organizations.
6. Always do a three-year business plan that shows how the program will begin and scale up during that time.
7. Keep an eye on the profit and loss.
8. Evaluate the program regularly—not only for participant satisfaction, but program efficiencies and financial viability.
9. Identify and sell the program to funders who want to reach the market you serve.
10. Look for ways to connect to other disciplines—reading, writing, and maybe even sports.

CHAPTER 28

Fifty Years of Museum School

KIT GOOLSBY AND CHARLIE WALTER

FOR MORE THAN HALF A CENTURY, MUSEUM SCHOOL, a program of the Fort Worth Museum of Science and History, in Texas, has been opening young minds to a world of natural wonder. Originally called the Frisky and Blossom Club, Museum School has provided life-changing experiences for over 200,000 children since 1949, and it is one of the first museum preschools in the nation to be accredited by the National Association for the Education of Young Children.

Museum School enhances a child's awareness of the world: sitting in the huge footprint of a plant-eating dinosaur, grinding corn inside a teepee, visiting the Noble Planetarium for an imaginary trip to the moon, and hundreds of objects from the Museum's collections make Museum School distinctive. The curriculum combines natural and physical sciences, history, and anthropology with art, music, and literature. The goal of Museum School is to provide age appropriate learning experiences that enrich children's lives. With programs ranging from a one-week science day camp adventure to a thirty-week preschool experience, Museum School is a valued asset of the north Texas community.

PROGRAM CONTENT

Museum School explores natural science, physical science, and culture in its various curriculum themes for preschool as well as school-age children.

The goals of Museum School are the following:

- Children will have firsthand experiences.
- Children will develop an appreciation for the natural world.
- Children will begin to construct their own knowledge.
- Children will gain skills in listening and language acquisition.
- Children will creatively express themselves through music, art, and drama.

The unique advantage in museum teaching is the availability of exhibits, specimens, and artifacts as teaching aids. The curriculum for three-year-olds introduces an individual animal each week. The four-year-old curriculum places the animal into a family of animals, such as a coyote into the dog family, along with the red and gray foxes and the child's own pet dog. The five-year-old, ready for the community concept, puts plants and animals together into habitat studies. Older children, entering grades 1 to 5, have a wide variety of program options ranging from the weeklong *Space Bound*, *Volts and Jolts*, *Coral Reef Adventure*, and *Art Exploration* programs.

The preschool curriculum is a thirty-eight week program that builds off the strengths of the museum's staff and collections. Using the first seven weeks of this curriculum as an example, the children explore topics as shown in table 28.1.

Table 28.1. Topics of Exploration

Week #	Three-Year-Olds	Four-Year-Olds	Five-Year-Olds
1	Rabbit	Small Rodents	Seashore
2	Skunk	Large Rodents	Seashore
3	Opossum	Cat Family	Seashore
4	Lizard	Dog Family	Space
5	Box Turtle	Earth Science	Space
6	Snake	Solar System	Planetarium
7	Spider	Planetarium	Pond

Week three of this program would have the three-year-olds studying the opossum. The teaching objectives of this class are to review the general characteristics of mammals, introduce new marsupial aspects, and explore the opossum's habitat, food, nocturnal nature, family life, predators, and defenses. Teaching aids for this class include opossum puppets, pictures, an opossum poem, plastic eggs, fruit, snake, and mouse to show food habits. A highlight of the day is a visit from a live opossum from the museum's live animal collection. A mounted specimen, skull, and magnifying glasses are also utilized. Children see a short film, *Mrs. Opossum*, and at reading time they hear stories from the books *Posie the Opossum*, *Henry Possum*, *Possum Baby*, and *Animals at Night*. Children imitate an opossum's sleep defense, and their creative project for the class is to build an opossum puppet, complete with prehensile tail, which they get to take home to talk about with their parents.

In week five the four-year-olds study Earth Science. The teaching objectives of this lesson are to explore

Figure 28.1. Budding naturalists, Fort Worth Museum of History and Science, Fort Worth, Texas. Photograph courtesy of the FWMSH.

properties of earth, properties of air and gravity, properties of the moon, sun as a gaseous star, and source of our energy, heat, and light. Materials utilized include mineral specimens from the collection, magnets to show gravity, straws to blow ping-pong balls, clay to make a core, mantle, and crust model of the earth, and candles and balloons to explore properties of air. Children also visit an earth model and rock and mineral displays in the museum's exhibit galleries. The book *Papa, Please Get the Moon for Me* and the records *The Earth Goes Around the Sun* and *Goodnight Moon* are utilized. For a creative project, children make a model of the earth using construction paper, paints, and cotton "clouds."

The five-year-olds are continuing their study of the Seashore in week two of their Museum School experience. In particular, students focus on Birds, swimmers and divers, and Mollusks. The classroom is rich with specimens from the museum's collection, including mounted birds (pelican, gull, black skimmer, cormorant, great blue heron, ibis, and curlew), eggs and shells in cases, and mollusks (live snails, conch shells, pen shells, whelk shells). Children listen to ocean sounds and the record *Birds Are Flying*, watch a National Geographic film, *The Seashore*, and are read *The Seashore Book*. As a creative project, children finger-paint waves on a table and use a piece of white paper to make a print of their pattern.

The museum's rich resource base of collections, exhibits, and staff expertise are the core from which all activities are developed. The Museum School provides a live animal collection and interpreters for children to have real hands-on experiences. Imagine the impact of holding the tiniest baby mouse or feeling the rings of a rat snake or watching the jaws of an alligator snapping turtle. While studying dinosaurs the Museum School students visit a re-created paleontology dig site where they can dig in the sand and uncover dinosaur bones (casts) and find real microfossils that they can take home. They also walk upstairs and see articulated skeletons and other cast models, which introduce them to the real science. Likewise, when these children are studying the solar system and planets, they visit the planetarium and take an imaginary trip to the moon.

Museum preschool students attend one day per week for two hours. School-age children attend classes on Saturdays and both age-groups may attend summer classes. Museum School students share experiences in the classroom and an exterior courtyard that serves as an outdoor classroom where all are explorers, naturalists, botanists, and dreamers. Children use their imaginations and are encouraged to let curiosity be a common guiding principle. Each Museum School class has a literature component that reinforces the topic of study for the day. When children are learning about rodents, a teacher may include *Under the Moon* or *Good Neighbors* in her storytelling time. The illustrations serve as a springboard for more discussion and the concepts that were introduced earlier in the day are now real and understandable. Many children attending Museum School live in urban settings where the natural world experiences are limited.

PROGRAM PARTICIPANTS

More than 1,800 preschoolers participate annually in this 38-week program. In addition, 2,000 six- through twelve-year-olds register for Museum School classes each year, many of these are museum preschool graduates. Museum School often serves as the initial introduction to our museum for young families in the Metroplex. The program currently serves children from eighty-two different zip codes: 72 percent from Fort Worth, 21 percent from Tarrant County, and 7 percent from other counties. During the summer sessions, Head Start centers, Summer Sunshine Club, an enrichment program for children living in a local homeless shelter, boys and girls in the I Have a Dream Foundation program and All Church Home have been actively involved in Museum School as scholarship recipients.

One grandmother writes often about her grandchildren and their experiences in Museum School.

As a child I never set foot in a Museum till I was 22 years old. . . . You see, my mother nor my father had any type of formal education. Neither one spoke English, so that alone made learning harder in school. My summers were spent picking cotton and picking other vegetables. The above are just some of the reasons that I am very grateful for the opportunity you have given to my grandchildren, Adrian 11, Arthur 9, Paul 6, and Ledia 5. I believe because of this, my grandchildren will be able to excel in school.

THE STAFF

Museum School staff is very carefully selected from candidates who possess academic backgrounds in child development, education, science, or related fields and who possess a passion for working with young children. An understanding of developmentally appropriate practices is paramount for the success of our program.

Figure 28.2. Girls with baby mice. Photograph courtesy of the Fort Worth Museum of Science and History.

The teaching team includes administrators, curators, undergraduate students, classroom teachers, early childhood specialists, scientists and artists (see included résumés). The following excerpt typifies a Museum School teacher:

Children are naturally inquisitive. Taking advantage of that attribute, I present a concept with clarity and illustrate its relevance to them. Direct experience, simulation, role-playing, and metaphors help students internalize an idea. The process of writing helps organize thoughts. To help students understand, I teach techniques such as drawing time lines, graphs, and sketches. Dramatic illustrations may be found in literature. Factual elements may be incorporated into art projects. Whatever the avenue, the new concept is introduced with a memorable experience and is consequently remembered.

In loving environments, children appreciate each other and themselves. In accepting environments, they feel free to ask questions and seek to understand life's intricacies. In enriching environments, children acquire a sense of their place in history and accumulate ideas of what they might accomplish in the future.

This is the environment Museum School provides through highly qualified teachers, chosen on the basis of educational background, experience, and aptitude for working with preschool-age children.

Museum School supports its teaching staff with professional development opportunities such as field trips and exploration workshops at the Museum's dinosaur excavation site.

Before teaching a wildflower theme or using wildflowers in an Artist's Paint Box class, teachers will go on a wildflower exploration with a science curator. Techniques in handling an opossum or reptile give staff confidence in hands-on interpretation that is shared with Museum School participants.

PROGRAM IMPACT

The primary evaluation tool utilized by Museum School is a parent evaluation that is filled out at the end of each session. This evaluation covers such items as program quality, richness of educational experience, and teacher-student relationship. Teaching staff meet regularly to reflect and refine teaching strategies. University interns (see Partnership section below) use Museum School as a learning laboratory and provide written feedback on their observations. Museum preschool is accredited by the National Association of Education of Young Children. As part of this accreditation process, two validators spend two to three days every three years on-site evaluating the program, facility, and staff. Prior to the on-site visit over 500 parent surveys are completed (validation requires at least 50 percent of program participants) and staff complete a self-evaluation. Several comments from these accreditation reports follow:

- The Commission commends the program for its continuing effort to provide a high-quality early childhood program for children and their families. The program is to be further congratulated for positive and frequent interactions with both children and parents.
- The Commission commends the teachers and assistants that fully meet the qualifications criteria.
- Congratulations for completing the self-study process. The Commission awards accreditation to your program.

Perhaps the greatest indications of the program's impact are its longevity and the ongoing stream of praise and encouragement by former students. The testimonies of Museum School alumni are filled with emotion and connectedness. Museum School currently serves second generations and has among its alumni many who share memories about the impact that Museum School had on their lives. During the school year 2000–2001, we enrolled the seventh grandchild in a family whose four children had attended in the 1960s. Former Congressman Pete Geren attended Museum School as a child and his two daughters attended recently. Another former Museum School student visited the Museum while home from college and wrote about vivid memories and the impact that Museum School had on his life. These personal recollections share a common thread—their lives were enriched and often changed by the opportunities that Museum School provided.

Looking back at what this institution has meant to me, I can honestly say its impact during my childhood benefits me to this day. It was through exhibits, the planetarium, and classes in Museum School that I first became aware of the larger world around me. As a child attending Museum School, I developed a fascination with rocks, bugs and stars that motivated me to read all I could about them.

What I learned then has carried over into the time I spend with my three daughters today. Our "quality time" together often involves an expedition to a nearby creek or river bed to look for fossils.

—Former Congressman Pete Geren

Before graduation from college last year, I took a friend to the museum during Christmas break to show her the scientific stomping grounds I played on long ago. I saw a group of Museum School students tromping to the dinosaur section as I did over fifteen years ago in that very same place. I paused from inspecting a Gila Monster exhibit and wondered if I was like that at one time: saucer-eyed, mouth gaping, and imagination certainly spinning in all directions.

The Museum School continued to teach me through hands-on learning. The day we constructed dark-glass viewers, which enabled us to see sunspots, was amazing. Compare this activity with a dry textbook and it is easy to see why the Museum School made such an impression on me. Today, our house is still stocked with countless books on astronomy and I can still remember facts from Museum School: Neptune and Pluto have intersecting orbits, angrily swirling poisonous gases envelop Venus, and Saturn has icy rings.

My Museum School experiences have not been forgotten. I regularly devour the Science and Technology section of the *Economist* magazine. I give credit to the Museum School for this scientific interest of mine. From a psychology research project I conducted at Rice on emotional (social) intelligence, I place a considerable value on the Museum School's often-unspoken instruction on how to interact and deal with others.

—Michael Chiu, former student

IMPORTANT PARTNERSHIPS

In addition to serving young children, Museum School serves as an important learning laboratory for university students preparing themselves as tomorrow's teachers. In an ongoing program that started in 1996, undergraduate students from Texas Christian University taking the course Introduction to Early Childhood Education spend 50 percent of their class time in Museum School. Ranae Stetson, a professor of early childhood development at TCU, says the Museum School is a vital part of her lesson plan. Stetson says her teachers-in-training are regulars at the museum to hone their real-life experiences with children in a learning environment. As part of this work, students are required to write an in-depth developmental profile of a child they have worked with that addresses the cognitive, social, emotional, and physical development of that child throughout the semester, as well as present recommendations for planning developmentally appropriate learning experiences and instructional activities for this child. All student reports are shared with Museum School teaching staff, creating an ongoing feedback loop of theory and practice.

THE NEXT FIFTY YEARS

For five decades, the community embraced Museum School with such devotion and loyalty that parents or grandparents would spend the night or a minimum of several hours waiting in line to register their children for classes. This registration process has changed most recently to include mail, fax, and telephone registration as well as a website acquisition of brochures and scheduling instruction. Technology is the tool of choice and responding to the needs of young families continues to generate capacity enrollments.

The Museum School audience has expanded from a neighborhood to the Metroplex. Demographics of the 1985 enrollment reflected 72 percent from city, 21 percent from county, and 7 percent out of county. The 2002 demographics reported 52 percent of enrollment from city, 18 percent from county, and 30 percent out of county.

These changes would normally demand creativity in advertising and promotion as well as an increased advertising budget. However, Museum School has decreased its advertising budget, increased its website presence, and maintained a capacity enrollment. Today we are serving multiple children of alumni. During a recent open house, many parents excitedly shared their own experiences as children in the same classrooms that they were exploring with their children. Museum School touches more than just its students. It reaches the whole family and extends into the community.

The response to change and the constant commitment to excellence have served Museum School well. While parents are pleased with the changes in registration procedures and information exchange, they continue to pursue a curriculum that is rich in firsthand learning and whose foundation is supported by the museum's collections. Natural and physical science presented in a developmentally appropriate classroom environment exceeds the expectations of families who are seeking enrichment for their children. The importance of helping children understand their world and their place in it will serve future generations of Museum School families. The joyful learning that engages young Museum School children and encourages creativity is only surpassed by the interaction with live animals, real specimens, and interpretive strategies that align with children's learning styles. Every child can succeed in Museum School. The parents recognize that this success builds a positive attitude toward learning—lifelong learning.

CHAPTER 29

Promoting Public Understanding of Research through Family Science Programs at MIT

BERYL ROSENTHAL

Research is the backbone of scientific and technological advancement. Moreover, both knowledge and applications in science and technology are advancing today at a rapid and unprecedented rate, a rate that continues to increase dramatically.

—(Boston's Museum of Science 2001, 2)

Two seminal conferences were held by major science centers in the United States regarding how to best address the issues of public understanding of research. The first, The Leading Edge, was held February 11–13, 2001, hosted by Boston's Museum of Science. The second, Museums, Media and the Public Understanding of Science, was held September 26–28, 2002, hosted by the Science Museum of Minnesota. The National Science Foundation (NSF) supported both conferences. The participants represented a variety of stakeholders, including research scientists, the media, and representatives from very small to very large science centers and museums in the U.S. and abroad. All had the same questions regarding how to present and interpret cutting-edge science and technology to the public. Some had addressed the issue using high-tech multimedia presentations, others held lectures or talks, and still others used live feeds from laboratories (and in one case, a surgical theater).

THE NEED FOR PUBLIC UNDERSTANDING OF RESEARCH

The MIT (Massachusetts Institute of Technology) Museum uses a variety of methods to address the issue. For the purposes of this article, I will concentrate on family programming. We chose a low-impact, relatively inexpensive method, one that took advantage of our greatest asset—the faculty itself.

A major function of museums and science centers is to assist the public in learning about the world around them. As informal learning institutions, we take a very different approach than do formal education settings and provide information and experiences that go beyond what is possible in a traditional classroom. This public expectation (our visitor charge) is that we will, particularly in the case of science and technology, provide insight into a world that is often viewed as beyond the scope of most people. The general public views science as difficult and that only highly trained scientists are capable of understanding it. There is a lack of public understanding of who does science and technology. The most commonly held and long-lived stereotype is Hollywood's mad (male) scientist in a lab coat, surrounded by chemicals and electrical equipment (although more recently, biological samples), divorced from ethical considerations or possibly even serving as the unwitting, naive tool of gangsters and evil politicians.

Finally, and perhaps most damaging, is the public's lack of understanding about *how* scientists and researchers do what they do. Scientists are supposed to have all the answers, and be omniscient, and when emerging data disputes previous beliefs, the public loses confidence in science and its practitioners. As I overheard in a restaurant recently, "These scientists can't even make up their minds about whether low fat or low carbohydrates is healthier, how am I supposed to know what to do?" The public does not understand that science is a self-correcting and cumulative process (and to be fair, a product of human endeavor and is thereby not always free of bias)—there is still a strong assumption that science is merely a chase for (correct) knowledge. Worse, the public has a poor grasp of the difference between basic science and ongoing contemporary research. Hyman Field and Patricia Powell (2001) describe this in detail. We can all name the number of projects that have been cancelled because of the lack of funding directly related to lack of public support (read: understanding). As noted in the report from the Boston conference,

> Most scientific research is highly technical, and therefore not entirely accessible to the general public. Indeed, research in one area of science is frequently so technical and specific that its details may not be fully accessible to researchers who have been trained or who work in other areas of science. Yet, the results of all this research have transformed society in a number of critical areas and continue to do so. They have also raised a number of new ethical and social questions that must be addressed by the public at large. It is therefore crucial that the public understand as much as possible about the new technologies, products, medical practices, environmental issues, communications media, social issues, ethical questions, and more, that result from ongoing scientific advancement. (Museum of Science 2001, 2–3)

In the present promotion of public understanding of research, the greatest programmatic challenge for science centers and museums is how to stay current, how to reflect the latest cutting-edge research:

> By its nature, current research is difficult to present in public venues. It is continually being modified according to the latest findings. Rather than presenting an established set of facts, educators must track a moving target and try to predict its trajectory. The ongoing nature of research does not lend itself to the means we traditionally use to inform the public.... The field of informal, public information is uniquely positioned to provide an orderly dissemination of the nature and scope of ongoing and emerging research to the general public, but it will require altering the way the field goes about conducting its business. (Field and Powell 2001, 421–22)

Professional conferences and listservs are peppered with concerns over how to best accomplish this very critical need. Museum and science center staff constantly seeks ways to join as a partner with local universities and researchers.

THE MIT MUSEUM'S RESPONSE TO THE PROBLEM

As a museum on the campus of one of the world's premier research institutions, the MIT Museum is fortunate to have world-class resources in this area. Four years ago, the MIT Museum began offering what became our (signature) weekend program for families, entitled F.A.S.T. (Family Adventures in Science and Technology.) The F.A.S.T. program provides family visitors with new and exciting opportunities to explore current science topics and technologies that include engagement and active conversation with MIT faculty, students, and researchers. It is our primary and most regular vehicle for demystifying science (and in particular, MIT) and in addressing the public understanding of research. The target audience is families, generally with middle school–aged children. Many of the children attending are exceptionally bright and are seeking enrichment they cannot get in formal classrooms. Still others are homeschoolers. It is, however, in no way limited to that audience, and large numbers of institute and other university students and single adults often participate. Many of the visitors are connected in some way to science and technology; others are just curious and interested in lifelong learning. Regardless, the common denominator is that they are seeking an understanding of the latest research.

The museum's public programs staff identifies projects that may be of interest to visitors (this is supported by visitor feedback regarding desired topics) and the associated groups or departments at MIT that help develop demonstrations that illustrate their current research and related activities that involve and engage families. The projects are identified through a number of means. Staff regularly reads institute newspapers, scans departmental websites, canvasses the news office, and relies on personal contacts (curators are particularly valuable partners in identifying researchers and in locating collections objects that support programming). Each F.A.S.T. offers a completely new set of learning experiences and creates an exciting learning atmosphere. The programs last two hours and are held on the last Sunday of the month (as determined by visitor feedback), and they have a low impact on scientist/engineer/researcher time (making it particularly appealing to extraordinarily busy researchers). The programs always have a hands-on component, but may also include a talk, a workshop, a demonstration, and so on. The type of research being highlighted generally determines the day's format. It also provides graduate students with an opportunity to communicate with the public in lay language (critical to both their general professional development and the need to effectively contribute to a public understanding of research), which the institute is supportive of. The program is free with museum admission.

The desire on the part of museums and science centers to join as partners with university faculty has spawned a new challenge—how to prepare the faculty/graduate students/researchers (who are used to working in formal settings) to present in an informal setting, where the expectations and dynamics are radically different from those they are familiar with. It should also be apparent that they are not necessarily naturally good conveyors of information to the public, and this needs to be addressed. In networking with educators from other museums and science centers that are building relationships with university faculty, we discovered that we were not alone in recognizing and responding to this challenge. Museum staff often holds conversations with the news office to determine which researchers are good public communicators and can discuss their research well. The potential presenter is approached and the program is explained. This first contact is usually done through e-mail (MIT's preferred and most effective communication device) or by telephone. A meeting is arranged, generally occurring in their laboratory, so that the public programs staff can observe the research setting and products and discuss successful translation and activity development with the researcher. The public programs staff then arranges a time when they can accompany the presenter on a tour of the museum, so that they can begin to get a feel for the setting, including the physical parameters (location and number of electrical plugs, access to water sources, and so on). They are also encouraged to visit museum public events, particularly F.A.S.T. events prior to theirs, in order to view firsthand the kind of visitors who are attracted to the program. There is an extended period that follows where the presenter and the public programs staff communicate by telephone, e-mail, and in person to discuss communication with the public, successful interpretation, and choosing appropriate and engaging activities and talks.

Museum staff has developed methods for expanding this experience into related programs. February School Vacation Week frequently coincides with National Engineers Week and is another prime time for us to engage with families and to meet MIT's goal of increasing community awareness. Essentially, the museum offers a F.A.S.T. experience every day during that week. In 2002, we began attracting quite a few visitors from the immediate neighborhood (a low-income, high English-as-a-second-language population), a particular goal of the museum. National Engineers Week set the context for the programming, and participating faculty were as excited as visitors. A bonus was the opportunity this offered to gain buy-in from MIT administration. The dean of the School of Engineering was contacted with a request for his aid in supporting our public outreach through F.A.S.T. and Engineers Week.

His office distributed a strong statement of support to department heads who distributed it to their students and faculty. This has been a highly effective recruitment tool. It has also raised awareness of the museum as a venue for fulfilling NSF proposal requirements for public dissemination of research.

In addition to the type of research conducted, holidays and anniversary celebrations may also provide a format for the program. In 2002, the March F.A.S.T. fell on Easter Sunday, and we seriously considered canceling the activity; however, recognizing the diversity of our community, we decided to go forward with the event, and focused our activities around the science of eggs. Visitors explored eggs through chemistry, biology, physics, and design engineering, and faculty and students involved took an exceptionally playful approach to the topic. This particular event drew a larger than normal local audience from the museum's low-income neighborhood, largely because of the topic's tie to the holiday and the post-church time slot. In 2003–2004, we implemented a yearlong celebration of the centennial of flight, and the focus of our F.A.S.T. programming was aviation. (This also resulted in increased programming specifically for adults.)

Attendance at the Museum's F.A.S.T. events is largely determined by topic, although there is always a general increase during the two-hour slot. Research on robotics typically attracts the largest crowds.

One could argue that MIT is an obvious location for very high-level focus on research using faculty as its

Figure 29.1. During the yearlong celebration of the Centennial of Flight, most F.A.S.T. events revolved around the principles and mechanics of aerodynamics. Visitors had the opportunity to examine models of innovative aircraft. MIT Museum, Cambridge, Mass.

primary resource; indeed the museum's primary audiences are adults and the MIT community, with families and school groups forming the secondary audiences. This does not, however, make the experience in any way inaccessible to a nontechnical audience. A major focus of the public programs department is to insure that the "what does it mean to me?" component is primary. Researchers are coached in how to describe their work in terms they would use in describing their work to their grandmother. They are further expected to include discussion and examples of practical use relevant to the general public. For example, a research group within MIT's Institute for Soldier Nanotechnology focused on how the smart fibers they were developing would be used in nonmilitary clothing applications, such as firefighting and police uniforms. Visitors were far better able to relate to such uses. This was confirmed anecdotally when floor staff noticed a correlation between these uses being raised and the verbal interest level of visitors—there were numerous affirmative responses, references to relatives in these fields and how it would make them safer, and so on.

ASSESSMENT

In 2003, a team of outside evaluators from Lesley University's Program Evaluation and Research Group assessed the program (Sandler 2003). The purpose was to assess (1) who was coming to the program and why, (2) what the visitor response to the programs is, (3) whether or not visitors were getting an increased understanding of science and technology, (4) where and how they heard about the program, and (5) whether and how the program meets the mission of the MIT Museum. Overall, we were interested in knowing if the program was successful in meeting its goals and determining what visitors wanted in the future. Findings were interesting in that although the visitors were excited and enjoyed the program, they often did not recognize that the researchers were from MIT. It became clear that the most common way participants learned about the program was because they were visiting the museum that day. Other sources included museum mailings, the internet (family activity oriented sites and the MIT Museum website), and word of mouth. The most significant attendance and awareness factor remains whether or not the local newspaper (the *Boston Globe*) picked up on our press releases.

Approximately one-third of the respondents had attended F.A.S.T. programs before. Some felt that they could not distinguish between the F.A.S.T. programming and other MIT Museum programming, given

that many of our programs involve MIT researchers. A key response to the reason for attending was a strong interest in science and technology and a desire to learn more, particularly about current research. Various responses included a desire to "feed" their children's interests, to have fun while learning, to engage in social family-based learning, and so on. Most responded that their expectations had been met, owing to clear descriptions in public relations materials or previous experience, and only a few expected something different. When asked about what they had learned, most responded with a science or technology topic. Others named new vocabulary, restated concepts, articulated scientific processes, noted career opportunities, and several noted the implications and application of new technology to their daily lives. There was overall positive response to the program. Regarding what they liked best about F.A.S.T., the four things most frequently mentioned were the hands-on experiments, the involvement of real scientists and researchers (particularly the opportunity to interact with them directly), the program format, and interaction with museum facilitators and floor staff. Other items included being exposed to new topics and the presence of female scientists. Visitors were also given the opportunity to identify topics they would like to see explored in future programming.

The fact that the museum is on the MIT campus grants it tremendous credibility. However, the data indicated that visitors did not always make the connection to research on the rest of the campus, *at MIT itself.* It was clear that the specific connection to MIT needed to be reinforced throughout the program. The survey also identified challenges, such as the need to choose researchers who can communicate their research effectively (that is, not rely on jargon) and the ability to quickly adjust to meet audience needs, environmental factors (creature comforts), and researcher commitment (on two occasions, researchers backed out at the last minute).

Changes made to the program as a result of the evaluation include offering very explicit introductions to the program, noting the specific connection to MIT, providing an agenda for the presentation topic, noting the appropriate audience, providing a map of the museum showing the location of activities, and labeling the activities specifically as part of the F.A.S.T. program. For the following year, F.A.S.T. is being reconfigured to have a more direct tie to the daily lives of the visitors and will be multidepartmental (that is, exploring a single topic in an interdisciplinary manner with the assistance of several departments and laboratories).

"WHAT DOES IT MEAN TO ME?" IMPLICATIONS FOR OTHER INSTITUTIONS

What we learned overall from the formal evaluation, anecdotal information, and our experiences, is that visitors have an expectation that museums and science centers are places where they can find out about new technologies in a satisfying way, through direct conversation with researchers, hands-on activities, and presentations geared toward the lay audience. We also learned that careful planning was needed for the recruitment of researchers (regardless of whether you are a university museum with built-in faculty or are a stand-alone science center with a local university partner), where to focus and how advertising programs takes place, and most important, how to listen to visitors, their expectations, and their perceptions.

But what is the broader applicability of this project? There are several key lessons that can be taken away from this:

1. Many participants came because they were already visiting the museum on the day of the event. This has implications for how one markets programs: how and where do you expend most of your efforts on recruiting visitors who were planning on visiting or on redirecting those whose visit was serendipitous?

2. Regarding the recruitment of researchers, we learned much about how *they* interpret public dissemination; many felt it meant formal lectures, and some were at first skeptical of informal settings, their importance, or their relative worth. One also begins to gain insight as to how museums and science centers are viewed by some researchers; those who felt that they were poor substitutes for the classroom were generally not the best presenters to engage.

3. While researchers are generally interested and intrigued by the opportunity, we learned that in some cases, it was important to sell the importance of informal settings. Such presentations are not a prior-

ity for faculty, and a concerted effort may be needed to convince them that it will not take up too much of their time and that their work is important to the public. They are generally unaware that meeting the scientist carries cachet with the public. This too, has recruiting and marketing potential.

4. Graduate students are often far more interested and available to participate in such programs. Generally speaking, they are often more comfortable with informal settings, and they also tend to be more excited about communicating their work.

5. It is important to know the audience and to listen carefully to them. It is critical to find out what is or is not of interest to visitors. In a later audience assessment project (Rosenthal and Rubin 2004) we began profiling adult visitors. They were very clear about what would or would not draw them. As expected, the "what does it mean to me?" personalization and daily applicability of the research were key. This information was carefully incorporated into subsequent programming.

6. It is also important to know where visitors hear about programs. In the later survey, we discovered that word of mouth was far more powerful than we had assumed. Programs of this type can be enormously successful in increasing the public understanding of research, but only if visitors are listened to carefully and programming is based on their needs and desires and presented by actual researchers who understand the importance and means of informal education. One need not be on the campus of a major institution to succeed.

REFERENCES

Field, Hyman, and Patricia Powell. 2001. Public understanding of science vs. public understanding of research. *Public Understanding of Science* 10:421–26.

Museum of Science, Boston. 2001. The leading edge: Enhancing the public understanding of research (report on the workshop at the Museum of Science, February 11–13).

Rosenthal, Beryl, and Jacqueline Rubin. 2004. "I'm glad we missed the first show at the movies": A survey of adult visitors to the MIT Museum, January 25 to April 25, 2004 (Power Point presentation, staff meeting, June 8).

Sandler, Jodi. 2003. MIT F.A.S.T. program evaluation report (unpublished ms.).

Part VII

WORKING IN A SCIENCE CENTER

CHAPTER 30

Leading and Implementing Innovation in Your Science Museum

RONEN MIR

I MOVED FROM ISRAEL TO THE UNITED STATES TO direct SciTech in 1999, following my science research (I am a particle physicist) and science education (as scientific director of the Clore Garden of Science) careers. I remember the shock of arriving in a lively museum, but with no money and too few visitors. I could see the great potential SciTech had, and the motivation both board and staff had to take the museum to the next level. Building on these strengths, we brainstormed and developed a master plan. SciTech was to expand by adding a mezzanine level, 100 new exhibit components, an outdoor science park, air conditioning, carpets, lights, and colorful paint. Everything but the historic building's external walls was to change.

Fund-raising, building permits, project management, and operating the museum during construction were real challenges. In parallel, we developed educational programs, formed partnerships, and increased attendance. Five years later, all pieces of the puzzle fell into place and the ambitious master plan has been accomplished. I would like to share with you some of what I learned along the way, hoping to encourage you to always dream big and achieve your vision.

VISION

- Develop a vision for your center and communicate that vision to others. If you don't, no one else will. You should set realistic and future goals for yourself and take into account the impact and value to the community, financial condition of the museum, facilities, exhibits, programs, staff, volunteers, and board. You should track your progress, as tracking is the surest way to achieve your goals.
- You are the standard bearer of the vision—that and fund-raising are your principal responsibilities. You should share the vision at staff and board meetings and at talks in front of civic groups.

SCIENCE

- Learn the science in your center. First, study each exhibit and attend each education program. If you can't understand or don't like what's going on, find out early. Ask your exhibits person to explain to you what the exhibits are all about. You do not need to be a scientist, rather try to understand the concepts behind your exhibits, as this forms the basis of all your work. Once you understand, try explaining your exhibits to a child. If you can explain the science concepts to a child, there is a chance an adult will understand them as well.
- Learn how to do science demonstrations. Choose a few demonstrations that you like, and practice using your staff as audience. As you get more comfortable with demos, use them in talks that you give to, for instance, the Rotary Club. As part of a commencement speech I was giving, I brought a small whistle, dipped its end in soap solution, and asked the college president to blow it: sound coming through a big bubble formed at the end of the whistle. This humorous show, demonstrating that sound is transmitted by waves through the air, is probably more memorable than anything else I said there.

TEAM DEVELOPMENT

- The best people to lead your departments may already be with you. They are people who have worked in your center for a while, know and love the center, and need to receive from you a lot of guidance and trust. When SciTech needs to hire new leaders, we prefer to promote insiders or hire people recommended by staff. But, having said this, each museum needs the infusion of new ideas and approaches that come with hiring new staff from the outside.
- Use senior staff to make important decisions and plan for the year and beyond. Encourage them to have their staff participate in department decision making. To create a successful team, treat staff like team members who have a say in what's going on.

Figure 30.1. View of SciTech's main exhibition hall, Aurora, Ill.

Figure 30.2. Using the Hourglass exhibit developed by students of the University of Chicago, SCOPE project.

- When a department head at SciTech asks for my advice, I try to hold it back. Instead I ask the department head to try and solve the challenge with another department head, and once they have come up with ideas, to implement them. This brings original solutions that are much better then mine.
- At SciTech, I throw a lot of tasks at the department heads, as science museums need to always change. Some people are comfortable with extreme multitasking. Others I ask to e-mail me back what they understand their tasks to be, and then I prioritize the list for them.

RESOURCES

- By definition a science museum never has enough money or visitors. Implement your department heads' ideas on increasing both money and visitors. To get more money, spend on development. To get more visitors, spend on marketing, but spend it wisely—use the power of free media coverage as often as you can, as opposed to purchasing ads.

- Bring in temporary exhibitions that will draw specific segments of the audience you're after and choose topics that your media and sponsors will support. SciTech recently received an offer we could not turn down to rent a 5,000-square-foot exhibition, Backyard Monsters, for three months at a significantly reduced rate. We had no budget allocated for this reduced rate, but together with the board we decided to take a bank loan. The increase in visitorship during the three months more than paid for the loan, and the public relations the exhibition generated was superb.
- Do a lot of outreach. Send your exhibits and programs to schools and festivals. Ask your staff to be in the field as much as possible. At SciTech, we reach over 50,000 people per year in the museum and an additional 35,000 each year through our outreach activities. These activities create awareness for our museum and reach out to audiences who would never consider visiting the museum.

PARTNERSHIPS AND COLLABORATIONS

- Develop and invest in collaborations with other museums, universities, and school systems. The collaborations may be local, regional, national, and international. The key for forming a successful partnership is identifying and working directly with one motivated individual who leads each institute in the partnership. The investment you make will bear fruit in unexpected ways over time. Here are some examples:

- SciTech works together with nine science museums in Anchorage, Alaska; Peoria, Rockford, Wauconda, Ill.; Evansville, Fort Wayne, Terre Haute, Ind.; Waterloo, Iowa; and Amarillo, Tex. The collaboration shares its expertise and knowledge to develop exhibits and outreach programs on Wild Weather and Space Exploration. Our collaboration gets significant funding from the National Science Foundation.
- SciTech formed a partnership with the Electronic Visualization Laboratory of the University of Illinois at Chicago. SciTech is the first museum to test the university's low-cost virtual reality GeoWall (www.geowall.org) technology and software. Our visitors rate shows such as My Heart and Lungs, Journey Through the Universe as "real cool!"
- SciTech formed a partnership with the University of Chicago. Together we train twelve graduate students each year in presenting University of Chicago, current science at SciTech. In both these museum-university partnerships, we provide dedicated staff time, opportunities to show the universities' research, and rapid implementation of the concepts developed.

ADVICE AND SUPPORT
- Ask for advice from other directors of science centers, large or small, as we all like to share. Talk with your floor explainers, who are most attentive to your audience's needs. Seek advice from your supporters—board members, community leaders—the more input, the better. Write down their suggestions, and once implemented, do let them know. Get your staff out to other museums. Rent a van, pack the staff in, and go on a field trip to another museum where your staff can meet with their peers.
- As director, you will soon find out that it is rather difficult and painfully slow to effect change in the museum. What you can do immediately is set a modest example by your personal behavior. Your staff will follow your actions much better than they will follow your words.

PROJECTS
- Think BIG, modular, and always high quality. Design projects that may be far-fetched as you can always implement them in stages. Implement a small pilot and evaluate it. The success of a pilot will pave the way for additional funding. When we started implementing SciTech's Outdoor Science Park, people doubted its feasibility because of the Chicago weather. The pilot's success brought additional funding. Now complete, the park is a marvelous at-traction functioning for eight months a year. (For more on outdoor science parks, see chapter 23.)

YOUR TRAINING
- I may have had the best preparation for directing a science center as a department chair. When I worked as second-in-command at the Clore Garden of Science, I learned so much from observation. This gave me the knowledge and motivation to succeed as director while (I hope) avoiding making some mistakes that I was subjected to.
- When you start your tenure as director, you will have a different vision than your predecessor. This is good. Each change of directors should bring about a course change, trying to retain what was especially good about the previous director's approach while trying new approaches. Nevertheless, this will cause tension, the more so if your predecessor was the founding director. The best you can do is stick to what you believe in and have the board absorb some of the ricochets. After all the board appointed you, and it is in their interest that you succeed in implementing your vision, which should be adopted by the board as their vision, too.

THE BOARD OF DIRECTORS/TRUSTEES
- "The Board" is made of individuals with differing agendas. To benefit from their potential you need to work with each one individually. Board committees need to be led by one board member who drives their work and is accountable for the results.
- Board members need to donate their own personal money to the museum. This makes them very credible when asking others to donate.
- Feed your board and staff well. Monthly staff development lunches increase morale. Board meetings are much more productive once board members, who come after a day of work, enjoy a light meal.

LAST NOTE
- In directing a science museum you may feel like a juggler with fifty projects spinning above your head. Delegating and prioritizing will help your team bring these projects to successful completion—and that's already the time to launch more! Good luck!
- It was so much fun writing this article! I thank Ed Sobey for critical review and thoughtful input. Thank you Cynthia Yao for coordinating this book and for giving me the opportunity to write with "no strings attached." Please e-mail me with your best hints! Ronen Mir, ronen@scitech.museum.

CHAPTER 31

Hiring, Supporting, and Developing Museum Educators for Your Science Center

Elsa B. Bailey

A friend of mine, a librarian, once asked me what do museum educators do exactly? I was surprised by this question as we'd known each other for years and our conversations often related to our work. In reflection, I realized that most people outside of the field of museums, even those in allied fields, have only a vague idea about what museum education is and what museum educators do. However, education is at the core of most science centers' missions and when possible it is best to hire museum education staff to lead these efforts.

This chapter will provide an overview of the work and culture of museum education; the hiring of museum education staff; what staff needs to know and learn to effectively conduct their activities; how these educators learn their practice; and how their learning can be supported by museum administrators.

THE WORK AND CULTURE OF MUSEUM EDUCATION

Ask a museum educator and they will say their work is about working with people. These people include families, school groups, teachers, and other adult groups. In a science center, education work supports the exhibits, and stimulates and builds people's interest and knowledge in science. Much of this work takes the form of programs that help visitors and others in the community connect to the ideas the museum endeavors to share. Museum programs take place both on-site at the museum and off-site when staff travel to other venues such as schools, community centers, festivals, and shopping malls. Work is often project based and may change as exhibits or program initiatives change. In some science centers, museum educators' responsibilities include involvement, in varying degrees, with exhibit development.

Museum education has its own unique culture. The museum educator works in a lively environment and the galleries and grounds are often noisy, bustling places. Most educators in small science centers wear many hats, as their work encompasses a variety of functions. Museums are often understaffed, especially if resources are scarce. Staff members tend to work long and nontraditional hours, such as evenings and weekends, to implement their programs.

The museum educator is affected by the context in which the institution functions. Institutional shifts, regional changes, and national and global developments in policy and politics profoundly influence museum education work. Finances also affect the culture of science center education. The national economy, accessibility of grants, the climate for private donations, and the financial health of the institution all play a part in influencing job security, staff turnover, and program continuity. To deal with limited resources, many science centers depend on volunteers to accomplish their education mission. An organization staffed to some degree by volunteer personnel has its own particular culture and management considerations.

HIRING MUSEUM EDUCATORS

When hiring museum education staff, the critical first steps are to think through what the position entails, to identify the specific responsibilities, and to clarify expectations for success. You should also think about the personal traits required and the professional values that your organization favors.

This description shapes ads placed in newspapers, professional publications, and on websites that will be seen by prospective candidates. In addition, communications with colleagues and local individuals by way of letters, e-mails, or word of mouth are useful methods for finding good candidates.

Science museum educators, as well as other science museum staff, often come to their positions through multiple pathways. A diversity of background and experience can considerably inform and enrich science museum work by providing alternate perspectives and sensitivities.

Having more than one person review the applications will strengthen the review process. You may elect to conduct phone interviews with the most promising applicants; and select from these candidates those you want to interview face to face. Final candidates should meet all staff. This procedure permits the candidates to get a feel for the organization and its culture and also provides staff an opportunity to gain impressions of the candidates that can be communicated to leadership.

As you review applications and conduct interviews, keep in mind the particular aptitudes, strengths, personal characteristics, and areas of knowledge and experience that are compatible and effective in the field of museum education.

MUSEUM EDUCATORS: THEIR QUALITIES AND LEARNING PROCESS

My own research (Bailey 2003), as well as that of others to date, regarding museum educators' professional growth (Leichter and Spock 1999; Matelic 2001; St. John and Hennan 1997) offer insights toward museum educators' qualities, professional needs, and styles and modes of learning.

QUALITIES OF SUCCESSFUL MUSEUM EDUCATORS

Interviews with museum educators suggested a number of qualities common to people drawn to and successful in museum education. First, many of these individuals are attracted toward unique working environments such as museums.

> Museum education draws people who . . . just march to [a] different drummer . . . People who are willing to do that will come to an environment like this.

Many can be described as risk takers, able to work in a fast-paced setting, and willing to work nontraditional schedules. They have specialized curiosity, a problem-solving bent, and a drive to keep on learning.

> I need to constantly be learning, or I go dead. I have to have something gnawing at me . . . I'm curious about everything. . . . I don't think you can go to work in a science museum if you're not that kind of person. It's the essence of science practice anyway. That's who I am.

Many assign a high value to working in a field where they are "making a difference," and achieving a "higher purpose." In some cases, these values have driven people's career choices; they have sacrificed earning higher salaries to follow these beliefs.

> I took a wage cut to come back. But you've got to be able to live with yourself. . . . Here I feel like maybe I'm making a difference. . . . You can see that you're planting seeds all over the place.

Some describe themselves as generalists, and museums are places where such an orientation is useful.

MUSEUM EDUCATORS AS LEARNERS

My research suggests that museum educators' on-the-job learning is primarily self-directed and self-motivated. A number of motivators drive their learning:

- A need to increase their professional capacity by developing specific and deeper content knowledge, and enhanced skills
- A need to have and disseminate accurate information
- A need to learn more about the culture of their specific audience
- A thirst for learning science
- A feeling that they should know something
- Developing an emotional connection with a topic
- A need to update or enhance their formal education background
- A need to learn more about a new working context or a new geographic region
- A desire to take on new professional roles
- A response to encouragement from an organization that supports staff development

WHAT DO MUSEUM EDUCATORS NEED TO LEARN?

- Building skill and knowledge in teaching and learning and comprehending how to apply these skills toward work with particular audiences is critical to museum education work.
- Educators must learn to develop effective science programs that incorporate museum exhibits, where participants learn by doing, and ideas and content are presented in manageable packages.
- Presentation skills are key to much that museum educators do on a day-to-day basis, thus, having a certain degree of stage presence, a unique presentation style, and the facility to infuse humor while sharing content knowledge; all are important abilities to develop.
- Substantial content knowledge in science is key to successfully implementing science programs. Even

those educators with a formal background in science must acquire additional knowledge as the situation and program merits.

- Museum educators need the ability and skills to initiate, develop, and coordinate projects, including the creative design as well as the nuts and bolts of project implementation.
- Some museum educators serve on exhibit development teams and must learn the concepts and vocabulary that are key to exhibit development.

HOW DO MUSEUM EDUCATORS LEARN?

Most museum educators express a preference toward learning through active, direct, and rich sensory experience, integrated with conversations and dialogue with others. Museum educators' learning has distinct modes and qualities. They report learning through trial and error, by being "thrown into the pool," by "the seat of their pants," through difficult experience, and through particular assignments.

> I learn . . . by making mistakes, trying things out, and going back and doing them again. It's useful to make mistakes because then you can do things better the next time. . . . So what I do is identify the glitches . . . and I try to anticipate the problems that might occur [and] . . . remedy the situation.
>
> I look back on that first summer now, and I shake my head sometimes because it was just so overwhelming. But yet, I learned from it, and I know—when times get tough, I say, "Okay, remember that first summer you were here? If you could get through that summer, you can get through anything."
>
> They . . . said, "We need a play to go along with [a] . . . dinosaur theme. Can you do a play?" And I said, "Sure." . . . And so I set out to learn about dinosaurs; and what I found out fascinated me. . . . So, I wrote the play, and the play was a big hit, and everybody thought, wow; he knows so much about paleontology! And I didn't know a thing when I started, except for the word dinosaur.

As part of their learning process, museum educators assess on a day-to-day basis the gaps in their knowledge.

> Because it's really been self-driven, a lot of it has been sort of hit and miss. So in conversations [when someone] starts telling me about . . . things, [I say] "Oh my, there is a hole the size of a Buick I need to fill." Just talking with people . . . when [they] start throwing out references, I start writing them down.

These educators tend to be very observant and employ observation as a tool in their own learning process.

I would notice those questions that people asked that I thought I had already answered, and that was one of the very strong ways that I learned to evaluate my own presentation. . . . I learned that people's questions were actually a way of assessing whether they were, one, engaged . . . and, two, what their understanding level [was].

Museum educators must develop an intuitive understanding about the particular styles to employ with each audience such as the appropriate language to incorporate or avoid and the climate to generate.

> There is a vocabulary that [teachers] . . . use, and if you want to be respected by them, you need to be on their wavelength and use their same terms.

Museum educators' learning is an ongoing process of experience, reflection, analysis, and subsequent actions.

> You know, you learn by doing . . . [through] the more people that you sit down and talk with, or the more questions that you ask . . . and you learn the [kind of] questions . . . that get the information that you need. . . . You build on everything that has gone on in the past that you've done. . . . Because we get hopefully better at what we do, because we try different things.

Learning through informal social interaction is a natural, and sometimes preferred way, for museum educators to learn. It appears to be critical to professional learning, providing a springboard for exploring work ideas and issues with colleagues and peers inside and outside their institution and people or groups from other communities.

> I have a partner that I work with here. . . . So I get to watch her in action and I learn from that experience. Even though I have taught with her for five years, I am still learning things from her, which is great.

Other significant social learning experiences include apprenticing with or seeking out experts and interacting with mentors.

> It was important to hear that other people had confidence in you. Sometimes they had more confidence in you than you do in yourself at the beginning. That mentoring was very important. Also important was doing that risk taking and having someone there to talk with you about it afterwards.

Museum educators participate in formal structured learning experiences such as workshops, symposia, conferences, and courses. They select these for reasons ranging from a specific learning focus, through a general exploration of topics relating to their work, to having a long-term goal such as a degree or certificate. In most cases, participation is voluntary, often self-directed, and motivated by a specific, job-related need to know.

> It was becoming more obvious that funders were really requiring us to be more accountable and do more evaluation and assessment for our programs and our grants. Since I was a grant administrator, that fell to me. It was clear . . . not too many people here knew much about assessment or evaluation; especially program evaluation. So, I thought, well, I'd better learn something.

Membership organizations such as the Association of Science-Technology Centers and American Association of Museums create collegial networks through conferences and meetings, are a source of pertinent resources, and often provide communication vehicles such as listservs, journals, and newsletters that sustain professional learning.

> I think the nice thing . . . is that people in this informal type of education are very open about sharing their successes but are also very willing to share the things that didn't work out, as well. And if someone's experience can save me from having that same problem, I am very appreciative of that.

Professional writing and presentations are reported as very useful to professional growth because they help to crystallize and clarify professional knowledge. Comprehension of the museum culture evolves over time as museum educators become integrated into the museum community of practice and increasingly identify as a member of that community.

THE ROLE OF MUSEUM ADMINISTRATORS IN SUPPORTING MUSEUM EDUCATORS' LEARNING

Museum administrators are important players in the process of supporting museum staff development. Their role is to provide a climate sensitive to the adult learning process and create systems and structures that will facilitate professional growth. Administrators need to be aware of the fact that adult professionals are constantly learning, and this learning is often self-directed. An institutional climate that scaffolds self-directed learning is one that encourages staff questions, dialogue, and requests for assistance. Many

museum educators report enjoying opportunities for creative freedom and innovation, explaining they work most effectively and enthusiastically in an organizational atmosphere that encourages risk taking.

> I think that in museums, it mostly has to do with the people on the upper levels, and if they want to bring you along or not. . . . If you have a director who wants to see people move forward and who wants to motivate people to do more, to do bigger, to do greater, then those people will . . . expand, and their horizons will expand and their knowledge base will expand. . . . A good director will . . . let you grow your own sort of way of doing things; your own path; your own flavor for something.

Some organizations establish structures that promote staff orientation and professional development.

> The museum philosophy is to tell you what you are expected to do and then say "Go do it and we will support you" . . . it's now become part of everybody's job description . . . to address their own professional development goals. . . . A certain amount of funding has been made available to permit people to take advantage of these opportunities.

In smaller institutions, priorities such as program coverage and funding can inhibit certain modes of professional development, particularly conference and seminar attendance. To deal with this issue, some smaller institutions offer formal on-site professional development activities for staff.

CREATING CONTEXTS FOR MUSEUM EDUCATORS' PROFESSIONAL LEARNING

Museum administrators should be sensitive to the fact that the context in which one works has an impact on staff professional growth and learning. The people and information to which one is exposed, the psychological influences of one's work surroundings, and the energy needed to perform the job are all connected to the physical work context. Distance can affect the information flow, and museum administrators must make sure that communications systems exist that allow all to be in the loop.

The ideal professional office context appears to be a space that both encourages and facilitates staff interaction but can also permit private time and space for tasks that require quiet and individual concentration. Informal structures such as common lunch and kitchen areas, copy centers, and resource centers also bring people together and provide opportunities for collegiality and sharing.

In conclusion, hiring, supporting, and developing museum educators are facilitated by building an understanding of, and sensitivity to, science museum education's culture; its required professional expertise; the learning styles of museum educators; and the influence of the day-to-day work environment on science museum educators' professional growth.

REFERENCES

Bailey, Elsa B. 2003. How museum educators build and carry out their profession: An examination of situated learning within practice. PhD diss. Cambridge, Mass.: Lesley University.

Leichter, H. J., and M. Spock. 1999. Learning from ourselves: Pivotal stories of museum professionals. In *Bridges to understanding children's museums*, ed. N. F. Gibans, 41–81. Cleveland, Ohio: Mandel Center for Nonprofit Organizations, Case Western Reserve University.

Matelic, C. T. 2001. Mentoring tradition. *Museum News* 80 (6): 44–49.

St. John, M., and B. Hennan. 1997. The teaching of first hand learning: A retrospective look at the ASTC Institute for teacher educators at science centers. Inverness, Calif.: Inverness Research Associates.

CHAPTER 32

Explainers: Youth Development at the Exploratorium

BRONWYN BEVAN AND DARLENE LIBRERO

THE PRIMARY FLOOR FACILITATION STAFF AT THE Exploratorium is a cadre of high school students who reflect and represent the cultural diversity of the San Francisco Bay Area. Their energy, commitment, and curiosity serve as a central driving force on the museum floor as they engage with exhibits, visitors, and museum staff.

These students are part of a youth development program that combines regular training sessions in science content and inquiry-based teaching with authentic (paid) work on the museum floor. On the floor, Explainers lead science demonstrations, support visitor interactions with the exhibits, and respond to visitors' questions and needs. They also develop a number of other projects or programs—from youth-oriented webcasts to newsletters and special events—on the basis of the interests of the Explainers and the needs of the museum programming. Explainers turn the exhibits on in the morning and turn them off in the evening. They greet visitors at the door and usher them out at the end of the day. In between, they are deeply engaged in learning about themselves, one another, science, teaching, and learning through a range of social interactions with visitors who come to the museum from all over the world. The program is thus a unique combination of youth development and core operations. This duality is both the essence of its power to provide transformative life experiences for participating youth and the kernel of its complexity for the museum organization and structure.

The program hires about 100 high school students each year. Explainers are hired for trimester-long terms, and at any given time, there are between thirty and forty actively on the payroll. Explainers work on weekends and after school, as well as full time during the summer months. During the school year, the museum operates a separate Field Trip Explainer program (with older adults) that works with field trip students. Since 1969,

the program has served more than 3,300 youth, about 200 of whom have come from foreign countries to work in the program for a summer.

Hiring and supporting high school youth as key museum staff can be a powerful source of community relationships for your museum. It can also connect you to local schools and after-school organizations. And it can create a legacy of committed alumni who can serve as advocates in the future. Giving youth a voice and a responsibility within your organization can vitalize the energy on your museum floor and, if supported well, can lead to a range of innovative programming that reflects the needs and interests of audiences your museum may seek to serve.

The Exploratorium was one of the first museums to create such a program, but other museums, under the YouthALIVE! initiative, similarly have created programs that put youth at the center of core museum activities. This article will describe how the Exploratorium's program is designed and operated to highlight features that need to be attended to, and benefits that accrue, for museums that commit to youth and a youth presence in their museums.

SOME BACKGROUND

The Exploratorium High School Explainer program was started when the museum opened its doors in 1969. It was the vision of Frank Oppenheimer. At that time, there were just five exhibits and one Explainer.

The Exploratorium was designed to provide visitors with firsthand experiences with science and scientific phenomena. The museum promulgated the idea that everybody could engage with science—regardless of their age, gender, education, or cultural backgrounds. Exhibits were designed to stimulate visitor curiosity and to provide the tools needed to investigate the phenomena. Driven by their own questions and investigations, all visitors could advance their own ideas, interest, and understanding with regard to the phenomena featured in the exhibits.

Their unfettered curiosity makes inquiry in a museum a fairly natural process for children. Moving to conceptual understanding is a different challenge, but plunging in and exploring phenomena is not a great obstacle for most young people. But this same level of freedom to inquire may not come as naturally for adults as it does for children—as we know, it is often drilled out of us in a world that values expertise over curiosity.

To support the engagement of adults, as well as children, the Exploratorium took the novel step of hiring expert questioners, in the form of high schoolers, to model inquiry and assist the active questioning of visitors. It thus made the enthusiastic novice a central learning resource for its visitors as well as for its staff.

As the central public facilitation staff, Explainers play an essential role in museum operations and the visitor experience. The authentic nature of their contributions to the museum is, in turn, a key aspect of the youth development program, along with participants' evolving experience as informal educators.

Explainers are encouraged to discover along with the visitors, rather than to provide answers to specific questions. When they start out, although they participate in trainings on a regular basis, they may know less about the museum and the science in the exhibits than many of the visitors. They learn on the job—not only about science, but about learning and about teaching. The program stimulates young people's interest in science and starts them on a path of noticing things around

them and looking for why things behave the way they do. The ongoing activity of noticing, contemplating, and sharing prompts reflection and discussion about their own learning.

They also learn right away that the museum depends on them—they are the frontline staff, they are there not only to interact with visitors at exhibits but also to reunite lost children and their parents, to deal with various issues as they come up on the museum floor, and to ensure that exhibits are operating properly. This responsibility is significant and monitored by a staff of three adults who work to supervise, train, and support Explainers as they grow into their responsibilities. Explainers who choose to return for subsequent semesters are given increasing levels of responsibility and opportunities to train and support their newer peers at the museum.

The program has been the subject of several qualitative studies that have found that it has had a lasting impact on the life choices and leadership development of its participants. A longitudinal study is currently underway.

SELECTING AND HIRING EXPLAINERS

The Exploratorium works with more than 100 high school–aged youth each year; these students come from African American, Caucasian, Hispanic, Chinese, Japanese, Cambodian, Filipino, Russian, Vietnamese, and East Indian backgrounds. Approximately 70 percent of the Explainers are people of color. Explainers come from all socioeconomic levels and reflect a full range of interests and academic performance. Over

Figure 32.1. Explainers at work.

60 percent of the participants come from underserved neighborhoods.

Explainers are hired by the program directors. There is not one type of person that the program seeks. In fact, the program deliberately seeks a range of people who together constitute a cohort of students who reflect a variety of skills, experiences, and backgrounds. For example, some may have stronger social skills than others; others may be more academically motivated. Because learning from one another is an important part of the youth development process, the balance of the group is perhaps the most critical aspect of hiring.

Explainers are recruited through high school teachers and counselors who refer students to the program. Some schools offer general credit to students in work experience classes. We also recruit from local community organizations that train students, look for youth employment opportunities, or are after-school centers for youth. Many Explainers learn about the program by word of mouth.

During the school year, about six to seven Explainers studying for their General Education Degrees (GEDs) are hired to work weekdays after school. Weekend and summer cohorts range from twenty to forty Explainers. Explainers stay involved with the program anywhere from one trimester to four years. The minimum commitment is four months. Every four months, a student must reestablish interest in the program by writing a letter to the program leaders. The letter is reviewed, and the students are re-interviewed. Students make new goals when they commit to the program another four months. This process serves as an important reflection and evaluation period for both the students and the program staff. It also helps students build their own reflective practices and articulation, in writing, of their strengths, interests, and intentions. It forces them to seriously address their job and their commitment to their job.

CONTENT AND PEDAGOGY DEVELOPMENT

Explainers participate in approximately seventy hours of training within a four-month period. These sessions, conducted by staff scientists, educators, and artists, include content and exhibit facilitation training—sessions focus on, for example, science content, exhibit facilitation, social skills involved in approaching strangers, mediating problems or disputes, and reflection sessions. During these training sessions, Explainers are encouraged to build from their own knowledge, experience, and instincts as budding educators. Initial orientation training provides Explainers with an overview of the main exhibit sections of the museum, museum operation procedures, and an introduction to exhibit facilitation skills.

Concurrent to the training, Explainers work on the floor with the public. They are either assigned to work at demonstrations (such as cow's-eye dissections; visual illusions; laser table; or flower dissections)—where they work in teams of two to three to provide essentially scripted demonstrations, with significant and open interactions and Q&A with the public—or assigned to roam through specific sections of the museum, usually in teams of two. Newer Explainers are teamed with more experienced Explainers and more often are scheduled to offer demonstrations, which are by their nature more controlled environments. At demos, visitors approach the tables on their own volition. Approaching strangers on the museum floor can be an excruciatingly difficult act—especially because visitors often are deeply engaged in reading or interacting with exhibits. Explainers must learn to accept a range of responses—from the enthusiastic to the bland to the hostile.

Learning to read people's readiness for engagement, or openness to interaction, is a part of the job, as well as part of the transformative part of the program for many youth. It can build tremendous levels of self-awareness and self-esteem, as well as social skills and adroitness. As they spend time on the floor, building their own capacity through the Explainer trainings, and developing a range of skills through discussions with peers, Explainers become increasingly more equipped to talk about the exhibits and to encourage visitors to explore exhibits.

Explainers also interact with the museum staff in a variety of ways. Informally, they provide critical feedback to the museum's exhibit design staff—discussing how visitors work with the exhibits, reporting questions that are raised and problems encountered. Formally, Explainers are involved in the design and development of floor demonstrations of a variety of scientific phenomena, depending on the new or special exhibitions that are developed by exhibit staff. They are also involved in the design and implementation of various youth-based and youth-oriented media projects.

LESSONS LEARNED

High school students bring with them—besides great enthusiasm, curiosity, and a wealth of ideas and innovations—high school student behaviors. For the most part, these behaviors are nothing short of responsible, energetic, and dedicated. Yet, for some staff, a core of retired volunteers might provide a more stable and predictable workforce for the key floor staff positions. In fact, the two are not incompatible, although the balance must be carefully maintained so as not to undercut the authenticity of the work and the responsibility for the participating youth. Being in charge of the floor is very different than playing a supporting role.

In running these programs, it is especially important to offer youth real-work opportunities and to listen to their ideas and issues with the genuine intention of making change. Giving youth real work motivates students to become more deeply engaged because in the end, they know they will be helping others—the institution, their peers, the visitors, and the program leaders. Program leaders and museum staff need to listen carefully to participants to develop positive relationships between the youth participant (teenager) and the institution.

In a program such as the Exploratorium High School Explainer program, which is complicated in its composition as youth development/visitor facilitator and which is tied so intimately to the museum's founding mission, time poses a challenge to its ongoing integration. Because the Explainer cohort is so varied and also changes from trimester to trimester, it has been important for the program to constantly work with the museum staff to educate them about the unique character of the program that leads to different expectations as well as outcomes as compared with more traditional visitor services programs. The following elements have proven essential to its successful integration and the maintenance of staff support for the program:

- Strong program leaders who can articulate the trade-offs and benefits related to such a hybrid program.
- Developing internal and external staff advocates by connecting them with the Explainer cohorts—through training and exhibit-centered work with staff.
- Creating inroads for Explainers to work with other museum projects, through internships and other means.
- Developing Explainer advocates through periodic reunions that bring back people to reunite, share, and testify about the pivotal role the program has played in their lives.
- Maintaining leadership status in the field of educational and museum operational programs for youth.

PROGRAM OUTCOMES

The outcome of the program is as varied as each student. Every trimester we have asked Explainers to comment on what they have learned. They respond saying that they have learned more about people than anything else. When we then ask, "What about science?" they respond with, "Yes, that too, but more about people." Our studies have shown that some students leave with an increased interest in science, while others decide they want to pursue a career in science or teaching. Many have pursued medical careers—which we speculate builds on their cultivated interest in both people and science.

The program is currently conducting a longitudinal study of a number of Explainers to track how the meaning of the experience changes over time. Short-term results of this study have shown that they all remark upon what they have learned by teaching. About 62 percent comment on learning methods for teaching subject matter; 33 percent stated that they learned how to teach; and 40 percent said that they learned about tolerance.

An earlier study conducted on the program in 1987 noted that while Explainers said that they learned directly from interactions with the museum staff, half of them noted that they learned from one another. About a quarter noted that it had developed their interest in science.

An Explainer alumnus comment summed up the goals of the program with this statement:

> The Exploratorium is what gave me the spark to wonder how things work and the joy when I'm finally able to figure it out. The Exploratorium had a lot to do with my decision to become an engineer, and I'll always be grateful for the foundation you guys gave me. The Explainer program to me was not just exposure to science but about exposure to new and different folks. P.M., 1985

What the institution gains is a group of young individuals who are shaping their ideas about the world, people, culture, and science. These youth provide tremendous service to the museum and its visitors. They remind the staff of that key part of our founding mission—that science is not for the experts, that it is a process of ongoing inquiry and learning, and that it welcomes and includes all.

Part VIII

AUDIENCE

CHAPTER 33

Understanding Your Audience

Kirsten M. Ellenbogen

There has been a dramatic growth in the museum audience over the last century (Lusaka and Strand 1998). Consider the following data: thirty years ago, one in ten people went to museums regularly. By the end of the twentieth century, three out of every five people visited a museum once a year. The increase in visitorship can be explained, in part, by the growth of new museums and the changes in the quality of exhibits and programming in existing museums. But the creation of new museums and improved exhibits and programming alone do not account for the increase in visitorship. Research (Falk and Dierking 2000) indicates that there is a shift toward making learning experiences, such as visiting museums, a part of everyday life.

The emphasis on learning experiences and growth in museum visitorship comes at a time when the mission of museums has shifted from a focus on collecting and preserving to one of educating the public. Stephen Weil (1999) describes this as a movement from being about something to being for somebody. This ideal confluence—increased museumgoing, increased attention to education in museums, and an increased interest in learning experiences in the general public—is promising for museums hoping to grow their audience.

This chapter discusses what we know about museum audiences and their motivations, and then it explores the ways in which this knowledge can be useful in creating satisfying learning experiences. It also focuses on why people go to museums in an effort to provide a perspective on museumgoing that can be used to increase audiences. Likewise, if we know what motivates people to learn in museums, we will be able to create exhibitions and programming that better support visitors' learning needs.

IDENTIFYING AUDIENCES

Research has found that museumgoing correlates with a number of demographic variables, such as education, income, occupation, race, and age (Falk 1998; Hood 1983, 1989). One important demographic variable to emerge from the data is age. Even when school trips are excluded, children between the ages of five and twelve still represent a significant percentage of all people who visit museums. Of course, these children visit museums with their parents, making the elementary school age-group and adults between the ages of twenty-five and forty-four the largest part of the museum audience.

Although this information is evidence that families are the largest part of the museum audience, it does not mean that demographic variables such as age are the best predictors of who will be most likely to visit a museum. For example, research by John Falk (1993) has shown that race is not an accurate predictor of museumgoing. Falk found that African American and white museum visitors were similar in terms of interest in science, occupation, annual income, and frequency of museumgoing. These findings suggest that psychographic variables, which describe people's psychological and motivational characteristics, are far more predictive than demographic ones. Despite varying demographics, museumgoers tend to fit a specific psychographic profile (Hood 1983): they value learning, seek the challenge of exploring new things, and place a high value on doing something worthwhile during their leisure time.

This psychographic profile correlates with the demographic data on museumgoers. For example, psychographic data indicate that museumgoers value education; demographic data indicate that museumgoers have a higher than average level of education. Falk (1998) cautions that this does not mean that a higher-than-average level of education is needed to value the museum experience. It is more likely that this correlation is because people who value learning are more likely to pursue higher education, such as college or graduate school. Psychographic data about museum

Table 33.1. Most Important Leisure Attributes

	Frequent Museumgoers	Occasional Museumgoers	Non-Museumgoers
Learning	X		
Challenging	X		
Worthwhile	X		
Social		X	X
Active		X	X
Family-Centered		X	X
Comfortable		X	X

audiences' values and beliefs may help predict museumgoing behavior, but like demographic data, they do not provide a complete explanation for why people go to museums.

Psychographic data have led to some important new generalizations about the museumgoing audience. Traditionally, the museum audience has been divided into two segments: museumgoers and non-museumgoers. Marilyn Hood's (1983, 1989) psychographic research led her to regroup the museum audience into three segments: frequent museumgoers (three or more visits per year), occasional museumgoers (one or two visits per year), and non-museumgoers.

Frequent museumgoers tend to prioritize three characteristics of leisure experiences unique to their group: the opportunity to learn, the challenge of new experiences, and the experience of doing something worthwhile in their leisure time. In contrast, non-museumgoers tend to give priority to experiences that are less important to frequent visitors: socializing, participating actively, and feeling comfortable. The characteristics of leisure experiences that frequent visitors rank high are ranked low by non-museumgoers. In addition, non-museumgoers typically have not gone to museums with their families when they were children.

Occasional museumgoers who visit museums once or twice a year have an unexpected psychographic profile. The traditional division between museumgoers and non-museumgoers assumes that all museumgoers, regardless of the frequency of their visits, would fit the same profile. Instead, studies show that occasional museumgoers resemble non-museumgoers more than they resemble frequent museumgoers. As children, occasional museumgoers are typically socialized into leisure activities that emphasize active participation and socialization. As adults, they continue to be involved in these sorts of leisure activities, including sports and visiting amusement parks.

Both non-museumgoers and occasional museumgoers place great emphasis on family-centered activities and value feeling at ease in leisure environments. The characteristic that distinguishes occasional museumgoers from non-museumgoers is that occasional museumgoers believe some of the characteristics they value in leisure experiences are available in museums, but not sufficiently to warrant frequent visits. Often, occasional museumgoers visit museums for special events, believing that activities planned during special events are more geared toward family interests and will best fulfill their needs.

AUDIENCE MOTIVATIONS

In order to better understand the needs of museum visitors, researchers (Anderson and Roe 1993; Falk, Moussouri, and Coulson 1998; Moussouri 1997; Rosenfeld 1980), have used in-depth observation and interview techniques to gather data on expectations and motivations. This research reveals that there are specific patterns in visitors' motivations for visiting museums. Although there is some diversity in the various patterns of visitor motivation that researchers have proposed, seven categories are consistently included in the findings:

- Entertainment: "We always have a good time here."
- Social: "I wanted to spend some quality time with my brother."
- Learning: "We always learn so much going to museums like this."
- Life Cycle: "I went to museums as a child and now I am bringing my children."
- Place: "You have to see this museum if it is your first time in town."
- Practical: "It was getting too hot so we decided to duck in."
- Context or Content: "I love butterfly gardens."

Museum visitors tend to mention more than one of these motivations when talking about why they go to museums, since many are overlapping and interconnected. Researchers argue that not only do visitors have specific motivations for their museum visits but also that these motivations directly influence what is learned during the experience. For example, Falk, Moussouri, and Coulson (1998) found that in a content knowledge test administered before and after visitors went through an exhibition, people with a high learning motivation scored significantly higher than people with a low learning motivation. More interestingly, people with a high learning motivation, regardless of their entertainment motivation, showed significant conceptual learning. Likewise, people with a high entertainment motivation, regardless of their learning motivation, showed significant vocabulary development and overall mastery of the topic. This result reinforces the belief that learning and entertainment are not two ends of a single continuum. It suggests that visitors do not distinguish between the value of learning and entertainment and that both are effective motivations in learning.

IDENTITY AND AUDIENCE

It remains difficult, however, to truly understand the museum audience on the sole basis of motivations. In many instances, it can be difficult for visitors to place themselves in one or more of the motivation categories. More problematic is that identity building, one of the major reasons for visiting a museum, cuts across the motivation categories above, can be extremely difficult for visitors to articulate, and tends to result in motivations that reflect people's general beliefs about who they are or who they want to be, rather than an immediate assessment of the reasons for that day's museum visit. This is arguably an indication that we must allow identity to take center stage in our efforts to understand the needs of our audiences.

Research has shown that there is great potential for people's existing and desired identity to influence their museum experience (Ellenbogen 2002; Ellenbogen, Luke, and Dierking 2004). Parents who want to develop a particular family identity are able to quickly adapt the general museum experience, as well as specific content, to reinforce the desired identity. Everything from behavior modification ("We don't bang on the computer screen like that.") to personal narrative episodes ("Do you remember the last time we saw one like that?") can quickly be used to reinforce the identity of the family. At times, this effort to reinforce family identity comes at the expense of the museum's intended experience or content goals. Museums often intentionally choose content and create experiences to support a specific theme, learning goal, or general mission. However, if visitors' efforts to develop their family identity do not coincide with the museum's intended goals, visitors may hijack the museum messages and reshape them to support identity building.

D. Hilke (1987) voiced a similar concern decades ago, suggesting that we focus not just on the museum as a resource for learning but also on the visitor as a resource for his or her own learning. This inverted point of view highlights the reality that visitors bring a great deal of resources with them, including previous experiences, concerns, and desires to act or be perceived in certain ways. All of these influence the way visitors interpret the museum experience. Visitors' motivations consequently have a strong impact on the learning that takes place in museums.

SATISFYING AUDIENCES

An understanding of the motivations and identity-building activities of audiences can be useful when creating successful museum learning experiences. Viewing exhibit and program experiences through the lens of specific visitor motivations can prompt alternative perspectives in all aspects of the visitor experience. In an effort to create a comprehensive checklist that positions visitors' needs and interests centrally, Judy Rand (2001) has created The Visitors' Bill of Rights. This document takes a holistic approach to understanding the visitor and demands careful attention. Eleven items are detailed:

- Comfort: Provide for visitors' bodily needs including food, restrooms, seating, and access.
- Orientation: Visitors should understand their options and be able to find out how to best act on these options.
- Welcoming or Belonging: Friendliness extends beyond the front desk to include representation within the staff and the exhibitions.
- Enjoyment: Visitors' fun is quickly derailed when they read confusing interpretation, or find exhibits in disrepair.
- Socializing: Access through positioning, size, seating, and content should accommodate social groupings.
- Respect: Inappropriate programming and labels are just two of the ways that visitors can feel excluded.
- Communication: All interpretation should be clear, accurate, and honest.
- Learning: Understand what visitors are interested in and how they learn.
- Choice and Control: Visitors should have some autonomy in their physical movement, directional freedom, and intellectual preferences.

- Challenge and Confidence: A range of activities will ensure that the visitor experience is neither boring nor frustrating.
- Revitalization: It is possible to engage visitors to such an extent that they feel a refreshing (flow) experience.

Attending to such an extensive checklist cannot happen overnight. Rand indicates that fully embracing these visitor rights will require extended attention across departments. What is most difficult is the initial and then sustained shift to approaching activities from the perspective of the needs and interests of the visitor rather than those of the museum. Although a shift from prioritizing the museum to prioritizing the visitor does not need to compromise the museum's mission, some of the most successful audience-centered activities may at times seem tangential to the museum's strategic activities. This change in focus positions the museum as context rather than content. The implication is that visitors are learning *in* museums, rather than *from* museums. In other words, the visitor controls the experience within a responsive learning environment. Understanding the museum audience thus serves as the first step toward empowering visitors to engage in meaningful learning experiences.

REFERENCES

Anderson, P., and B. C. Roe. 1993. The museum impact evaluation study: Roles of affect in the museum visit and ways of assessing them. Summary, vol. 1. Chicago: Museum of Science and Industry.

Ellenbogen, K. M. 2002. Museums in family life: An ethnographic case study. In *Learning conversations: Explanation and identity in museums*, eds. G. Leinhardt and K. Crowley, 81–101. Mahwah, N.J.: Erlbaum.

Ellenbogen, K., J. Luke, and L. Dierking. 2004. Family learning research in museums: An emerging disciplinary matrix? *Science Education* 88, Suppl. no. 1: S48–S58.

Falk, J. H. 1993. *Leisure decisions influencing African American use of museums*. Washington, D.C.: American Association of Museums.

Falk, J. H. 1998. Visitors: Who does, who doesn't, and why. *Museum News* 77 (2): 38–43.

Falk, J. H., and L. D. Dierking. 2000. *Learning from museums: Visitor experiences and the making of meaning*. Walnut Creek, Calif.: AltaMira Press.

Falk, J. H., T. Moussouri, and D. Coulson. 1998. The effect of visitors' agendas on museum learning. *Curator* 41 (2): 107–20.

Hilke, D. D. 1987. Museums as resources for family learning: Turning the question around. *The Museologist* 50 (175): 14–15.

Hood, M. G. 1983. Staying away: Why people choose not to visit museums. *Museum News* 61 (4): 50–57.

Hood, M. 1989. Leisure criteria of family participation and non-participation in museums. In *Museum visits and activities for family life enrichment*, eds. B. Butler and M. Sussman, 151–69. New York: Haworth Press.

Lusaka, J., and J. Strand. 1998. The boom—And what to do about it: Strategies for dealing with an expanding field. *Museum News* 77 (6): 54–60.

Moussouri, T. 1997. Family agendas and family learning in hands-on museums. PhD diss. Leicester, England: University of Leicester.

Rand, J. 2001. The 227-mile museum, or a visitors' bill of rights. *Curator*, 44 (1): 7–14.

Rosenfeld, S. B. 1980. Informal learning in zoos: Naturalistic studies of family groups. *Dissertation Abstracts International* 41 (7). University Microfilms No. AAT80-29566.

Weil, S. E. 1999. From being about something to being for somebody: The ongoing transformation of the American museum. *Daedalus* 128 (3): 229–58.

CHAPTER 34

Girls, Boys, Moms, and Dads: Learning about Their Different Needs in Science Museums

DAVID TAYLOR (1953–2005)

WHAT INSPIRED MY PASSION

I spent more than thirty years working in public radio, planetariums, and science centers before returning to the university setting to learn more about the visitors we were serving. Over the years, I had been involved in the ongoing discussion about what visitors were really taking away from their experience with our exhibitions and programs. Were they learning content material? Probably a minimum amount. Were they going away excited about a subject they hadn't thought about? Maybe on occasion. Were they having experiences they would later be able to integrate with school-like learning? How was their use of exhibits and programs influenced by the social nature of these experiences? Are there differences in how children and adults, males and females, use our exhibitions and programs, and are the differences in what they take away influenced by their age or gender? Everyone seemed to have an opinion but most of the discussion was conjecture. I wanted to know more...to satisfy my own curiosity and to possibly provide some more insight for those of us in the museum business, so I returned to the university to pursue these questions through a PhD program.

My interest was first piqued in the late 1970s when Carole Kubota, an education manager at Pacific Science Center, was working on her PhD. Kubota was studying children's use of Science Playground, one of the exhibitions I had helped develop. She was investigating whether the use of a slide show about the exhibition shown in the classroom prior to the kids' visit would make a difference in how students used the exhibit when they arrived. Kubota found a difference in how the slide show affected boys and girls. In her control group, girls would tend to come into the exhibition and move to the first thing that grabbed their attention. Boys on the other hand, would come into the exhibition and spend the first few minutes running around and looking at almost everything before settling down to interact with a single exhibit. When kids were shown the slide show first, there was less of this general running around for boys. She hypothesized that the boys were building a mental map of the exhibition and all their choices before choosing specific experiences to interact with. The girls did not seem to need to go through this map-building process. This finding struck me as really interesting and made me start to think that we need to understand that our visitors have different needs and experiences with the things we develop.

Over the years I tried to learn more and more about visitors from folks who were studying visitors including, Chan Screven, Minda Borun, Ted Ansbacher, Elsa Feher, John Falk, and Lynn Dierking. Through conversations with them and by reading their work I have discovered new insights, but I have also raised a thousand new questions.

Sometime in the early 1990s, I started attending a session on girls and science centers that Dale McCreedy of the Franklin Institute was offering annually at the Association of Science-Technology Centers (ASTC) conference. Each year she brought together researchers, folks developing and researching programs for girls, and interested others in discussions about reaching girls in our programs. This got me wondering about why single-sex programs sometimes seem to be more effective.

Since much of my emphasis in the museum field has been in managing exhibit development projects, I tried to incorporate a bit of what I had learned from each of these efforts into my work. It's important to realize that we aren't developing exhibits for one type of visitor, there are lots of individual differences that we need to try to provide for. Does an exhibition have something for a small child as well as an adult and a senior adult? Families are unlikely to stay in an adult-oriented exhibition if the child cannot find anything to do, and even in children's museum exhibits, there should be something with which adults can become engaged. Does the exhibit allow for use by individuals as well as groups? Is there something there for the novice as well as the knowledgeable?

One of the most successful examples that I have seen of how an exhibition can be effective with experts and novices was an astronomy exhibition at Pacific Science Center in the 1990s. We included a working cloud chamber to show the atomic particles from space that are constantly passing around us. This exhibit was great because it was about a real phenomena happening in real time and visitors could experience this phenomena directly by watching the path the particles left as they passed through the vapor in the box. It was beautiful and it was graceful. It caused people of all kinds of backgrounds to make comments back and forth to one another and to ask questions. At the opening reception for the exhibit, I happened to be standing near the exhibit when several university astronomy professors were watching it. It was wonderful to hear one of them turn to the others and say, "I've taught about all these things for years, and theorized about them in my research, but this is the first time I've actually gotten to experience them." This kind of comment as well as comments such as "Mommy, this is great. Come take a look," are what we all strive for.

WHAT I LEARNED FROM MY RESEARCH

During the last five years, I have undertaken more than a dozen studies of kids and families in museum exhibits and camp programs. Each participant and each family has been different, but here are some generalizations about groups of people drawn from four studies. Many of the ideas, and particularly interpretations of the data, are based on small sample sizes and hunches. Each participant and each family has been different, but here are some thoughts about what I've observed from an experienced exhibition developer, now trying to gain a more research-based perspective. Please remember that the challenge with much of the research on humans is that one study often contradicts another for a variety of reasons including how the researcher framed the study, who is in the sample, and so on, so my observations may be situational to the setting in which I did the research, as well as related to the size of my samples. Other researchers may draw different conclusions from the same data, and I have tried to point out those differences.

(1) Families Visiting Exhibits

I have found that in families visiting museums, kids do the exhibits, dads do the exhibits if they are interested, and moms make sure everyone gets a turn and hold the coats. During a study I did about interaction with a bicycle wheel gyroscope exhibit, I observed this pattern over and over in this setting. Boys would usually do the exhibit first (even if it meant pushing their sister out of the way) and once finished would run off to the next thing. Girls would do the exhibit but usually stay with the rest of the family. Dads might or might not choose to do the exhibit next. Moms seldom did the exhibit themselves when visiting in the family group. By the time everyone else had a chance to use the exhibit and was ready to move on, mom would give up her turn for the good of the family. Moms also were almost always the ones holding all the coats. These observations left me wondering about what these behaviors say to boys and girls about their gender roles as they grow up. When the daughter sees the mother giving up her turn and carrying the coats, is she learning that science is for her or that a woman's role is a supportive one? When boys see this, what will their expectations be for their future wives and children? It is important to note that this is an area in which the research is very equivocal with not all researchers observing these same behaviors (Dierking and Falk 1994).

(2) Same Gender Visiting Exhibits Together

In an informal study of the Body Works exhibition, I videotaped more than 400 visitors moving through the exhibition for two hours a day, over a ten-day period. I reviewed these tapes to try to look for patterns of how visitors used exhibits as they moved through the exhibition. In reviewing the tapes, I tallied group size, group composition, how individuals in the groups interacted with individual exhibit elements, and how they moved through the overall exhibition. I became more interested in the differences I was seeing between how all-male groups and all-female groups interacted with and used the exhibits. The same patterns seemed to apply whether these groups were a parent and child, or adolescent friends, or adult friends.

Two or more males visiting together seemed to move through the exhibition more independently, each choosing exhibits to interact with and then from time to time coming back together to note exhibits they especially liked or what score they had achieved on a particular exhibit. Statements such as "that one was fun" or "you should try that one" or "I got a 92 on that one" were common for the males.

All-female groups (friendship groups or moms and daughters) tended to move through the exhibition as a unit, generally staying together, using exhibits together, or waiting for one another to have their turn at a specific exhibit. They also tended to do a lot more talking at the exhibits and laughed a lot more. My take on these findings is that female groups use the exhibits as part of their social interactions, while males do the exhibits in a more task-oriented way, wanting to know what to do, how well they performed on the exhibit, and how they compared with others. Women behaved differently when in family groups than when in all-female groups, which is confirmed in some early research by Paulette McManus (1987).

(3) Kids in Camps

During the summers of 2002 and 2003, I studied fourth through to sixth grade kids in four weeklong science camps. Two of these camps were mixed boys and girls and two were for girls only. I was looking at differences in how the kids "did" camp and in girls' opinions on all-female and mixed-gender camps. Campers in these programs, such as those in most museum programs, are not representative of the total population. They tend to come from families with higher education levels and higher incomes, and they are usually kids who do reasonably well in school, so it is difficult to expand my findings to the total population

Given these qualifications, what I find is that the older girls tended to be more aware of gender differences and more supportive of the choice of same-sex camps. There were many girls who felt that there was no real difference for them in a camp that was all girls from one that was mixed boys and girls, but many felt that it was easier for them to learn and speak out in the all-girls camps because there were no boys to be disruptive or be distracted by. This fits with some of the research literature that indicates that girls tend to get less airtime in class because they are more polite and wait for a break in the conversation before speaking out. Boys on the other hand tend to interrupt and break into the conversation so they are better able to control the amount of airtime they get. Even in the mixed-gender camps, I found that the subgroups that formed were between members of the same sex. In one of the mixed-gender camps where there were only a few girls and many more boys, one girl described it as "we needed to band together to protect ourselves from the boys."

Numerous studies have shown that as girls move into puberty and beyond, their interest in science and technology tends to wane. The amount of nature versus nurture involved is not clear, but the underlying messages that science and technology are "guy" areas continue to be heard by young women. Are all-girls science camps, camp-ins, clubs, and programs such as Girls in

the Center, directed by Dale McCreedy of the Franklin Institute, effective in breaking through this stereotype or are we attracting those girls who are already more interested in science and technology? Probably a bit of each.

(4) Interviews with Women

As part of my research, I have interviewed more than a dozen women with science degrees to find out their views on why they did or didn't stay involved with science or technology. While this is not a very large sample, some patterns seemed to reoccur in the reports of these women and their experiences.

One of the contexts for this interest was my high school chemistry teacher. She is now in her eighties and we have kept in contact over the years. In an interview I conducted with her, she tells of falling in love with chemistry and being single-minded in her pursuit of the subject. Her parents were supportive, but when she graduated from college in the 1940s and went around looking for a research chemist position, she was told by companies such as Bell Labs, "we don't hire women, we'll put lots of time into training you and then you'll just leave us to have babies." As a result she ended up in what was considered a women's profession—teaching. I benefited, but I think the field of chemistry suffered as a result. Since the advent of the women's rights movement in the 1960s and laws forbidding discrimination against women, some things have changed, but for many young women, the message still comes across that science and engineering are not areas that welcome them. Several of the women I interviewed tell of being one of a handful of women in classes with many more men. They speak of having almost all-male teachers and few female role models and of being encouraged to go into the marketing or educational end of science and technology rather than basic research. In fields such as medicine, biology, and psychology where the percentage of women is more than 25 percent, there tend to be fewer of these disincentives for women entering the professions.

A pattern seemed to emerge from these interviews, which may assist the museum field in having a greater impact in their efforts to reach girls. It appears that having early hands-on experiences with science and technology is very important. Many of the women who remained in the sciences had been encouraged from an early age and had been involved actively in using science and technology. Girls who helped their fathers work on the car or change a light switch or repair something, seemed to be more empowered and got the message that they were capable of doing science and technology stuff. Those girls who didn't have these early experiences were less likely to have a self-image later in life where they saw themselves involved in science and/or technology. This is true for boys as well, but the percentage of girls who have not had these early experiences has traditionally been larger. So, in our museums, we should provide some of these empowering experiences that will help girls and boys later in life. Many of us do this already, but we should consciously think about it as we develop new exhibits, demonstrations, classes, camps, and programs.

I still have a question of how well exhibits with their short interaction duration and generally limited kinds of interactions can accomplish this. Of course, what is important to note is that a brief time at an exhibit is building on prior experience and knowledge and will be built on subsequent experience and knowledge so museums and science centers are one component of what it is hoped is a systemic approach (school, home, and community) to encourage gender equity for girls in science. Having said that, how can we make exhibits more effective at this? Workshops, carts, and classes generally allow greater time on task that may greatly aid the areas we are trying to get across in exhibits.

WHERE DO WE GO FROM HERE?

As I mentioned at the beginning of this chapter, the challenge with much social science research is that one study can contradict another for a variety of reasons including how the researcher framed the study, the nature of and size of the sample, the setting in which the research was conducted, the meanings which researchers drew from similar data, and so on. Appreciating these limitations, here are a few thoughts on how my observations might influence the work of museum or science center professionals:

- Train staff to think about all the differences in your visitor population and to realize that different people are coming with different motivations, backgrounds, and ways of learning. There is no one way a visitor should visit the museum—the staff's job is to make it possible for all visitors to find something they can relate to.
- Provide real experiences for boys and girls at an early age to help them build a self-image that says they can do these things.
- Provide extended experiences where more complete interactions can occur and where a facilitator can help visitors be successful and gain understanding.

- Encourage parents to use good parenting skills to support their sons and daughters to help them be successful and to let them know that opportunities are there for people with all kinds of differences, including them.
- Provide some experiences that are geared to just reaching girls. Do this especially for girls who are nearing and passing through puberty when gender is a strong and often contradictory influence and many stereotypical behaviors can shape their belief systems.
- Encourage mother and daughter days at the museum. It's important for girls to have some social time seeing their moms outside the traditional family setting.
- Finally, read as much as you can—you're reading this book, which contains thoughts of many who have been involved in the field. Read more and I would recommend reading some of the research by those I cited at the beginning of this piece.

Learning is a great passion!

SUGGESTED READING

Borun, M. 1989. What research says about learning in science museums: Naïve notions and the design of science museum exhibits. *ASTC Dimensions* 9:3–5.

Csikszentmihalyi, M., K. Hermanson, and U. Schiefele. 1995. Intrinsic motivation in museums: Why does one want to learn? In *Public institutions for personal learning: Establishing a research agenda*, eds. J. H. Falk and L. D. Dierking, 67–77. Washington, D.C.: American Association of Museums.

Diamond, J. 1986. The behavior of family groups in science museums. *Curator* 29 (2): 139–54.

Dierking, L. D., and J. H. Falk. 1994. Family behavior and learning in informal science settings: A review of the research. *Science Education* 78 (1): 57–72.

Falk, J. H., and L. D. Dierking. 2002. *Learning from museums: Visitor experience and the making of meaning.* Walnut Creek, Calif.: AltaMira Press.

Honey, M., B. Moleller, C. Brunner, D. Bennett, P. Clemens, and J. Hawkins. 1991. Girls and design: Exploring the question of technological imagination. Center for Children and Technology, report no. 17. www2.edc.org/cct/publishions_report_summary.asp?numPubId=48.

McManus, P. 1987. It's the company you keep: The social determination of learning-related behavior in a science museum. *International Journal of Museum Management and Curatorship* 53:43–50.

Rosser, S. 1997. *Re-engineering female friendly science.* New York: Teachers College Press.

Serrell, B. 1997. Paying attention: The duration and allocation of visitor's time in museum exhibitions. *Curator* 40 (2): 108–25.

CHAPTER 35

Evaluation 101

Elsa B. Bailey and George E. Hein

WHY EVALUATE?

The present emphasis on accountability requires museums, like other social and education agencies, to demonstrate what they accomplish. The current term *outcome evaluation* suggests that the proper assessment of any activity is not whether it was actually carried out, that is, its "outputs," but the extent to which that effort has made some difference for its users or participants (IMLS n.d.). However, acknowledging these realities, the most compelling reason for carrying out evaluation should be that if conducted throughout the process it can help us better understand the audiences we plan to serve, thus shaping an effort that will meet an identified need or set of needs, and can actually improve the chances that our goals will be achieved. We can only learn from what we have done, can only improve our practice—or, to put it in practical terms, learn from the inevitable shortcomings and have a reasonable record of what's been accomplished, if we document what we've done.

Rewards of conducting evaluation also include the satisfaction of better understanding the strengths and limitations of your institution and the professionals with whom you work, as well as providing the beginning of planning for the next project; evaluation provides a rational basis for future action. The first task of any evaluation, to determine what outcomes are intended for any activity, requires careful thought and discussion and can guide all project planning. Clearly articulated goals, objectives, and desired outcomes provide a road map for what you actually intend to accomplish in a project.

EVALUATION DEFINED

Evaluation is the systematic documentation of the development and outcome(s) from an educational effort, be it a program or exhibit or some other activity (website, and so on). Each of these italicized words requires some elaboration in order to understand the key components of an evaluation.

The most important aspect of any evaluation effort is that it be systematic. You can't assume that evaluation will happen (in fact it usually doesn't) unless there is a deliberate and systematic effort to carry it out. Unfortunately, this requires time, commitment and resources—usually both staff time and some financial input. The essential first step is to make the commitment, no matter how small, to acknowledge that evaluation is an essential task in the conception of a program or exhibit and that it will take some hours and days, money, and, possibly, managing the work of others. Not surprisingly, like most aspects of museum work, the limiting factor frequently is available resources. How much evaluation actually occurs (and consequently how much you learn from it) will depend on the resources committed. Not surprisingly, professional evaluators tend to estimate evaluation budgets on the high side, while project managers, worried about the cost of everything, tend to calculate low. Realistic estimates hover around at least 10 percent of total project costs, including both time and money. If the maximum you can commit is significantly less than 5 percent, it may not be worth the effort, and you might be better off just doing the project, hoping you can recreate the best aspects and avoid the worst problems the next time around.

Systematic also means that evaluation should be planned into a project *from the beginning*. This maxim, so frequently ignored, can't be emphasized enough. Evaluation can only assist in planning if it is started early. Also, one crucial feature is to know where you started; having some baseline data. Unless you have some idea of the situation in your museum—current visitorship, what people do, what activities they appreciate and which they avoid, what they get out of their visits, and so on—you'll have difficulty demonstrating particular impacts attributable to the new program or exhibit. Even a simple example illustrates this point. If you want to claim that a new exhibit boosts attendance, then you need to count visitors before the new activity occurs and take into account seasonal and/or daily variations. (You won't be able to convince anyone that a special school vacation-week program brought in many visitors, if your only comparative data is the quarterly or annual average visitation.) So you need to think about evaluation long before the actual program occurs. If you have more ambitious goals, that visitors learn something, change their attitudes or behavior or demonstrate other outcomes, then you obviously need to incorporate activities into your evaluation plan that will enable you to demonstrate what did or didn't occur in your institution before the new program.

Third, systematic means a plan to systematically collect data. This doesn't have to be elaborate, but it has to be consistent and reasonably objective. Even the relatively simple example above, that of counting visitors, a task that can be done by ticket takers, volunteers, or a guard with a counter requires planning, a pilot trial, and like all data collection, some decisions. Are all visitors counted or only those who pay? What about school groups? What about someone who only goes in and right out? Usually, there are no right or wrong answers to such questions; what's important is that you set up a consistent system for data collection across the various relevant conditions (before and after, for example) and stick to it. More complex questions require more thorough analysis of how and what data will be collected (interviews in person or later by phone, observations, unobtrusive measures, and so on), but issues and potential problems are similar.

Finally, systematic means deciding clearly what you value and want to find out from an evaluation. This important step is also frequently overlooked, especially in evaluations carried out internally without the assistance of professional evaluators. Just giving visitors a questionnaire or asking teachers to fill out an evaluation form at the end of a workshop usually isn't sufficient. Unless you're clear about what you need to find out, it may be a complete waste of time. Part of being systematic is clearly matching your needs with your deeds: collecting the data that can best answer the questions important to your institution.

WHAT DOES EVALUATION LOOK LIKE?

Once you've accepted the need and challenge of incorporating evaluation into your professional responsibilities and are willing to integrate it into your planning and carry it out systematically, then what?

Evaluation needs to be on the agenda at every planning meeting, or constantly on your personal to-do list if you're doing a one-person project. There may be times during a project or exhibit development when evaluation will not be necessary, but these times are rare; more often, you need to think about (sometimes only briefly) whether you're following your evaluation plan (as well as your implementation plan; the two are interrelated but not identical), what needs to happen next, and whether any changes need to be made.

Decide what you want to know; what your evaluation questions are, and the available resources; there will probably be more questions about all aspects of the project than you can afford to pursue. You may be able to carry out only a fraction of what is recommended below.

EVALUATION STAGES

Traditionally, evaluation work is described as applying to three stages of any program or exhibition: *front-end* (or preliminary), *formative*, and *summative*.

Front-end Evaluation

Front-end evaluation refers to information gathering before you begin the program or decide to build the exhibit. What skills should you offer teachers? How will the public react to an exhibit on AIDS? What do they already know or think about the topic? Since such questions need to be answered before significant investment in the program or exhibit, they usually involve oral or written responses from a potential audience. It's possible to develop prompts or tasks that provide clues to what visitors know, think or believe, but all these are obviously not exactly like the actual program or exhibit to be offered or built and therefore require some educated estimates of their likely relation to the final product. Visitors may say that an exhibition on AIDS offends them, but inclusion of the topic in a larger exhibition on public health issues may seem appropriate to the public. Instead of simply asking visitors questions, they can be shown objects or pictures of possible exhibit components, asked to participate in an activity, or to converse in groups on a topic. Program needs can be assessed through interactions with community groups and school administrators; observations at meetings; or by noting trends in local or national policy (Dierking and Pollock 1998).

Formative Evaluation

Formative evaluation focuses on improving a program or exhibit while it's being developed. The purpose is to try out parts or the whole activity to see how well they will work. It's the systematic consideration of responsible practice: good developers routinely try out prototypes of exhibit components on visitors. If you hire outside developers, you can insist that they take this step. Ideally, the evaluation should be carried out in your institution under the conditions that will prevail when the exhibit is installed. Most visitors are delighted to be asked to interact with an exhibit in progress; they like to be part of something special and to feel, correctly, that they are assisting the museum. You need to remember not to think it's a shortcoming of the visitors when they don't understand something or can't use the interactive as it was intended. The point is to use the formative evaluation to change what you put on the floor so it will be useable and understandable for the intended audience. It's also important (and required both ethically and legally) that you inform visitors that you're using them in an experiment. Formative evaluation is often quite informal; bits of exhibits may be tested on small numbers of visitors and few permanent records kept of the results. But even very small trials can help you avoid costly errors. If you're going to do this on a regular basis, it will be more efficient and the institution will learn more if you develop a system for formative evaluation and collect data following some standardized procedures. Results will gradually lead to a richer understanding of what all visitors to your institution do and don't understand. Even if you have limited time to allot to formative evaluation, it's better to do something rather than nothing. Small steps toward initiating an evaluation process can serve as pilot studies, build staff's experience with evaluation, and help determine the methods, styles, and approaches that are a good fit for your particular institution and staff. Also, like an exercise program, once you get used to the routine, you wonder how you ever carried out work without it.

Summative Evaluation

Finally, *summative evaluation* is work done after an exhibit or program is in place. Does it accomplish what it was intended to do? How do visitors use it and to what extent? Usually this is the most elaborate evaluation, most likely the type required by funders, and most often carried out by professional evaluators. If the evaluation process was started when the project or exhibition planning began, then you've already identified both the outputs (what you intended to do) and the outcomes (what effect your activities will have on participants). Summative evaluation will document the extent to which these goals were achieved.

It's also important to organize a summative evaluation so that unanticipated outcomes are acknowledged. Educational activities often result in unexpected outcomes—both desirable and less welcome. The means you use should be flexible and broad enough to allow you to find out what actually happened as a result of your activities. The matrix described below illustrates how our research group addressed these challenges.

CONDUCTING AN EVALUATION

At the Program Evaluation and Research Group (PERG) at Lesley University, Brenda Engel and George Hein developed a matrix format useful both for the planning and implementation of program and exhibition evaluation, and they and their colleagues have utilized it with numerous evaluation studies over two decades. The matrix describes the activities and outcomes of the program to be evaluated and matches these with a list of evaluation methods that can be used to gather information (Engel and Hein 1981; Hein 1995). A sample blank evaluation matrix is provided in figure 35.1, and examples of actual matrices that were used are in both references above. Preparing the matrix as a joint effort by evaluators and program staff provides a road map for the evaluation and a planning tool for keeping track of project activities. It requires a careful description of all the proposed actions and products (outputs) as well as the expected outcomes. If it is developed early enough in a project, it can become the basis for front-end and formative evaluation activities as well as, later, serve as a guide for summative evaluation. In multiyear projects, the matrix often is revised annually, since as a project develops, both activities and goals inevitably change.

During the initial articulation of the major tasks required to carry out the project, it's important to examine not just outcomes but also what actually occurred. In order to argue that outcomes are the result of a program, it's necessary to be able to describe what that program actually is. In practice, this turns out to be a harder task than project staff usually assumes. Although a project may appear to be straightforward in a proposal description, the reality of an operational project (frequently with a reduced budget) may require reformulations and clarifications. Equally important and significantly harder is the next step, to determine

	METHODS						
	Observation	Interview	Document review	Survey	Program products		
Activities							
Outcomes							

Developed at the Program Evaluation and Research Group
Lesley University

Figure 35.1. Evaluation 101 Matrix.

what outcomes are to be evaluated. This requires that staff actually verbalize in concrete terms what impact they hope to have on users or participants, and what the educational goals for their exhibit or program are. We have found that this is the most difficult step in any evaluation.

Once project activities and intended outcomes are clear, then—and only then—is it time to think of ways to collect data and to list all the possible methods that could be used. There are actually only three categories of methods: observation, language (spoken or written responses), or some product that results from the experience that can be examined. But these three categories can be exploited in a wide range of ways: besides the obvious interview forms or questionnaires, visitors can be asked to draw, build something, or look at and sort pictures. Evaluators have brought carts full of objects onto a museum floor and talked to visitors using the objects as prompts, asked them to take photographs of things they liked (or didn't), to think aloud while in an exhibition, and devised other ingenious ways of trying to find out what people know, think, or feel. The most important component of any method used is that it be done systematically, with some reference point (for example, carried out both before and after an experience, or on visitors that did or didn't have the experience) to help interpret the results.

Usually, a matrix will include many activities and several outcomes. The limit to what is evaluated is a compromise between what is most important and what data can be gathered in the time and with the resources available. Outcomes such as long-term changes in visitors' attitudes and behavior (although often desirable and frequently included in program goals) are seldom evaluated since they would require data collection over many years. In many evaluations, the questions addressed are only surrogates for what one would actually like to know, but they frequently represent the best that can reasonably be measured.

Once a matrix is completed, the other components of the evaluation plan are decided. A useful exercise is to explore all the possible categories of people who may be able to provide information. Will museum security guards or frontline staff know something about visitor outcomes? Is it likely that school administrators have knowledge of a program for school children? Is it worth attempting to telephone parents of children who participated in a family program a week or a month after the program? The potential list of sources is often surprisingly large and provides rich opportunities for data collection.

Finally, the logistics of the evaluation need to be addressed. Who is going to collect data and when, who will be responsible for analysis and for writing reports? These are decisions that often need to be made very early, or precious opportunities slip by. For example, if a school-vacation program needs to be evaluated, then data has to be collected during that week. But, that's often a time when there are heavy demands on staff for other responsibilities and the reliable, well-organized part-time employee who does excellent interviews may be planning a family vacation that week. The secret to addressing these kinds of issues is to plan as carefully for the evaluation as you do for the program or exhibit— and then realize that the same unforeseen problems and last-minute crises will arise during the evaluation as they do for every other activity. Guidelines for evaluations can be found in two AAM resource reports (Korn and Sowd 1990; Diamond 1999), useful primers on evaluation in museums.

WORKING WITH AN OUTSIDE EVALUATOR

If you plan to use an outside evaluator, before you select him or her, it is wise to think first about your evaluation needs and expectations. By doing this initially, you are in a better position to consider whether or not your needs and expectations align with what a particular evaluator or evaluation team will do for you. First, seek references from colleagues about evaluators they have used. Hearing about people's firsthand experiences with an evaluator will give you valuable insights from a user perspective that might otherwise be more difficult to attain. Second, speak to potential candidates directly, and determine if you are comfortable with their communication styles, the services they offer, and their approach to the evaluation process. Ask potential candidates if it is possible for you to see examples of reports they've written for others. The clearer you can be in explaining your evaluation needs to the evaluators during the selection process, the more likely you will be to find a good match.

There are a number of traditions in both research and evaluation, and finding a person or group with whom you are theoretically and philosophically compatible is essential. Some evaluators emphasize a quantitative approach, some a qualitative approach, and some incorporate both in their work. *Quantitative* methods can provide straightforward answers to questions, such as what is happening, where it is happening, and how often it occurs, and are useful for reporting things that need to be counted, listed, and compared. *Qualitative* (or *Naturalistic*) methods are more appropriate for reporting how things are happening and why they are happening, and they can provide in-depth insights into people's attitudes and understandings. Data that inform qualitative findings are often gathered through interviews with people or by asking people to respond to open-ended questions in either a verbal or written format.

Budgets may determine much of what can be accomplished in any evaluation. Many people have unrealistic expectations as to what an evaluator can do within budgetary constraints. Organizational personnel may not be aware of what is involved to achieve what they request. Professional evaluators know how many hours and days a quality evaluation requires. Conversations up front with the evaluation team can help to alleviate future misunderstandings and encourage a constructive dialogue. For instance, if time allotted for data collection emerges as an issue, outside evaluators may be able to resolve this through developing a plan to utilize (and train) staff to do the data collection, assist in developing embedded evaluation tools, or suggest a sampling strategy that provides the needed information while collecting less data.

EMBEDDING EVALUATION INTO YOUR PROGRAMMING

If it's impossible to hire an outside evaluator, in-house evaluation is an alternative. Although conducting in-house evaluation requires an expenditure of organizational time and resources, evaluation that is done in-house is likely to provide the organization with a number of benefits. Embedding evaluation into project development offers important richness to the program or exhibit planning. It encourages staff to think "evaluatively." When project planners are encouraged to grapple with the key question discussed earlier—"What do we want or need to know?"—it pushes them to describe the program objectives and goals early in the project development process, and it identifies places where evidence for achieving those objectives may become visible, be captured, documented, and analyzed. Typically, formative evaluation is the type most effectively conducted from within the institution, with some guidance from a professional evaluator, if possible.

An example of embedded data collection in an exhibit is a component that encourages visitors to respond to a question and record their responses through writing, recording (video or audio), or another collection device such as punch cards. This data collection tool can be framed to elicit visitor attitudes and/or specific knowledge. An embedded data collection example in a program might have participants keep journals during the program (after they are informed up front that they will be read by staff). Such journals are useful in assessing participant learning. They also offer staff an opportunity to observe participants' metacognitive process (their thinking about their thinking), and they provide data about participants' learning experiences.

Depending upon available staff, another way to collect data about an exhibition and/or a program is to assign a particular member of the staff to serve as an observer and note-taker as visitors interact with exhibits or during program events. This not only provides documentation about the exhibit or program but also provides a professional development opportunity for staff. Making video and audiotapes of program activities allows program implementers to revisit program

events at a later date, supports the debriefing process, and encourages reflection, dialogue, and perhaps revision around the program they are implementing. Videos and audiotapes should be made only with participants' knowledge and consent.

And finally, should the resources become available to permit hiring professional outside evaluators, the experience gained in conducting in-house evaluation activities can better prepare the organization and its staff members for the professional evaluation process.

UTILIZING EVALUATION

No matter how thorough and competent an evaluation, whether performed by staff or a professional evaluator, it is only of value if it is used (Patton 1996). Evaluation studies need to be discussed, circulated among staff, and acted upon in order for an institution to learn from them. Surprisingly, both funding agencies and organizations that invest heavily in evaluation frequently shelve the reports without much discussion or reflection of their conclusions. A plan for dissemination and discussion with staff at all levels should be integral and the final required step for any evaluation.

REFERENCES

American Association of Museums. 1999. *Introduction to museum evaluation*. Washington, D.C.: American Association of Museums.

Diamond, J. 1999. *Practical evaluation guide: Tools for museums and other informal educational settings*. Walnut Creek, Calif.: AltaMira Press.

Dierking, L. D., and W. Pollock. 1998. *Questioning assumptions*. Washington, D.C.: Association of Science-Technology Centers.

Engel, B. S., and G. E. Hein. 1981. Qualitative evaluation of cultural institution/school education programs, in *Museum school partnerships: Plans and programs*, eds. S. N. Lehman and K. Igoe, 39–45. Washington, D.C.: Center for Museum Education.

Hein, G. E. 1995. Evaluating teaching and learning in museums. In *Museums, media, message*, ed. E. Hooper-Greenhill, 189–203. London: Routledge. (Fig. 35.1 is available at www.lesley.edu/faculty/ghein/papers_online/Leicester_1993/Hein_Leicester_1993.html.)

Institute for Museum and Library Services. n.d. *Perspectives on outcome based evaluation*. Washington, D.C.: Institute for Museum and Library Services.

Korn, R., and L. Sowd. 1990. *Visitor surveys: A user's manual*. Washington, D.C.: American Association of Museums.

Patton, M. Q. 1996. *Utilization-focused evaluation: The new century text*. Walnut Creek, Calif.: Sage Publishing.

Part IX

GOVERNANCE

CHAPTER 36

Working Model: A Mechanism for the Effective Board

Harold and Susan Skramstad

PEOPLE DECIDE TO JOIN A BOARD OF TRUSTEES FOR many reasons—some good, some bad. Service on a museum board gives a person a certain prestige within the community—an identity as someone with whom others wish to associate—and acquaintance with interesting people and activities. For some, it may be a nice addition to a resume or a way to make personal contacts with people who are influential in the community. For others, board membership may be primarily social, an opportunity to interact with like-minded people and be associated with an organization that is known and valued. For these groups, regular attendance at meetings and a deep interest in the concerns and challenges of the museum are secondary. But for most, service on a museum board is an important volunteer activity that offers an opportunity to make a difference.

A good board doesn't just happen, it must be crafted carefully. Recruiting a good board begins with a real understanding of the museum's purpose, value, and audiences, as well as its plans, priorities, and needs. It also requires a self-examination designed to identify board strengths and weaknesses as they relate to the museum's current needs.

Once the board has recruited the right people to serve as trustees, it must work to keep them. Nothing will alienate a new trustee faster than a culture of hostility or passivity among the trustees, or board meetings that are long, boring, and about unimportant issues. If the board does its job right, new trustees will enter board service with energy and enthusiasm—they'll want to get on with it. It is up to the board to ensure that this happens.

Board development is a continual and continuing process. Times change, people change, and boards, too, must change. The basics of board development should never be far from the minds of current board members. Naturally, a board is only as good as its members and leadership, its understanding of the museum and its priorities, and its understanding of the mission and how it is carried out. But organizing and disciplining the work of the board and the individual trustees will position any board to be the best that it can be—guaranteed.

We believe there are 10 areas that trustees should consider when working to improve board performance—a mission for the board, the board meeting, keeping board members busy with real work, committees and task forces, the culture of the board, motivating and retaining good trustees, terms of service, succession planning, the exit interview, and evaluating board performance. Several of these items are discussed below.

THE MISSION OF THE BOARD

While the magnet for all museum activities and for all staff and board activities should be the institution's mission, the board should have its own mission. A mission provides a rationale and a charge for the board's work and may help to prevent misunderstandings among trustees. The board's mission should clarify three things:

- what the board does
- the outcome of that activity
- and the value of that activity

What the board states in its mission has real meaning and should serve as a guide for the work of the trustees. Mission statements are not easy to write, and board members should not expect to develop the perfect one after a few hours of discussion. We suggest that the development of a mission be on the agenda for three consecutive board meetings, beginning with a general conversation about the work of the board and ending with closure on the statement. The process by which the board arrives at its mission and the discussions that lead to its development are almost as important as the mission itself.

THE BOARD MEETING

The board meeting, sometimes impolitely called "the bored meeting," is where the real work of the trustees is carried out. That is where full discussion occurs, votes are taken, and courses of action approved. And yet, all too often, trustees either dread the thought of a board meeting—expecting an endless and boring series of reports—or look forward to a nice social get-together.

The board chair and the CEO, who plan and run the meeting, should strive to ensure that trustees look forward to the meeting and arrive ready to give their best. Questions the CEO and chair should consider during the planning process include:

- What decisions will be needed at this meeting?
- How will we handle board members who do not come prepared?
- How will we ensure that everyone has a chance to speak?
- Who will follow-up on board assignments to ensure that they have been completed?

A lively board discussion in which everyone participates, problems are solved, decisions are reached, and assignments are made will go a long way toward keeping the board engaged. The best way for the chair to discover whether the trustees find meetings effective is to ask them. We have found that a short survey following each board meeting—with questions ranging from "did the meeting begin on time?" to "did you feel comfortable speaking up?"—often aids the planning of future meetings. The goal is to use the time of board members to the best advantage; the moment they feel their time is being wasted, they're gone.

COMMITTEES AND TASK FORCES

Committees can keep an organization moving in the period between scheduled meetings of the board and allow trustees to use their time wisely. No matter what the board size, certain functions—namely, oversight of finances, development (raising funds), and board development (training and education)—are usually best delegated to standing committees. Many museums have a plethora of other standing committees, covering such functions as collections, exhibits/educational programs, and buildings/grounds, etc. Whether such standing committees will be effective tools for improving museum performance will depend on the size, membership, resources, and stage of the institution's development.

In a small museum, the board may be called upon to take on or extend the skills that staff would be responsible for in a larger organization. In this case, committees function as volunteers to support, in a practical sense, and sometimes even do some of the staff's work. However, while there is no question that board committees can bring a common-sense perspective and special skills to the oversight of staff functions, sometimes that oversight leads to micromanagement and counterproductive second-guessing of the staff. That, in turn, can cause severe tensions and morale problems. In thinking about what committees to establish, the board must be able to answer the following questions:

- Do we need this committee, and why?
- How will it improve the museum, or board, performance?

If there is no ready and positive answer to these questions, then there is probably no need for the committee.

An alternative to the semi-permanent committee structure is the task force, a fluid body that has a specific charge and a specific deadline—at which time it goes out of business. The advantage of this system is the mechanism for self-destruction that is built into the very nature of the work—a focused agenda, a limited time, and a particular product. The urgency of dealing with a pressing issue of importance to the future of the museum can be a great stimulus for good board members and can help spread the workload of the board.

BOARD CULTURE

Every board has a culture—or what we might call "the personality of the board." Think about your own board for a moment. Is it stimulating and exciting; functional and efficient; dysfunctional and chaotic; or just not very interesting? Are board members lazy or are they high performers? Do all trustees participate in important board discussions or only a few? Is the board cliquish or culturally inclusive, reflecting the community that the museum serves? Do trustees engage in lively debate or passive acceptance? Do staff or board opinions dominate, or is there a productive exchange of ideas and good partnership between the two? These are all culture issues. What is important is to ensure that the board's culture—the way in which it routinely does its business—is a positive one in which the right work is being done in the right way.

It is difficult to explain how a negative culture becomes embedded in an organization. It often begins with one or two very strong personalities who want to control the discussion, deal with staff as little as possible, communicate nothing, and run the museum, in some cases by going around the CEO. Or sometimes it begins with a weak chair; little information before meetings, which consist solely of staff and committee reports; tolerated absences and lateness; and very little of significance accomplished. The interesting thing is that after the strong or weak trustees have left the board, the culture remains as an unwelcome legacy. And people just accept it.

There is almost always at least one person in the room who knows when things have gone awry. More often than not that person remains silent. That is why bad things don't change; no one raises the flag; no one pushes; no one says, this isn't right. But while negative perceptions of the museum or its board must remain within the museum—trustees, above all others, must speak only positively of the institution to the outside world—they must be discussed openly by the board. Listening to other opinions is the only way to grow.

SUCCESSION

The constant motion of trustees on and off the board makes it very hard to plan effectively for succession. Consistency and continuity are extremely important to the effective functioning of the board, however, and boards should think more about succession planning than they do. There is no value in the haphazard elective process that often brings in a new chair, and certainly none in the practice of a short term for the chair. How would boards like it if the museum's CEO left every year, or every two years, the most common terms for board chairs? That kind of "herky-jerky" leadership makes it hard to take the museum where it needs and wants to go.

The chairmanship of a museum board of trustees is a powerful leadership position and should not be given to someone who is unprepared; the leadership must be passed smoothly so the work of the board can go on without missing a beat. That argues for a formal succession process, built into the procedures of the board itself. In addition, every member should be on the lookout for potential leaders from the first day a

new trustee appears on the board. It is the responsibility of the chair to develop leadership when he sees it; it is the responsibility of the other trustees to acknowledge and promote it.

EXIT INTERVIEWS

Don't miss the opportunity to interview board members when they leave the board, whether they are leaving because their term is up, for personal or professional reasons, or because they are being removed. The exit interview will provide you with invaluable information that will help ensure that the board is operating to its potential. A retiring, resigning, or fired board member may have a great deal to say, which, because of the board culture or his own personality, was difficult to say before. Take advantage of this opportunity; the exit interview should be an ongoing activity of the board, closely associated with board development.

It also will provide you with an opportunity to encourage the trustees' continued involvement in the museum. Former board members should be considered a part of the institution's family, kept informed of museum activities, and involved in the actual work of the board, when appropriate. It goes against common sense to let all that experience drift away.

EVALUATING BOARD PERFORMANCE

Every board, from the smallest to the largest, from the newest to the most sophisticated, should evaluate its own performance annually. This is something few boards actually do. Trustees know that they are required to review the museum's director annually and examine the performance and progress of the museum. And, for the most part, they are eager to do this.

Many boards, however, are reluctant to establish goals for themselves, analyze their own performance, and look for areas that can be improved. The idea that the museum's director and others outside the organization might give a candid assessment of its service seems to strike terror into the collective heart of even the most seasoned board. But if the annual review of the museum director is an opportunity for an honest exchange of ideas, an opportunity to learn and grow, why would such a review be a less significant opportunity for the growth of the board?

CONCLUSION

Service on a museum's board of trustees is a privilege that carries great responsibility. It provides an opportunity to make a real difference to the museum, to the people who use the museum, to the community in which the museum is located, and—if the board and staff are truly successful in developing an innovative and engaging museum experience—to the museum field.

The people who agree to serve on museum boards of trustees generally have a deep sense of civic and cultural responsibility; they understand and value the trust that is given to them and begin their tenure willing and eager to serve. But all too often this eagerness turns to disappointment, and new trustees become disillusioned as the real culture of the board is revealed. Some choose to leave the board and use their talents in other ways; some stay on, frustrated but accepting, until their terms end. In neither case do these people stay involved with the museum after they leave the board.

It is the responsibility of all trustees to ensure that this description does not represent their board. Our basic message is to use the time of the trustees wisely and well, but, above all, to use it. Trustees have agreed to serve; let them serve.

We encourage you to evaluate board performance and individual trustee performance on an annual basis.

It is the only way to know whether the board's service to the museum is organized and efficient, and, more important, whether it is valuable to the museum and its CEO. That requires an act of faith on the part of the CEO and the board, but without trust between these two parties, not much will happen. Given that there is trust, an open and honest evaluation of the board's performance will be extremely helpful to the board.

Trustees expect the CEO to get the best out of the staff; the CEO should expect the best from the trustees. And the trustees should expect the best of themselves.

Harold Skramstad, president emeritus, Henry Ford Museum & Greenfield Village, Dearborn, Mich., and Susan Skramstad, former vice chancellor for institutional advancement at the University of Michigan-Dearborn, are consultants to nonprofits on issues of change, planning, fund raising, and board and staff development. This article is adapted from their book, *A Handbook for Museum Trustees* (American Association of Museums, 2003).

CHAPTER 37

Code of Ethics for Museums

AMERICAN ASSOCIATION OF MUSEUMS

INTRODUCTION

Ethical codes evolve in response to changing conditions, values, and ideas. A professional code of ethics must, therefore, be periodically updated. It must also rest upon widely shared values. Although the operating environment of museums grows more complex each year, the root value for museums, the tie that connects all of us together despite our diversity, is the commitment to serving people, both present and future generations. This value guided the creation of and remains the most fundamental principle in the following Code of Ethics for Museums.

CODE OF ETHICS FOR MUSEUMS

Museums make their unique contribution to the public by collecting, preserving, and interpreting the things of this world. Historically, they have owned and used natural objects, living and nonliving, and all manner of human artifacts to advance knowledge and nourish the human spirit. Today, the range of their special interests reflects the scope of human vision. Their missions include collecting and preserving, as well as exhibiting and educating with materials not only owned but also borrowed and fabricated for these ends. Their numbers include both governmental and private museums of anthropology, art history and natural history, aquariums, arboreta, art centers, botanical gardens, children's museums, historic sites, nature centers, planetariums, science and technology centers, and zoos. The museum universe in the United States includes both collecting and noncollecting institutions. Although diverse in their missions, they have in common their nonprofit form of organization and a commitment of service to the public. Their collections and/or the objects they borrow or fabricate are the basis for research, exhibits, and programs that invite public participation.

Taken as a whole, museum collections and exhibition materials represent the world's natural and cultural common wealth. As stewards of that wealth, museums are compelled to advance an understanding of all natural forms and of the human experience. It is incumbent on museums to be resources for humankind and in all their activities to foster an informed appreciation of the rich and diverse world we have inherited. It is also incumbent upon them to preserve that inheritance for posterity.

Museums in the United States are grounded in the tradition of public service. They are organized as public trusts, holding their collections and information as a benefit for those they were established to serve. Members of their governing authority, employees, and volunteers are committed to the interests of these beneficiaries. The law provides the basic framework for museum operations. As nonprofit institutions, museums comply with applicable local, state, and federal laws and international conventions, as well as with the specific legal standards governing trust responsibilities. This Code of Ethics for Museums takes that compliance as given. But legal standards are a minimum. Museums and those responsible for them must do more than avoid legal liability, they must take affirmative steps to maintain their integrity so as to warrant public confidence. They must act not only legally but also ethically. This Code of Ethics for Museums, therefore, outlines ethical standards that frequently exceed legal minimums.

Loyalty to the mission of the museum and to the public it serves is the essence of museum work, whether volunteer or paid. Where conflicts of interest arise—actual, potential, or perceived—the duty of loyalty must never be compromised. No individual may use his or her position in a museum for personal gain or to benefit another at the expense of the museum, its mission, its reputation, and the society it serves.

For museums, public service is paramount. To affirm that ethic and to elaborate its application to their governance, collections, and programs, the American Association of Museums promulgates this Code of Ethics for Museums. In subscribing to this code, museums assume responsibility for the actions of members of their governing authority, employees, and volunteers in the performance of museum-related duties. Museums, thereby, affirm their chartered purpose, ensure the prudent application of their resources, enhance their effectiveness, and maintain public confidence. This collective endeavor strengthens museum work and the contributions of museums to society—present and future.

GOVERNANCE

Museum governance in its various forms is a public trust responsible for the institution's service to society. The governing authority protects and enhances the museum's collections and programs and its physical, human, and financial resources. It ensures that all these resources support the museum's mission, respond to the pluralism of society, and respect the diversity of the natural and cultural common wealth.

Thus, the governing authority ensures that:

- all those who work for or on behalf of a museum understand and support its mission and public trust responsibilities
- its members understand and fulfill their trusteeship and act corporately, not as individuals
- the museum's collections and programs and its physical, human, and financial resources are protected, maintained, and developed in support of the museum's mission
- it is responsive to and represents the interests of society
- it maintains the relationship with staff in which shared roles are recognized and separate responsibilities respected
- working relationships among trustees, employees, and volunteers are based on equity and mutual respect
- professional standards and practices inform and guide museum operations
- policies are articulated and prudent oversight is practiced
- governance promotes the public good rather than individual financial gain.

COLLECTIONS

The distinctive character of museum ethics derives from the ownership, care, and use of objects, specimens, and living collections representing the world's natural and cultural common wealth. This stewardship of collections entails the highest public trust and carries with it the presumption of rightful ownership, permanence, care, documentation, accessibility, and responsible disposal.

Thus, the museum ensures that:

- collections in its custody support its mission and public trust responsibilities
- collections in its custody are lawfully held, protected, secure, unencumbered, cared for, and preserved
- collections in its custody are accounted for and documented
- access to the collections and related information is permitted and regulated
- acquisition, disposal, and loan activities are conducted in a manner that respects the protection and preservation of natural and cultural resources and discourages illicit trade in such materials
- acquisition, disposal, and loan activities conform to its mission and public trust responsibilities
- disposal of collections through sale, trade, or research activities is solely for the advancement of the museum's mission. Proceeds from the sale of nonliving collections are to be used consistent with the established standards of the museum's discipline, but in no event shall they be used for anything other than acquisition or direct care of collections.
- the unique and special nature of human remains and funerary and sacred objects is recognized as the basis of all decisions concerning such collections
- collections-related activities promote the public good rather than individual financial gain
- competing claims of ownership that may be asserted in connection with objects in its custody should be handled openly, seriously, responsively and with respect for the dignity of all parties involved.

PROGRAMS

Museums serve society by advancing an understanding and appreciation of the natural and cultural common wealth through exhibitions, research, scholarship, publications, and educational activities. These programs further the museum's mission and are responsive to the concerns, interests, and needs of society.

Thus, the museum ensures that:

- programs support its mission and public trust responsibilities
- programs are founded on scholarship and marked by intellectual integrity

- programs are accessible and encourage participation of the widest possible audience consistent with its mission and resources
- programs respect pluralistic values, traditions, and concerns
- revenue-producing activities and activities that involve relationships with external entities are compatible with the museum's mission and support its public trust responsibilities
- programs promote the public good rather than individual financial gain.

PROMULGATION

This Code of Ethics for Museums was adopted by the Board of Directors of the American Association of Museums on November 12, 1993. The AAM Board of Directors recommends that each nonprofit museum member of the American Association of Museums adopt and promulgate its separate code of ethics, applying the Code of Ethics for Museums to its own institutional setting.

A Committee on Ethics, nominated by the president of the AAM and confirmed by the Board of Directors, will be charged with two responsibilities:

- establishing programs of information, education, and assistance to guide museums in developing their own codes of ethics
- reviewing the Code of Ethics for Museums and periodically recommending refinements and revisions to the Board of Directors.

AFTERWORD

In 1987 the Council of the American Association of Museums determined to revise the association's 1978 statement on ethics. The impetus for revision was recognition throughout the American museum community that the statement needed to be refined and strengthened in light of the expanded role of museums in society and a heightened awareness that the collection, preservation, and interpretation of natural and cultural heritages involve issues of significant concern to the American people.

Following a series of group discussions and commentary by members of the AAM Council, the Accreditation Commission, and museum leaders throughout the country, the president of AAM appointed an Ethics Task Force to prepare a code of ethics. In its work, the Ethics Task Force was committed to codifying the common understanding of ethics in the museum profession and to establishing a framework within which each institution could develop its own code. For guidance, the task force looked to the tradition of museum ethics and

drew inspiration from AAM's first code of ethics, published in 1925 as Code of Ethics for Museum Workers, which states in its preface:

Museums, in the broadest sense, are institutions which hold their possessions in trust for mankind and for the future welfare of the [human] race. Their value is in direct proportion to the service they render the emotional and intellectual life of the people. The life of a museum worker is essentially one of service.

This commitment to service derived from nineteenth-century notions of the advancement and dissemination of knowledge that informed the founding documents of America's museums. George Brown Goode, a noted zoologist and first head of the United States National Museum, declared in 1889:

The museums of the future in this democratic land should be adapted to the needs of the mechanic, the factory operator, the day laborer, the salesman, and the clerk, as much as to those of the professional man and the man of leisure.... In short, the public museum is, first of all, for the benefit of the public.

John Cotton Dana, an early twentieth-century museum leader and director of the Newark Museum, promoted the concept of museum work as public service in essays with titles such as "Increasing the Usefulness of Museums" and "A Museum of Service." Dana believed that museums did not exist solely to gather and preserve collections. For him, they were important centers of enlightenment.

By the 1940s, Theodore Low, a strong proponent of museum education, detected a new concentration in the museum profession on scholarship and methodology. These concerns are reflected in *Museum Ethics*, published by AAM in 1978, which elaborated on relationships among staff, management, and governing authority.

During the 1980s, Americans grew increasingly sensitive to the nation's cultural pluralism, concerned about the global environment, and vigilant regarding the public institutions. Rapid technological change, new public policies relating to nonprofit corporations, a troubled educational system, shifting patterns of private and public wealth, and increased financial pressures all called for a sharper delineation of museums' ethical responsibilities. In 1984 AAM's Commission on Museums for a New Century placed renewed emphasis on public service and education, and in 1986 the code of ethics adopted by the International Council of Museums (ICOM) put service to society at the center of museum responsibilities. ICOM defines museums as institutions "in the service of society and of its development" and holds that "employment by a museum, whether publicly or privately supported, is a public trust involving great responsibility."

Building upon this history, the Ethics Task Force produced several drafts of a Code of Ethics for Museums. These drafts were shared with the AAM Executive Committee and Board of Directors, and twice referred to the field for comment. Hundreds of individuals and representatives of professional organizations and museums of all types and sizes submitted thoughtful critiques. These critiques were instrumental in shaping the document submitted to the AAM Board of Directors, which adopted the code on May 18, 1991. However, despite the review process, when the adopted code was circulated, it soon became clear that the diversity of the museum field prevented immediate consensus on every point.

Therefore, at its November 1991 meeting, the AAM Board of Directors voted to postpone implementation of the Code of Ethics for at least one year. At the same meeting an Ethics Commission nominated by the AAM president was confirmed. The newly appointed commission—in addition to its other charges of establishing educational programs to guide museums in developing their own code of ethics and establishing procedures for addressing alleged violations of the code—was asked to review the code and recommend to the Board changes in either the code or its implementation.

The new Ethics Commission spent its first year reviewing the code and the hundreds of communications it had generated, and initiating additional dialogue. AAM institutional members were invited to comment further on the issues that were most divisive—the mode of implementation and the restrictions placed on funds from deaccessioned objects. Ethics Commission members also met in person with their colleagues at the annual and regional meetings, and an ad hoc meeting of museum directors was convened by the board president to examine the code's language regarding deaccessioning.

This process of review produced two alternatives for the board to consider at its May meeting: (1) to accept a new code developed by the Ethics Commission, or (2) to rewrite the sections of the 1991 code relating to use of funds from deaccessioning and mode of implementation. Following a very lively and involved discussion, the motion to reinstate the 1991 code with modified language was passed and a small committee met separately to make the necessary changes.

In addition, it was voted that the Ethics Commission be renamed the Committee on Ethics with responsibilities for establishing information and educational programs and reviewing the Code of Ethics for Museums and making periodic recommendations for revisions to the board. These final changes were approved by the board in November 1993 and are incorporated into this document, which is the AAM Code of Ethics for Museums.

Each nonprofit museum member of the American Association of Museums should subscribe to the AAM Code of Ethics for Museums. Subsequently, these museums should set about framing their own institutional codes of ethics, which should be in conformance with the AAM code and should expand on it through the elaboration of specific practices. This recommendation is made to these member institutions in the belief that engaging the governing authority, staff, and volunteers in applying the AAM code to institutional settings will stimulate the development and maintenance of sound policies and procedures necessary to understanding and ensuring ethical behavior by institutions and by all who work for them or on their behalf.

With these steps, the American museum community expands its continuing effort to advance museum work through self-regulation. The Code of Ethics for Museums serves the interests of museums, their constituencies, and society. The primary goal of AAM is to encourage institutions to regulate the ethical behavior of members of their governing authority, employees, and volunteers. Formal adoption of an institutional code promotes higher and more consistent ethical standards. To this end, the Committee on Ethics will develop workshops, model codes, and publications. These and other forms of technical assistance will stimulate a dialogue about ethics throughout the museum community and provide guidance to museums in developing their institutional codes.

Part X

TRANSITIONS

CHAPTER 38

Building a Sustainable Future

Thomas Krakauer

THE SCIENCE AND CHILDREN'S MUSEUM FIELD continues to grow meteorically; new projects arise daily, with facilities opening continuously. I know how easy it is for a board, community, and CEO to get carried away by an exciting vision that grows daily on the drawing board. I recently retired as president of the North Carolina Museum of Life and Science in Durham, N.C. My thirty-year career was spent planning, designing, building, and expanding such facilities and ultimately living with the consequences of the drawing board plans. As a student of the field, I would like to share some observations on the practices that lead to successful growth, as well as the pitfalls that can interfere with an institution becoming sustainable. In 1997, I convened an ad hoc Association of Science-Technology Centers (ASTC) Committee on Museum Expansion and Sustainable Growth,[1] so this chapter also reflects and expands on the lessons learned in that brief study.

The rules to follow are pretty obvious. Can the funds to build the facility be raised? Can the operating budget be generated? What is the minimum it will take to maintain the facility as a trust for the community?

The most significant variable is based upon an estimate—how many people will come through the door after opening? All too frequently, the visitation estimates drift upward and assumptions made at the beginning of the project are not reexamined in spite of changes. What are some of the critical variables?

USING MODELS FOR FEASIBILITY STUDIES

There is some question about the best models with which to estimate attendance. More often with large projects than small, many institutions hire a marketing firm to conduct a feasibility study that estimates attendance using "shopping mall capture numbers" that estimate visitation on the basis of some expected percentage of population.

In my opinion, facility size is the most significant variable, based on a study conducted in the 1980s by the National Science Center in Georgia. They measured facility (indoor space), economic, and demographic characteristics at forty-six science centers in the eastern United States.[2] These data were analyzed to identify the importance of different variables; the variable with the smallest impact was population. The greatest was total exhibit square footage. Subsequent testing of the data supported the finding.[3] Considering both size and population, it is most important to compare attendance estimates with those of institutions of similar size rather than those from similar-size communities, though the truth is probably that the best estimate takes into account both population and facility size.

ESTIMATING COSTS PER VISITOR

Next, turn your attention to the expense side of the ledger. Divide your proposed budget by the number of visitors that you project.

A recent publication by the American Association of Museums (AAM) presents the following data on the costs per visitor for 2002:[4]

Discipline (all data are median):	
Children's/Youth Museum	$8.33/visitor
Nature Center	$11.88/visitor
Science/Technology Center/Museum	$16.27/visitor

When data were combined, the very smallest institutions (budgets under $180,000), spent significantly less per visitor. However, since costs per visitor vary significantly between disciplines, one cannot assume the same size to costs-per-visitor relationships apply. The sample size of the AAM study did not allow costs per visitor by size to be calculated.

If your estimated costs (exclusive of collections and research) are significantly above those medians, you might have difficulty meeting expenses. If your costs-per-visitor estimates are low, it is also problematic; you may be overestimating visitation or underestimating expenses. Both will have severe implications.

Also, while earned income reflects more than ticket price, the median cost per visitor usually far exceeds anticipated admission charges and other earned revenue. You will have to develop a plan for unearned income, for example, donations and governmental support.

GROWING BEYOND YOUR FINANCIAL ASSUMPTIONS

There are numerous examples where a project team develops an exciting concept, the project is sold to community leaders, and excitement grows. The project then takes off. After all, the leaders who can make a project happen seldom dream small dreams and few are willing to caution those making leadership gifts. Yet, without that word of caution, a project can become unrealistic, growing beyond its financial assumptions. It is vitally important to have a full and open exchange with your board; otherwise a project will be developed for which revenues don't meet expenditures.

PROJECT DELAYS

Time is money. Project delays result in obvious inflationary costs. Equally dangerous are the implications of budgeting staff expenditures out of one-time money. If there are delays, staff is still needed and the center hasn't opened when expected, so earned income does not increase to absorb staff costs. Without midcourse corrections, funds intended for bricks and mortar must cover staff expenditures instead.

Of equal concern, delay-caused overruns often occur when the project is under construction and midcourse corrections are impossible to make; the capital campaign is completed, the center has received contributions from its major donors, but it does not have the capacity to raise additional money. A center can cut size or quality, but cutting size influences visitation, and since the community expected a certain look and feel, no one wants to disappoint donors or visitors. This is another time when communication between staff and board is vital.

PROJECT COSTS

Construction budgets often are not held in check. Finger pointing is the name of the game when construction overruns are examined. Whatever the cause(s), it is vitally important to make midcourse corrections to reduce costs, or find additional money.

OVERESTIMATING REVENUES AND UNDERESTIMATING EXPENSES

This is almost a given in every project: fewer people will attend, and operating costs will be higher. These may be stand-alone budget errors, the result of incorrect planning assumptions, midcourse changes in scope or not anticipating the attendance decrease that often occurs the year after opening.

ANTICIPATING SECOND-YEAR SLUMP

It is usual for attendance to decrease the year after opening. Very high attendance during the opening frenzy may contribute to second-year slump, but there are other issues. Financial trauma in year one means less money for year two. To balance the budget in year two, you may have to cut staff and reduce marketing expenditures.

OPENING ON BORROWED FUNDS

There have been a number of high-profile institutions that completed their construction by borrowing money.

Don't.

Few institutions are able to generate cash from their operations to pay back debt. The situation is exacerbated because overruns are frequently large relative to the institution's operating budget, easily as high as 25 percent. Such overruns can be staggering. For example, if your construction budget was $80 million, but you spent $100 million, you are faced with debt of $20 million. Five percent interest on that $20 million is $1 million. The operating budget can not support an additional $1 million expense let alone begin to pay the principal.

Scale this example down to a smaller project. The concept is the same since the capital budget and its overruns must be supported from a smaller operating budget. Most in the museum profession recoil at the concept of large-scale debt. A few, whose institutions have lived through it say, "The museum did not close. Somehow the debt was paid, and the community has a larger, better institution." The jury is out on this issue, and more study is needed. But, be very careful.

LIFE AFTER OPENING

A science center has managed to open and avoid the pitfalls; how does it craft a sustainable future?

Sources of Revenue

How will the center generate revenue? Keep in mind that limitations on earned and unearned income are different. As discussed, the attendance model showed that visitation (that is, earned income) is primarily a function of facility size not population. Unearned income, for example, governmental support and contributions, is influenced by population. The higher the contribution of earned income, the greater the institution's independence. So how does the size and nature of the community play a role? What limits the size of a science center in the real world?

Super Science Center (SSC) has a total operating budget of $1 million and is fairly self-reliant; $750,000 is earned from admissions, memberships, classes, and other activities, and $250,000 from government support and charitable contributions. If SSC expands and its budget doubles, then earned income goes to $1.5 million and the community needs to provide $500,000. In contrast, if SSC raises only 50 percent of its budget through earned income, then when its budget increases from $1 million to $2 million, it will have to earn $1 million from government support and contributions.

Is that realistic? The vision for the expanded facility often is sold to political leaders because of their expectation that there will be less dependence on tax dollars. In fact, that seldom occurs, and a sustainable institution depends on its size and upon the political will and private resources of the community. A politically conservative community may not be willing to invest tax dollars in a museum; however, the private and corporate contributions of that same community may be extremely generous. Questions to keep in mind: What is the center's capacity? Will growth result in new sources and amounts of unearned income? Be realistic in the amount of unearned income available and who is willing to give; that income is needed in good times and bad.

Also be aware that the budget problems an institution faces because it failed to heed growth cautions can negatively impact its unearned income. Once the media learns of budget problems, there will be a media feeding frenzy, often resulting in changed leadership, which may make the community cautious. However, there are other examples where a community rallied to put an institution back on track under new leadership.

Cost Increases Don't Take Vacations

Let's assume a science center with a $1-million budget operating in an environment with a modest 3 percent cost of living increase. The center needs to find a new $30,000 in revenue just to maintain itself. Since personnel costs are apt to be about 60 percent of that budget, they have a disproportionate impact, particularly at a time when health care costs are skyrocketing. A center cannot cut its budget each year, so it must take a strategic approach, keeping in mind that as its budget grows, the absolute value of the new money needed every year also increases.

Keeping Things New and Fresh

The day a new science center opens, everything sparkles and all the big ticket items have fresh warranties. Fast-forward a decade and the center needs to have a source of funding to replace big ticket items, for example, a roof or HVAC compressors. Meanwhile, the visitors' feet wear out the carpet and their hands smudge the paint. The exhibits, built with grants or funds from a capital campaign, are beginning to wear out. What is a realistic, or should I say optimistic, replacement schedule? Ideally, interactive exhibits need to be replaced every seven years, specimen-based exhibits every twelve to fifteen years.

In addition, since many guests visit science centers for social or entertainment value, upkeep and cleanliness are critical to maintaining the perceived value. It is that perceived value that preserves the positive word-of-mouth reputation and supports strong attendance. It is also possible to keep an institution fresh by creating strong and changing programs.

CONCLUSIONS

So, how do you make this all work? It is the people stupid! It takes the right people "on the bus,"[5] that is, your board, executive team and staff, and volunteers, before you decide where you are going to drive your institution. I have been fortunate to work with extraordinary individuals in my career. Their strengths have permitted me the time to step back from the day-to-day work to think through and develop some fundamental rules, allowing me to grow the small institution that I inherited into a strong, large museum. I'll pick a few examples of where the application of the rules, at the right time, proved to be critical.

1. The request to create a downtown facility led to questions about how that would impact attendance and operations, which led to the development of the attendance model. Our analytical approach served the museum well.
2. The Museum of Life and Science operates on a 70-acre campus with eleven buildings and extensive outdoor facilities. When the realities of drifting schedules and insufficient construction budgets became apparent during project planning, we

were able to phase our growth. By dividing a big project into smaller ones we avoided the majority of the second-year slump issues. This was possible because the museum was 80 percent self-funded so the growth generated sufficient new earned and unearned revenue to maintain financial viability.

3. Keeping it fresh has been a challenge. We have been successful at being included in voter-approved general obligation bond issues, and have balanced those funds with capital campaigns and federal grants for exhibits. We have been able to grow unearned income somewhat, but it took more than a decade to persuade local government that their investment in facilities also implied annual capital expenditures. They now see the costs to the community and to the museum of deferred maintenance, and they are willing to step up to the plate.

4. Less dramatic, but equally important, the museum now schedules a museum work day in September, not a busy time for us. We close to the public and all staff members work on projects that cannot be done easily within a seven-day-a-week operation and with very small facility staff. Storage spaces get emptied and public spaces get cleaned, rehabilitated, and painted. It improves the appearance for our visitors and helps prepare for facility changes. It is amazing what sixty staff members, and about the same number of volunteers, can achieve—imagine about five industrial-size dumpsters being filled in a single day.

5. The museum has been strong in generating visitor revenue and competing successfully for federal grants. We have not done as good a job at building patron support. We needed to "get the right people on the bus" and give them enough time to change that critical balance of earned and unearned income. Having articulated the issue made it an institutional priority; however, it took a staff reorganization and clear priorities for board selection to begin to taste success.

SOME KEY POINTS

When planning a new facility or an expansion, the best sources of comparison are similar-size facilities.

Figure 38.1. North Carolina Museum of Life and Science, Durham, N.C.

Conduct a thorough set of financial predictions early in the project and make sure to update them with every change. Be very conservative with predictions and avoid a heavy debt service.

Know your community provides the best information to understand the ability to generate unearned income.

Good luck and enjoy what you provide to the community!

NOTES

1. VanDorn, Bonnie. 1996. Growing pains. ASTC Newsletter 1, 13.
2. Krakauer, Thomas. 1990. Science museum attendance: A desktop computer model. Washington, D.C.: Association of Science-Technology Centers. (Out-of-print diskette and manual)
3. This model is available from the author via e-mail as an Excel spreadsheet, tkrakauer@mindspring.com.
4. Merritt, Elizabeth E., ed. 2003. *Museum financial information*, Book 142. Washington, D.C.: American Association of Museums.
5. Collins, Jim. 2001. *Good to great*. New York: HarperBusiness.

CHAPTER 39

Renovation as Innovation: SciWorks, the Science Center and Environmental Park of Forsyth County

BEVERLY S. SANFORD

THE CREATION AND DEVELOPMENT OF SCIENCE museums across the country have come in all shapes and sizes. Many have brand new facilities designed specifically for the purpose of being a science museum. Others have found themselves in sites that were originally built for different purposes and have been renovated into science museums and nature centers. I know of some in old schools and one located in an old car dealership. SciWorks is located in an old hospital and county home. As I share our story, I think that you will see how a successful renovation can create an innovative science center with a unique personality and place in its community.

HISTORY

SciWorks, the Science Center and Environmental Park of Forsyth County, N. C., formerly the Nature Science Center, is an institution that was founded in 1964, has been located at two different sites, and has never built its own "house." The first site was the old barn and silo located in Reynolda Village, part of the estate of R. J. Reynolds. The building had only 4,000 square feet of public space, a classroom, and a couple of closets. The unheated silo served as the first planetarium. Unlike many of the other nature centers built in the 1950s and 1960s, the Nature Science Center started in a renovated facility. In the late 1960s, it became apparent that the Center was rapidly outgrowing its facilities. Reynolda Village was donated to Wake Forest University, which had other plans for the property, and an expansion of the Nature Science Center was not one of them.

Thus, the search was on for a new site. According to news stories from the early 1970s, the board of directors looked at over twenty-five sites. After much discussion with officials of Forsyth County, the board of directors negotiated a lease for the 27-acre Forsyth County Home and Nursing Center. The county home

closed in 1966 and was being used as a storage facility for county operations. The property contained several buildings that dramatically increased the amount of square footage available for the museum's collections, enhanced exhibits, and programs or classes. The acreage provided the outdoor space the museum wanted to increase its natural science programs, nature trails, and live animal exhibits. The property was leased for a nominal fee and the museum assumed responsibility for the upkeep and improvements to the property. In the thirty years since the Nature Science Center moved, it has totally redefined the property and the physical plant in ways that meet the programmatic needs of a contemporary science center.

The renovation of SciWorks has taken thirty years. In 1973, money was raised for the move from Reynolda Village by preparing a couple of the hospital wings for the museum. Five years later, in 1978, a capital campaign raised money to construct a 5,000-square-foot exhibition gallery with a loading dock between two of the hospital wings. For many years, this gallery housed the science exhibits and a small planetarium. Live animal displays were kept in outdoor areas and in other rooms within the buildings. One really positive aspect of the location is that with all of these empty hospital wings, the museum had plenty of storage!

In 1989, the museum launched a capital campaign that raised $4 million in bond money and $2 million from the private sector and foundations. The outcome of the campaign resulted in new gallery space, a new entrance, a new planetarium, a renovated interior, and a redefined nature park and outdoor space. In 1992, SciWorks—the new Nature Science Center—was born. The museum made a huge leap from its homegrown atmosphere to one of a contemporary science center. The physical changes to the grounds and facilities were outstanding. However, we ran out of money before we developed many exhibits.

Figure 39.1.   SciWorks Exterior.

Figure 39.2.   Children's Camp at SciWorks.

The year 2000 brought another capital campaign, SciWorks 2001. Three million dollars—$1 million in two federal grants and $2 million from the community—was raised. The project added a multipurpose eating area, two outdoor patios, a 1,000-square-foot collections gallery, two renovated classrooms, and 10,000 square feet of renovated interior space. Most important, the museum added exhibits. Three years later, in the summer of 2003, all of the exhibit areas of SciWorks were full. The journey from 1964 to 2004 has been a long one with a successful conclusion. SciWorks has adequate public spaces, filled exhibit spaces, and classrooms that will allow the museum to serve the community for many years to come, and we haven't even started on the outdoor area!

THE LESSONS

SciWorks is, in many ways, an emerging science center even with its forty-year history. It has had six executive directors and three interim directors. With each change in leadership, there has been a shift in the vision for the center. However, the expectations of the community have not changed significantly from wanting to create a nature center with outdoor nature trails and a museum where adults and children alike can learn astronomy, botany, zoology, geology, and the physical sciences.

Therefore, *lesson one* is to know the community, what it expects from a science center, and what it is willing to support. I think that this is particularly important for small science centers. Most small centers are in small or medium-size towns and depend on some degree of public support. The financial viability of the small center is often put at risk when public support decreases. In my opinion, the 1992 expansion and renovation of SciWorks were almost too much. Going from 10,000 square feet of exhibit space to 25,000 square feet

significantly increased the operating expenses of the facility and the community was not prepared to support this increase in costs.

Lesson two is to create, plan, and document the long-range plan and vision for the center for your successor. We all know executive directors, and a new one will always want to put his or her mark on a project. However, we all need a place to start and a good plan can provide direction. We all need to remember when going into established institutions that someone before us had a vision that the community supported. In the case of SciWorks, the 1992 project was the defining plan. The educational concepts had been tested in the community—the museum just ran out of money before the full plan was implemented. The 2000 project was built on the previous one. In fact, not all of the projects were funded. We still have a long-range vision for the facility and grounds.

Lessons one and two apply to most situations. *Lesson three* (more applicable to renovation projects) is to learn all you can about your site and its history. Learning about the history of the site can be a lot of fun. As stated earlier, SciWorks is located on the site of the Forsyth County Home and Nursing Center. In fact, the folks who stayed here in the 1930s, 1940s, and 1950s helped to operate the Forsyth Dairy. There was also a prison unit here. That building has been renovated into the exhibits fabrication shop and collections storage. In 1998, the final five structures of the old home were demolished. Two of the buildings were houses, used by the superintendent and assistant superintendent, and an eight-car garage. Also removed were the final two hospital wings—one that was used for patients with tuberculosis, and the mental ward, complete with barred doors. For many years, the Jaycees (Junior Chamber of Commerce) and the museum staff turned the mental ward into a haunted house each Halloween. The building was a little scary under normal circumstances but was really scary during Halloween! Finally, many former staff members of the late 1970s and 1980s are convinced that the site is haunted. Up to this point, I have not had any supernatural encounters, but the staff tells some very convincing tales.

Lesson four is to be realistic about the use of an old facility. I think that adapting an existing space to a science center is truly an imaginative, innovative process. First of all, the characteristics that you might like in a building may not be there, for example, high ceilings. In the former hospital wings at SciWorks, the ceilings are 10 feet high. There are many windows creating challenges with natural light. On the other hand, these old buildings have a lot of character, which can be considered and incorporated into the architectural and the exhibit design. Many times the old buildings are community icons. Adapting the building to another use often builds community support. Bringing an old facility up to contemporary codes is a challenge and often an expensive one.

My final lesson learned is one that I keep learning over and over again. *Lesson five* is to keep your sense of humor. I have long believed in the premise that if it can go wrong, it will. As with any construction project, renovating a facility is a task that requires the involvement of many people and oftentimes you are the last to know of the issues and challenges of the project. At times you will feel out of control and can't show it. Old buildings and sites have lots of surprises—old storage containers, old plumbing, underground wires. Just knowing that you should expect the unexpected can help the renovation move along. In our last project, we needed to knock through what we thought was a Sheetrock wall to create a new entrance to an exhibit hall. This project needed to be completed so that we could close down part of the building and still have a traffic pattern through the facility. The Sheetrock wall turned out to be reinforced concrete, and the five-day project turned into a twenty-day project. No one was more surprised than the construction worker who swung the sledgehammer at the wall.

RENOVATION/INNOVATION

When does renovation become innovation? I think that is when the project is completed. People touring SciWorks are really interested in the SciWorks story and in seeing the old parts of the facilities. Those who visited many years before are amazed at the changes and are quick to share their early experiences.

The innovation is in the comprehensive project plan that provides a long-term look at the redevelopment, coupled with creative use of space and a design plan that complements the existing facility.

CHAPTER 40

Collaborations: From Sharing a Museum Site to Winning NSF Grants

SARAH WOLF

THERE HAVE BEEN MANY ARTICLES, EVEN BOOKS, written about collaborations, and I think we all understand the benefits: shared resources; different perspectives; synergy of shared ideas and skills and donor responsiveness; and shared risks and rewards. Our museum was founded by a collaborative effort, and the major benchmarks in our history have all occurred through collaborations. Some that I will describe have been sought by our museum and some have been imposed. In seeking partners for collaboration, we generally have had shared goals and desired outcomes. We have looked for partners with similar ideas who could bring new skills and perspectives to the partnership. When Discovery Center Museum had the collaboration thrust upon us, the shared goals were not the motivator and our experience was not always a positive one.

Discovery Center Museum is a nonprofit, hands-on science museum located in Rockford, Ill. The mission of the museum is to operate a participatory, family-oriented museum designed to promote learning by doing: a place to explore, to experiment, and to experience the arts and sciences.

In 1980, the Rockford Arts Council and the Junior League formed a collaboration to research and create a hands-on family-oriented museum. It was housed in a former post office building, and the rent was subsidized by a unit of local government. In the first four years of operation the museum was run entirely by Junior League volunteers. The operating budget was provided by the two founding organizations, and a small amount of revenue was generated from earned income.

The exhibits were built by volunteers following many of the ideas in the Exploratorium Cookbooks. A planetarium was donated by the public library. In the early years, the programs and traveling exhibits were designed to attract repeat visitors, both school groups and the general public. As awareness of the center grew and the quality of programs and exhibits improved,

our attendance, member base, and awareness in our community also increased. After the first four years, sufficient income became available to hire staff, and the museum incorporated and formed a board of directors. After serving on the founding committee, I was hired to be the first paid executive director.

Two years later, the building that Discovery Center Museum had called home was put up for sale, and a search for a more permanent home began. A MAP 1 (a museum assessment grant-funded program through the American Association of Museums) study conducted in 1987 helped the museum determine the critical elements for sustainability: ongoing operating support funds, a permanent location, and community and political support. Our board was committed to a downtown location that was accessible to a diverse audience. Much of the growth in our community was moving east toward Chicago, and the central city was being abandoned by some businesses. Our goal was to keep the downtown vital as a cultural area. We wanted to keep our admission price affordable and offer free parking adjacent to our museum. Another goal was to have all new, professionally designed and fabricated exhibits and outdoor space for a science park.

A state grant from the Illinois Arts Council helped support the hiring of a team of consultants to research potential sources for ongoing support funds and the location for a new home. This study involved many local community leaders, and an eighteen-month search began for the best location. The study resulted in the commitment from our area state legislators to work together to find funding for capital dollars for a new Discovery Center Museum facility. This $300,000 grant was the first significant funding our museum had ever received.

At the same time that our search was beginning, Sears, Roebuck and Company donated a 120,000-square-foot former Sears store in the downtown

Rockford area to the Rockford Art Museum. The gift of this building, located on 6 acres of land on the Rock River, supplied the museum with a large permanent facility, which they later learned was larger than its needs or resources. Initial meetings with the art museum to explore the possibility of a shared home were not productive. Although the art museum had a building and the Discovery Center Museum had a commitment of state funds, the timing was just not right for the art museum to consider a partner in their new facility, and our museum continued the search. The art museum realized that being the sole occupant of the large facility was not financially viable and this led to further discussions a year later.

The art museum and Discovery Center Museum entered into a partnership that eventually included four additional cultural organizations that had also been searching for permanent homes. The Rockford Symphony Orchestra, Rockford Dance Company, Storefront Cinema, and Northern Public Radio joined with the two museums to refurbish the building and named it Riverfront Museum Park.

The State of Illinois allows park districts that operate museums to levy tax to help support the operations of the museums. Our park district already supported three other museums. Our newly formed group approached the Rockford Park District and asked to become part of the park district museum tax levy funding. The park district imposed three requirements that needed to be met before they would accept ownership of the building, which would provide the six organizations free space and building maintenance. We had to get the state to increase the amount that the park district could levy, we had to raise sufficient funds to renovate the facility, and we had to establish a method of governing the building. We had a relatively short amount of time to reach these goals, which caused later problems. Because of the positive relationships Discovery Center Museum had established with the legislators through our building search, they were already committed to helping with the project. They worked very hard to get the state law changed.

Because there were several organizations working together to improve the quality of life in our community, it was easier to launch a successful capital campaign. A capital campaign committee chaired by two community leaders was formed and $6 million was raised to renovate the building to meet the needs of the six organizations. After the completion of a fourteen-month $6.2-million campaign and after the state approved the change in the law to increase the levy, the Rockford Park District voted to accept the project and to accept title to the building and grounds upon completion of the renovation. Riverfront Museum Park would become the fourth museum receiving an annual grant from museum tax levy funds for operating expenses of the building.

A team of architects was hired and the renovation began. The two museums (Rockford Art Museum and Discovery Center Museum) occupy the largest part of the building, the symphony has offices and storage space, the dance company has studios and office space, the public radio station has a satellite studio and the cinema has a ninety-two-seat theater. The cinema went out of business and Discovery Center Museum and Rockford Art Museum now share the theater space. The art museum's total square footage is 40 percent of the building, the Discovery Center occupies approximately 36 percent of the total, dance company has 14 percent, symphony 2 percent, and public radio 5 percent. The remaining space is shared space and includes a 5,000-square-foot space for programs or traveling exhibits and several classrooms. These common spaces are scheduled through the building manager.

The project was completed and opened to the public in 1991. Discovery Center Museum did achieve our goal to raise sufficient funds to design and purchase all new exhibits. Our new exhibition space totaled 18,000 square feet, and space for the outdoor park added an additional 8,000 square feet, with additional space for classrooms, gift shop, party room, and a large storage and workshop space. Discovery Center Museum had also planned a major outdoor science park scheduled to open four months after the building opening. This partnership was with the Junior League and the Rockford Park District. The Junior League raised funds (more than for any other project they had ever undertaken) and provided a steering committee. The Rockford Park District provided manpower and expertise. Robert Leathers and Assoc. was selected to design the park and oversee the construction and thus began the first community-built science park in the nation. This project brought over 4,000 area residents together in a community effort unmatched in our region. The park was built in just over three weeks, with people working from dawn to dusk. We still have visitors who enjoy pointing out the particular part of the park they created—part of the cave, a special gate, a ball tower—they have fond memories and want to share them with their family. The park continues to be one of our most unique and popular attractions.

Figure 40.1. Dino Dig, Discovery Center Museum, Rockford, Ill.

The housing of six organizations under one roof was, and is, still an innovative concept. There were a few others in the country at the time our project began. We contacted several of them seeking information on fund-raising, governance, and space allocation. Center in the Square in Roanoke, Va., and a complex in Midland, Mich., were two that provided us with some information. Just as we researched complexes like ours, cities and museums from other parts of the country have visited our complex, gathering ideas and information in the twelve years that we have operated. The community embraced the concept because it was viewed as a progressive, positive sharing of resources and an efficient, effective way to enrich our citizens. The hope was that collaborative programming would emerge and the spirit of cooperation would prevail.

The tax levy funds of $500,000 a year pay salaries of the building manager and assistant, the custodians, the security guards, and building maintenance. There is an additional capital improvement budget from the levy, which helps support major improvements such as the new HVAC system installed two years ago.

The organizations each have their own board of directors and function autonomously. Each organization has full responsibility for raising its own operating dollars. Our operating budget has grown from $225,000 in 1991 to $1.4 million in 2003. Each board has two members who serve on a governing board for the building. There are also community representatives on this board. I am the only director of one of the six organizations who has remained the same during the twelve years that we have coexisted.

As new directors have taken charge of their organizations, the philosophies of the organizations and their priorities have shifted or changed. This has impacted our ability to work together for joint programming and event coordination.

We naively thought that once the hard work of raising the funds was behind us, we would coexist in our new facility and be able to function in much the same way we had in our previous location. The reality was we had become part of a layered bureaucracy, a difficult transition from the independent, entrepreneurial structure of our earlier days. We anticipated that we would be sacrificing some of the independence but this seemed to be a good trade for the advantages of sharing a facility that met most of our other objectives. It was in the downtown, had an operating subsidy, free parking, and space for an outdoor science park.

Although all of the organizations expressed the desire to remain autonomous, there were many new restrictions placed on our marketing and branding. We were told that all of our press releases, printed materials, and public service announcements had to include the name of our building and the park district and that all printed materials had to be approved before we could print them. This not only took considerable time but also imposed restrictions on our ability to react quickly to opportunities for promotion. It became clear that the building management and governing board for the building were more interested in marketing the building rather than the organizations within it. Their energies seemed misdirected. When we wanted to hang large banners on the outside of the building to promote a major blockbuster exhibit, our request was voted down; when we greeted our school groups we were told that we could not say, "Welcome to Discovery Center," but rather "Welcome to Riverfront Museum Park." These are just a few examples of the struggles to retain our identity within a multiuse facility.

Discovery Center Museum needed weekend custodial services, but the other organizations mainly functioned on weekdays, and it took two years to convince the management of our need for cleaning on weekends.

The difficulties with our organizational structure might have been avoided or lessened if we had held firm to our convictions during the negotiation period before we moved into the facility. As with many projects, our highest priority was the capital campaign, raising the funds for the renovation and new exhibits. The governance and administration took a back seat. When six groups try to plan together everyone works hard to be conciliatory, and verbal commitments are made between board members and staff and the chairpersons for the campaign are striving for a positive community image. The emphasis on unity was important for the success of the project. The momentum was very important and our attempts to dig into the realities of how this building and the organizations would function once the project was complete were dismissed, or submerged for the good of the whole. As we learned, the devil is in the details. Had we worked through the specifics before any of us moved in, and developed a structure that had safeguards built into it, the last twelve years could have been more productive and the struggles, at least in part, avoided.

In retrospect, all of this should have been negotiated prior to moving in, the verbal assurances should have been put in writing, and more time should have been spent in resolving the concerns that we voiced. Written bylaws or governing rules needed to be adopted early in the conception of the project. Clear job descriptions for the building personnel should have been established. But we were inexperienced and had few models to follow.

In thinking about the keys to successful collaborations I would advocate the following:

• A mutually agreed upon mission
• Defined operating procedures, roles, and responsibilities
• Active involvement in defining goals and objectives
• A communication system both formal and informal
• A sense of common ground and ownership
• Skilled leadership

After struggling with the governance structure for the past twelve years, the leaders of the organizations are now attempting to restructure and make alterations, which will, we hope, improve the governance, and thus, the productivity of the organizations and the benefits they can provide to the community. Just recently, one of the organizations, the cinema, has closed their operation because of lack of funds to pay the yearly assessments. These assessments are charged on the basis of square footage and in the past few years have become part of the budget because the park district levy funds no longer are sufficient to cover the operating costs of the building.

All this being said, are we better off here in this location than we would have been on our own? We were able to meet most of our goals for sustainability set forth when we began the search for a new home. This facility has provided us with ongoing overhead operating support, free parking, expanded exhibit space, and better visibility. Our attendance has grown, to serve over 100,000 visitors every year. Our exhibits and programs are high quality, and we have had many successful partnerships within our community, allowing us to serve the diverse audiences we hoped to reach. The challenges of coexisting, however, have taken a toll on our staff and board. Some staff members left because of the ongoing negative interactions with the building manager. Some board members refused to serve on the governing board because the meetings were so confrontational. We do think about the negative energy that has been expended and what additional wonderful things might have been accomplished if we were not bogged down within the structure. Fortunately, our community is not aware of the inner turmoil and problems; our staff has been resilient and our board has worked diligently to improve the structure. Last year, *Child* magazine, ranked Discovery Center Museum the fourth best children's museum in the nation; the outdoor science park we built has received several awards and national recognition. There have been some terrific joint programs with the Rockford Art Museum and other organizations working together. So, yes, we are better off and so is our community. The project has been a success, and yes, our community embraces our museum and yes, we have learned a lot.

As a director, I have learned what to look for in other collaborations and partnerships. Our museum has been fortunate to extend partnerships beyond our community. One of our first national endeavors has been with the very successful TEAMS collaboration. This partnership began with five museum directors who knew each other from the Association of Science-Technology Centers (ASTC) meetings and other affiliations. The museums were of similar size, both in budget and square footage, and the directors saw a common need for quality small traveling exhibits that were affordable for museums of our size. We successfully submitted a grant to the National Science Foundation (NSF) and received funding to create five exhibits that would travel to each other's museums and then travel through the ASTC traveling exhibition service. This collaboration had successes at many levels. It provided opportunities for our staffs to grow professionally and learn evaluation techniques from the experts who were hired through the grant, established communication networks for exhibit and education staff that have continued to thrive, and provided our museums with a variety of quality exhibits. The partnership was so successful that it continued and more museums were added in the second round; the original museums serving as mentors for the new museums added. In the second round, in addition to creating new exhibits, we also learned more about creating quality exhibits for people with a variety of disabilities. The third round of TEAMS has submitted a proposal to NSF and we have high hopes of it being funded. New research on sociocultural impacts on exhibits is the learning theme for this grant. All of this knowledge on creating quality exhibits for a diverse audience has been extremely valuable to our staff and ultimately to the general public and could not have occurred if we were functioning in isolation. The strength and commitment of all of the museums working together has made it possible to reach new levels of expertise.

Discovery Center Museum has also participated in another collaboration with a more centralized geographic location. Again, we were fortunate to work with other science museums in the midwest that had a common goal of creating exhibits that could travel to schools in our general areas, focusing on the wild weather of the midwest. This collaboration was funded through NSF and resulted in each of our nine museums researching, prototyping, and fabricating exhibits on the weather topic that we then take to schools. We jointly developed programming and training for our staff to serve as facilitators for the exhibits when they are in schools. Again, there was a professional evaluator for the entire project, and the staff received excellent training and developed new skills. Networking with the other museums and learning from one another is a valuable benefit to staff in a museum of our size. We have only nine full-time staff, but through these collaborations, we have many staff members from our partners whom we trust to work with, who give suggestions when we are struggling with an issue, or who just help in problem solving.

Collaborations in our community have gained much support from city leaders, corporations, and foundations. We have found that several organizations working together to provide more opportunities for our citizens is perceived as the type of projects that people are most likely to support, both through funding and participation.

By encouraging your staff and board to seek the benefits of collaboration, you are allowing new ideas, new methods of accomplishing goals, and new opportunities for growth for your museum.

Certainly, there are challenges associated with collaborations, and I would encourage you to seek out partners who share your goals. If you are in a situation where a collaboration is imposed on your museum then proceed carefully and forge agreements in advance that will protect your museum, your board, and your staff.

CHAPTER 41

Exploration Place: Science Center and Children's Museum Combined

Al DeSena

In April 2000, Exploration Place in Wichita, Kans., opened its new facility—a 100,000-square-foot science center and children's museum combined. This project was the culmination of years of discussion and planning by the community (in 2005, a four-county metropolitan area of 580,000), which had operated a small (25,000-square-foot) city-run planetarium and science center since 1976 and a small (12,750-square-foot) privately managed children's museum since 1984. I joined the organization in fall 1993 as president, specifically to help the board work out the details of developing and operating Exploration Place. Over the past decade, the organization has had to grapple with questions about size, mission, and audience in a region with a relatively small market. The experiment is an ongoing one. However, some insights have emerged, along with a lot of questions.

MIXING AND MATCHING

Many museums are organized by academic disciplines such as science, art, or history. As in academia, a variety of forces and interests can lead museums in communities to a generalized or a specialized focus. Thus we may treat combinations of science centers and children's museums as a subcategory of museum mixing and matching.

Over the past twenty-five years, the numbers of both science centers and children's museums have increased substantially around the country and the world. Also, for years there has been a partial overlap of institutional membership in the respective professional associations—the Association of Science-Technology Centers and the Association of Children's Museums. The mixing and matching of science centers and children's museums has been handled in a variety of ways. In some cases, community leadership has opted to develop only one or the other kind of institution. In many other instances, both science centers and children's museums were established and have operated as independent organizations in a city or region. And in several places, science centers have enlarged the scope of their exhibit and program services to include young children. Combining the two institutional types into one organization's structure has been a rarer phenomenon, although this alternative appears to be more seriously considered today.

Looking at the world from the children's museums' point of view, other forces are at play that affect how they position themselves and their services. Two trends are important to note. First, over time, more and more children's museums have expanded their scope beyond arts and culture, which was once their primary focus, and they now include science, technology, and mathematics. Second, and more recently, not only science centers but also art, history, and other museums are increasingly providing at least some hands-on experiences for young children. It seems that just about everyone has discovered the children and family market and wants to build audience awareness, market share, and loyalty at earlier and earlier ages.

What motivations have led Wichita and other cities to decide to combine a science center and children's museum? What makes such a mixture possible at all? What are some important advantages and disadvantages of this strategic positioning? And how has its implementation fared at Exploration Place? Can this combination serve as an institutional model for others?

MOTIVATIONS

To my knowledge, no systematic study has been conducted to evaluate the variety of motivations and interests that lead communities down different paths when deciding how science centers and children's museums might coexist. However, on the basis of our experience in Wichita and other cities, two primary motivations appear to be at work.

Figure 41.1. Exploration Place, Wichita, Kans. Moshe Safde, architect.

The first is economic. A single, combined entity should be able to realize certain economies of scale and higher levels of market penetration. For example, one would expect that in smaller cities in particular, it could be more efficient to have one institution providing services, raising funds, and trying to maximize the size of its audience in a limited market region. It is reasonable to assume that this motivation is the one that drives certain communities to mix and match museums of art, science, and history under one roof.

The second motivation is philosophical. In some instances, leadership holds the view that the organization's purpose is to help develop "the whole person" or to show the links between the arts, sciences, and humanities. Noteworthy in this regard is that children's museums are fundamentally dedicated to this goal of developing the whole child.

TIES THAT BIND, FORCES THAT SEPARATE

Their being compatible in certain respects facilitates the possibility of combining science centers and children's museums. By and large, professionals in both worlds share strong, common beliefs in the importance of process-centered, hands-on experiences, in the core value of informal or free-choice learning as a major component of the organization's mission, and in being populist, family-oriented operations. In addition, while most children's museums are inherently multidisciplinary, more science centers over the years have experimented with expanding the range of their exhibits and programs to include the arts and humanities.

However, certain forces also make the science center and children's museum combination a complex maneuver. For example, their audiences and missions do not overlap completely. Even though children and adults visit both kinds of institutions, children's museums obviously position themselves as serving children and families, whereas many science center offerings are designed to appeal, on average, to an older audience. Science centers tend to promote their role in improving the public's level of science literacy, while children's museums see themselves as advocates for child development and a variety of children's interests. To some degree, though there continues to be movement toward a middle ground, science center professionals still tend to emphasize science content more so than do their children's museum counterparts, who are more likely to stress the value of unstructured play in a child's growth. An observable effect of these differences is apparent, for example, when one compares the number of signs, instructions, and explanations incorporated into exhibits at the two kinds of institutions—science centers have a considerable amount, while children's museums have much less.

Another factor in a joint organization is the interaction of the different professional backgrounds and cultures of staff. In general, the differences in staff background and orientation are similar to those one finds between elementary school teachers and middle and high school teachers. This applies both to the distinction between a generalist and a specialist, as well as the implied hierarchy of those who know a subject in depth and those who don't. Finally, children's museums are much more likely to be smaller institutions in all respects and to take hold in smaller towns and cities, not just in large metropolitan areas.

Given the interacting dynamics of all these variables, what guidance might one give having been through the process?

RECIPES FOR THE CREATIVE, RISK-TAKING COOK

There are no recipes for assessing the merits of combining science centers and children's museums. Anyone considering the options would be advised to take on the spirit of a creative, risk-taking cook and to look less for best practices and more for good questions, processes, and ingredients.

Key ingredients in the pot include mission, brand positioning, market size and segmentation, community support, organizational culture, and the relationship of these to the proposed project's scale and scope. Assuming that the community's goal is to neither underachieve nor overachieve, but to provide the maximum, sustainable level of museum resources for its residents. Then the process-oriented question could be formulated in this way: What factors should a community consider as it determines what a sustainable scale and scope are for the population it wants to serve?

The implications for this broad question are interrelated and can be broken down as follows:

MISSION: Is it more important to the community leadership to have an institution that is dedicated primarily to science and that serves all ages, or one that serves children with multidisciplinary experiences, or both, or some new mix?

MARKET: Is the potential audience likely to be larger if one focuses more on the science, or more on children, or both?

COMMUNITY SUPPORT: Is it more likely that the capital and operating costs can be secured if the institution's mission is science, or children, or both?

BRAND POSITIONING: If the community decides to combine the two kinds of organizations, can the institution effectively communicate what it is and provide enough offerings to satisfy its market mix?

ORGANIZATIONAL CULTURE: If the institution is a combined science center and children's museum, can the organization create the appropriate culture to implement its mission?

WICHITA'S MOVING TARGET EXPERIENCE

In Wichita, Exploration Place combines a small science center and a small children's museum into one single, larger entity. What this means, how to do it, and what to change after some years of experience, have been ongoing questions over the six years of planning and the five years of operation.

MISSION

Before I joined the project in 1993, the community had already come to believe that it would be better to have one, rather than two museums. Some were more inclined to develop a full-fledged science center; others took up the flag of the children's museum. Eventually, a task force determined that the two should be combined for both financial and philosophical reasons; that the new entity should serve all ages; and that the content should be organized thematically across disciplinary boundaries.

Here are some of the mission-related questions that had to be raised and addressed: What should the name of the organization be if it combines a science center and children's museum? How should the needs of different audiences (who don't always want to be in the same space with one another) be accommodated in exhibit, theater, and program spaces? How much exhibit and other space should be created so various audience segments will feel there is enough value for them for the price? It became obvious that the name and tag line (Discover the Explorer in You) had to be generic and the logo design needed to convey excitement for all ages and interests. Answers to the other two questions were more problematic. Of the six indoor exhibit areas, two were developed for younger children—one for toddlers and another for preschool children. Equivalent spaces are available outdoors. The organization also experimented with putting a few exhibits for younger children in other exhibit spaces. However, visitors can freely move through all spaces, resulting in some issues with mixing younger and older children and adults without children. Staff sometimes receive complaints from adults that there is not enough for them.

One advantage of being a combined science center and children's museum has been the extended range of exhibits and programs that the organization have developed and hosted. For example, during 2004, the spread of offerings spanned exhibits and programs on Dr. Seuss for young children to cosmology for high school Advance Placement physics students.

MARKET

From the outset, the underlying assumption about the market for a combined science center and children's museum in Wichita had been that the two former institutions were underachieving with respect to attendance. Thus, the main question was, if so, by how much? After market research and strategic planning for the facility and operation, it was decided to build a larger facility than the sum of the two former parts. The complete answer to the question is not yet in. To put the numbers into perspective, the combined annual attendance of the Wichita Omnisphere and Children's Museum was about 80,000, of which about 30,000 were school children. The 1994 master plan for the new institution called for an average annual attendance of about 250,000. This was much exceeded in year one and achieved in year two. The third year's (2002–2003) total attendance was 218,000 or 2.7 times the combined attendance of the former institutions (but below the target of 250,000). Obviously, the conservative view on attendance has proven to be off the mark. However, the organization now has to come to grips with how to interpret this more recent experience—a lagging economy, a settling in toward more realistic attendance patterns, falling short on attracting repeat visits, and new audiences.

COMMUNITY SUPPORT

Community support for the capital campaign was considerable with over $62 million raised. In my assessment, the motivations for this extraordinary level of commitment to the project fall into the following categories: first, a general belief that the city needed a significant, high-quality, and architecturally prominent project that would increase opportunities for public leisure-time activity and community pride; second, an interest in promoting certain local industries and causes; and third, an interest in the welfare of children and families. I do not believe the project could have achieved the success it has if all three motivations were not at work. Nor do I believe that the project would have been successful if the rallying cry had been centered exclusively on the need for either science literacy or child development.

BRAND POSITIONING

Positioning Exploration Place in the minds of potential users has had its challenges, but 2003 market research indicated that it is heading in the desired direction. Of course, when an organization tries to appeal to the needs and interests of such a wide audience, it is not surprising that public perceptions vary. The good news is that after five years of operation, the organization has learned that a majority of the visitors like what is offered and describe Exploration Place positively—learning, fun for the whole family, something of interest for everyone. Designing public communications is no easy task when the audiences to be reached and the offerings provided span such a wide range.

Just before Exploration Place opened in 2000, we had the notion that perhaps we should position ourselves more generically as a center for creative learning—one for all ages that is multidisciplinary, and that helps visitors develop creative process skills. If a children's museum seeks to develop the whole child, why not have a creative learning center that does the same for teens and adults as well? Over the years, staff has worked with creative process consultants, such as Robert Root-Bernstein, PhD, and Michele Root-Bernstein, PhD, and Todd Siler, PhD, to help elaborate on this concept. As staff continue to explore these ideas internally, presenting such an image to the community has been difficult. It is much easier to conform to and tweak existing stereotypes than to create new categories.

Another aspect of the decision making on branding was the design of the facility. The trustees desired to have a landmark building for the city. It was clear that the design had to appeal to a wide audience and not be seen as exclusively a children's place or one that conveyed only science and technology. It is interesting, therefore, to compare the result with some of the more striking children's museum facilities, such as the Children's Museum of Houston and the Creative Discovery Museum in Chattanooga, Tenn., or with science centers constructed around the same time, such as COSI in Columbus, Ohio, and the Science Museum of Minnesota in St. Paul. The design of the children's museums cries out "children" with their fanciful exteriors and bright colors. The science centers present a much more technology-oriented image.

ORGANIZATIONAL CULTURE

Organizational culture, patterns, and traditions can be a mammoth stumbling block when trying to combine institutional types. One advantage Exploration Place had during its early development was the small size of staff at the two former museums. This allowed the organization to grow with fewer strongly held convictions about what kind of institution we were or were not. Staff were hired who were deemed enthusiastic about serving all ages, being multidisciplinary in scope, and being an "experiment in the making." For the most part, the organization has learned how to expand its thinking, take risks, and benefit from experience. Staff are becoming better, more creative cooks.

REFERENCES

Root-Bernstein, Robert, and Michele Root-Bernstein. 1999. *Sparks of genius.* Boston: Houghton Mifflin Company.

Siler, Todd. 1996. *Think like a genius.* New York: Bantam Books.

CHAPTER 42

From Little Acorns

Chuck O'Connor

So what is a 320,000-square-foot science center in Columbus, Ohio, with a $16.5 million annual budget (after depreciation) doing in a book for small science centers? COSI didn't start out the size it is today. In 1957, a small band of visionaries, led by founding director S. N. "Sandy" Hallock, were determined to have a science center in Columbus. After five years of planning, they convinced the Franklin County commissioners to lease the vacant fifty-year-old Memorial Hall to the Franklin County Historical Society for $1 per year. The historical society became the parent organization and COSI opened in 1964 with six staff members and just over 40,000 square feet of sparsely populated exhibition space. The first year attendance was 136,000. When I joined COSI in 1966, none of the staff, including myself, had ever worked in a museum before, so we didn't know what not to do. We tried lots of new ideas and COSI grew rapidly. Soon a volunteer core of teenagers and a women's association far outnumbered the COSI staff.

Today, COSI relies on over a 1,000 volunteers as board members, program and workshop assistants, guest service team, office help, and a band of retired craftsmen who can fix or build just about anything. Throughout the years, COSI has been able to attract a talented and dedicated team of stars. The team, consisting of both paid and volunteer members, is COSI's most valuable asset and the primary reason for its long-term success. COSI celebrated its fortieth anniversary in 2004. In-building exhibits and educational programs have touched over 17 million lives. Three generations of families have explored COSI. Just about everyone I meet has a favorite story about his or her experiences at COSI. More than a few tell of how COSI changed the course of their lives. Those happy memories create friends, donors, and community support. No organization can instantly create that kind of following; it is built one day, one guest at a time.

It is essential to have broad community support and diversity of funding to build and sustain your operation. Unless of course, you have unlimited financial resources, but then you probably would not be reading this book.

WHAT'S IN A NAME?

When COSI opened, our official name was the Center of Science & Industry of the Franklin County Historical Society. How's that for a catchy moniker? The acronym COSI had virtually no public recognition and did not depict what COSI was. It meant nothing. I lost count of the number of times I would introduce myself as working at COSI only to have the response be a blank stare, followed by, "What's COSI?" I'm pleased to report that that has changed dramatically over the years. Now the response evokes a broad smile and people tell me how fortunate I am to have worked there. I fully agree.

Over the years, COSI expanded both programmatically and physically. Beginning in the late 1980s, COSI was seeing a large percentage of attendance from all across Ohio. Our outreach program, COSI on Wheels, travels to all of Ohio's eighty-eight counties. Our annual Ohio State Science Workshop program sent science activity kits into every classroom in the state. Campers come from all over Ohio and surrounding states to stay overnight and participate in our Camp-In program. Camp-In was invented at COSI and has been replicated in many museums and zoos across the country. To reflect our statewide impact, we decided to change our name. We became Ohio's Center of Science and Industry. (As a die-hard Irishman, I suggested we change the COSI acronym to O'COSI, but no one took the idea seriously!) Our name change helped position COSI for a $50 million capital appropriation from the State of Ohio toward our new facility in Columbus.

In 1997, we opened a sister museum in Toledo. To differentiate ourselves, we became COSI Toledo and

Figure 42.1.   COSI, Columbus, Ohio.

COSI Columbus, our current legal names. Now just about everyone in Ohio recognizes COSI as a place for science, learning, and fun. However, when someone asks me what do the letters in COSI stand for, they are in for a long story.

My advice on naming a new center is to pick a unique, short, memorable name that depicts who you are and what you do. The Exploratorium and Science Place are good examples. Most important, whatever your name is, protect it, and use it consistently.

COSI COLUMBUS

Over the years, two additions and a temporary tent structure had grown the original COSI facility to 150,000 square feet. Annual attendance of 700,000 was taxing the East Broad Street location. A decision was made to expand or relocate. We evaluated expanding at the original site and explored several potential new locations. In 1994, we were able to secure, from the city, a 14-acre site that housed the former Central High School. The site is located on the west side of the Scioto River with a magnificent view of the Columbus skyline. We sit on a 42-acre peninsula that is being developed into a public park.

Museum professionals rarely participate in more than one major capital program in their career. I have had the opportunity to be involved in two additions to our original building, a 10,000-square-foot temporary structure, the COSI Toledo facility, and the development of the new COSI Columbus, which is over twice the size of our original building. Our team produced a detailed and well-documented Architectural Program that provided invaluable guidance to the architects during the planning of the new Columbus facility. This was one of the keys to the success of the project, which came in on time, under budget, and without any debt.

Perhaps the most difficult aspect of the building projects was the challenge of dividing our time between maintaining the ongoing museum operations while overseeing the construction project. Fortunately, we were able to assemble a group of experienced consultants who helped keep the project on course.

COSI TOLEDO

I think it was in the early 1990s that a group of citizens from Toledo, Ohio, first approached us about helping them start a science center. Discussions soon developed into creating a COSI in Toledo. The plans began to mesh when the City of Toledo made the recently defunct downtown Portside Shopping Mall, on the shore of the Maumee River, available to the science center planners. The statewide reputation of COSI helped secure the property, state capital funding, and private support to make COSI Toledo a reality. The total cost for remodeling the 90,000-square-foot building and creating seven major interactive exhibit areas (see Hi Wire Cycle in frontispiece) was $12 million. The two centers operate independently, with separate boards, budgets, and programs. There is carefully coordinated consistency in our brand, and COSI Columbus provides some administrative services to Toledo. COSI Ohio is the governing board and consists of trustees from both institutions. (Note: In 2005, COSI Toledo was awarded the National Award of Museum Service, the highest honor for the museum field. Its former and founding president, William Booth, deserves credit for this prestigious award.)

If I had it to do over again I would have preferred to be relieved of all daily operating responsibilities to focus on the building projects. A museum staff entering into a building expansion should not underestimate the time commitment needed to do it right. It was not

232   CHUCK O'CONNOR

unusual for COSI team members to work eighty or more hours a week as the project progressed. I also learned that any construction project will take longer and cost more than originally estimated. The opening of a new or larger museum requires an increase in staff. This is particularly difficult for a new museum that has no existing operating staff. I cannot stress enough the importance of hiring key staff early in the project so they become invested in the effort. The frontline team needs to be well trained before opening day. First impressions of your new operation will last a long time, so they better be good.

Opening your building is just the beginning. At COSI, the one constant is change: a never-ending succession of new ideas, exhibits, and programs. In just the last few years, we have produced, a new permanent exhibition on Space, a traveling exhibition on Speed, launched a very successful inquiry-based professional development program for teachers, beamed electronic outreach programs into classrooms around the world, established an award-winning surgical suite where students can talk to the surgical team during live open-heart and knee surgery, and a new after-school program for inner-city students.

COSI is not exempt from difficult economic times that more than once have required staff and other cutbacks. However, as a valued community resource, COSI has come through these periods stronger and wiser. For more information on current COSI Toledo and COSI Columbus activities, see our website at www.cosi.org.

The two most important assets of any museum are its team members and public trust. Take exceptional care of these two possessions and everything else will work out fine. Regardless of your size, pay close attention to the details of every guest's experience and do not neglect to keep looking over the horizon to what you can become tomorrow. Have Fun!

CHAPTER 43

The Tech: The Challenges of Growing from Small to Large

PETER B. GILES AND MAUREEN E. KENNEDY

HISTORICAL BACKGROUND

The idea for The Tech Museum of Innovation developed from the need for public understanding of the technological revolution that was occurring in the Bay Area. A member of the Junior League of Palo Alto had visited the Chicago Museum of Science and Industry and noted that a similar institution was needed in Silicon Valley to explain the impact of technology advances. In 1978, the Junior League assumed the challenge, formed an advisory committee of technology leaders, and completed a feasibility study. The envisioned museum would appeal to all ages and interests, explain the technological advances of Silicon Valley, explore the process of innovation, and illustrate the impact of technology on everyday lives. On the basis of positive results from the feasibility study, The Tech was formally organized in 1983 with a small board of directors composed of leaders from technology companies and participants from the Junior League to pursue what was envisioned as a five-year project.

Initial efforts focused on fund-raising to support the growing organization and securing a site for the museum. Silicon Valley had expanded very rapidly, well ahead of any coordinated planning. Several cities claimed leadership in Silicon Valley and desired to locate The Tech in their cities, assuming that its location would define the center of Silicon Valley. A competition among the cities resulted, which allowed The Tech to fulfill its objectives: secure the land, the building, and annual operating support. The City of San Jose was selected for the museum site, benefiting both San Jose and The Tech. San Jose included The Tech in its major downtown redevelopment plans and supported The Tech during its planning efforts. The Tech provided the city with a needed destination that would attract and lengthen the stay of visitors to the city. To solidify the partnership, the mayor of San Jose joined the

museum's board of directors, becoming an invaluable asset to the board.

The organization continued to develop, attracting Peter Giles, who became president and CEO, and key staff who continued through the opening of the final institution, providing both stability and knowledge. Progress, however, depended on raising the capital funds for the exhibits, which proved to be very difficult. Silicon Valley had not yet established a tradition of philanthropy, and there were no examples of a technology museum. San Jose, while the largest city, was not acknowledged as the cultural and commercial center of this fragmented region. When the capital fund-raising did not meet its goals, it was decided to follow the Silicon Valley model and begin small, like a start-up, and create a prototype of the ultimate museum. San Jose had a convention facility that was available for conversion into a prototype museum. During the next two years, a board member who was a retired corporate executive led a small, dedicated staff with the assistance of outside consultants to complete the building renovation and exhibit design construction. In 1990, The Garage, named after the birthplace of a successful global technology company, opened with a 20,000-square-foot prototype facility, including 6,000 square feet of exhibits, two program spaces, store, and café at a cost of $5 million. In its first year of operation, The Garage attracted 94,000 visitors with 28 staff members, 250 volunteer docents, and a $2 million operating budget.

At the same time The Garage's facility was being planned, program development and implementation began in order to build a presence in the community and develop a core group of supporters. The major programming efforts were Tech Talks, a speaker series of well-known innovators presenting their stories of how they addressed tough problems using technology; Tech Challenge, a team competition to solve interesting

Figure 43.1. Old Tech Museum, San Jose, Calif.

problems, such as repairing a vehicle in space; and a Teacher's Institute, workshops providing teachers with hands-on experiences supporting interactive science learning. With the opening of its prototype museum and implementing of its programs, The Tech began to fulfill its dual mission to be an educational resource established to engage people of all ages and backgrounds in exploring and experiencing technologies affecting their lives, and to inspire the young to become innovators in the technologies of the future.

INITIAL OPERATIONS: 1990–1998

After opening the prototype facility, The Tech focused on planning for its permanent facility. The Garage continued to operate with dedicated paid and volunteer staffing, while a new group was formed to plan the future Tech. The agreement with the San Jose Redevelopment Agency called for the agency to fund, design, and manage the construction of the building with approval from The Tech. The Tech was responsible to fund, develop, and manage the fabrication of the exhibits with approval from the agency. The Tech was also required to reach fund-raising milestones before the construction of the building could begin. This initiated a six-year capital campaign that raised $32 million cash and $25 million in-kind to fund the exhibits, capital campaign, marketing, and startup expenses for the new facility. Silicon Valley corporations and their leaders were asked to participate in exhibit definition, provide technical expertise in exhibit design, donate equipment, and assist with funding. Corporations were responsible for one-third of the cash and all of the in-kind contributions while individuals, many of whom were technology leaders, accounted for one-half of the cash contributions. During the capital campaign, an-

nual fund-raising campaigns continued each year to support the ongoing operations of The Garage.

The exhibit process began by defining the exhibit concept. This phase required many iterations and considerable time but ended with the concept of People and Technology, emphasizing the relationship between people and technology rather than technology in isolation. The exhibit space was divided into four regular galleries and a temporary gallery. A team consisting of an exhibit developer and designer took responsibility for each gallery and developed the look, feel, and content for their gallery. The reviews of exhibit plans required for funding by the redevelopment agency resulted in many revisions to these plans and an improved product.

While the new building and exhibits were being designed and constructed, The Garage provided opportunities to prototype new processes and procedures for the new museum, such as the admissions system and retail operations. The Garage exhibits exceeded their five-year design life before the new museum was completed. In order to close on a positive note, most exhibits were replaced with a traveling exhibition, which brought in record attendance. The Garage closed, allowing three months for final preparation of the new museum.

OPENING THE NEW MUSEUM

In 1998, fifteen years after incorporation in a Los Altos home, The Tech Museum of Innovation opened the doors of its new facility in downtown San Jose. The 112,000 square-foot building included four permanent exhibit galleries—Innovation: Silicon Valley and Beyond; Life Tech; Communication; and Exploration— as well as a changing gallery, Center of the Edge. The new facility included an IMAX Dome Theater and the

Figure 43.2. New Tech Museum, San Jose, Calif.

Center for Learning (a teacher development area), a café, store, a large public gathering space, and entrance lobby.

In its first year of operation, The Tech attracted 809,000 visitors, about 150,000 more than the projected capacity. The large attendance during year one resulted from extensive media coverage and advertising campaigns planned to position The Tech as a major new Bay Area attraction and educational resource. The target audiences were defined with priority given to parents with children ages nine through to fourteen. In the summer prior to the opening, a new look was presented aimed at both building on The Garage's positive identity and establishing the new facility as a major science and technology center. In a competitive review, a new advertising agency was selected and offered its services pro bono. The $1 million advertising and public relations campaign for opening was funded through the capital campaign. The highlight of the campaign was a special section published by the San Jose *Mercury News* in their Sunday edition on the opening of The Tech.

Concurrent with the advertising and public relations campaign, the number of volunteers and staff increased dramatically. Between 1997 and 1998, staffing levels grew from 64 to 110 FTEs (full time equivalents), and the number of volunteers increased from 185 to 740, requiring an enormous amount of training and assimilation. Along with the mission statement, a core ideology, "to inspire the innovator in everyone," was developed. This phrase, easier than the full mission statement to remember, became the slogan and centerpiece of staff and volunteer training. The Tech, with its theme of "people and technology," was more about inspiring visitors with the incredible range of technological applications and their impact than instructing visitors in the history and workings of technology. A statement of values was defined to move the largely informal culture of The Tech to something more tangible that could be expanded and transmitted to new players. These values provided a basis upon which the behavior of the expanded staff could be molded around a shared vision and ideals without inhibiting the passionate sense of mission that impelled the establishment of The Tech from the first days of its founding.

As The Tech staff prepared to open the doors of a new, expanded facility, the entire process of planning and budgeting as communication and accountability tools was reexamined. The Tech's board chairman shared the administrative processes used by his company to ensure that a larger organization could set objectives and act in unison in achieving them. The Tech adapted these processes and terminology to its needs, introducing critical action plans, measurable goals, and plans in support of larger strategic imperatives that cut across the lines of multiple reporting units. Cumbersome at first, this process has proven useful in establishing shared vocabulary, common and well-communicated goals, and the format for quarterly performance accounting against numeric measures.

In the rapid-fire steps leading up to the 1998 opening, inevitable trade-offs between quality and deadlines were made and processes were reexamined. The network infrastructure and the MIS overlay were fragile and required extensive remediation, rebuilding, and documentation of procedures. In the period of a year, The Tech evolved from a "barnstorming" and entrepreneurial phase to a permanent, institutional phase depending to a greater degree upon processes and teamwork rather than individual talent and enterprise. With assistance from the Center for Excellence in Nonprofits, process improvement techniques were applied to many of the processes that depended upon individuals from different departments working together, such as membership (membership, finance, guest services) and exhibit maintenance (exhibits, engineering, gallery managers). Human resources and facilities, where practice had been to respond to problems and opportunities, adopted systematic, longer-range processes. Maintenance schedules, staff training, and performance reviews are just three areas where The Tech moved from an ad hoc approach to a "building to last" mentality.

CHALLENGES FACED

There were many challenges in growing The Tech from a small science center to a mid-size center.

KEEPING THE SMALL SCIENCE CENTER (THE GARAGE) FRESH AND ATTRACTIVE FOR TWO YEARS LONGER THAN PLANNED, WHEN THE ATTENTION WAS ON IMPLEMENTING THE NEW FACILITY:

A critical decision was made to divert scarce engineering resources from the new museum implementation to bring in a traveling exhibition.

INTEGRATING THE EXHIBIT SCHEDULE, CONTROLLED BY THE TECH, WITH THE FACILITY SCHEDULE CONTROLLED BY THE REDEVELOPMENT AGENCY:

The building had to be "clean" for installation of the exhibits. At the end, the schedule for both the exhibit and the building construction pushed inordinate amounts of work into the same time period. This required delicate balancing of exhibition installation with construction activity, all the while staying committed to a late-October opening.

COORDINATING AND COMMUNICATING BETWEEN STAFFS IN TWO LOCATIONS:

The members of the exhibition project team were located in donated space, keeping them apart from other critical staff, such as engineering and marketing. Inadequate opportunities for face-to-face interaction led to misunderstandings that grew and festered. Two cultures developed in one organization, each apprehensive of the motives and good faith of the other.

SUPPORTING BOTH AN AMBITIOUS CAPITAL CAMPAIGN WHILE RAISING $1.5 TO $2.0 MILLION A YEAR IN ANNUAL SUPPORT:

Concurrent efforts had to be made to invite donors to continue their support because of its merit, and be on the "ground floor" of making something much more significant happen.

MOBILIZING STAFF TO OPERATE A SMALL SCIENCE CENTER WHILE GEARING UP TO RUN A MUCH LARGER ONE:

A key to managing this dual focus was a loaned executive who served as the launch coordinator. The addition of this position allowed the executive staff to give priority to their first duties without losing the administrative oversight necessary to bring all parties together, on schedule, and on budget.

INSTALLING EXHIBITS EN MASSE WITHOUT ADEQUATE PROTOTYPING AND AUDIENCE TESTING:

Because of the technical sophistication of the exhibits, inadequate engineering testing and specifications that depended on a distant fabricator led to excessive exhibit downtime. This occurred at a time when crowds were heavy, further aggravating the situation. A thorough exhibit evaluation engaging all staff led to a systematic process of remediation and an institution-wide receptivity toward critical feedback.

BALANCING PRIORITIES CALLED FOR BY THE TECH'S MISSION WITH THE NEED FOR EARNED REVENUES:

Seeking earned-income, The Tech placed a conscious priority on corporate clients for the use of the building rather than giving priority to important community relationships. The income from the large number of corporate events made The Tech dependent on a source of revenue that was highly sensitive to economic fluctuations.

BROADENING THE TECH'S ACCESS TO THE COMMUNITY TO ATTRACT A MORE DIVERSIFIED AUDIENCE:

Prior to opening, a group of community advisors was convened to involve community groups in the opening ceremonies and to establish discount admission programs. Over time, The Tech has placed increased emphasis on engaging lower-income individuals through partnerships with organizations serving these individuals.

FACTORS OF SUCCESS

The partnership between The Tech and the City of San Jose was essential to the successful evolution of The Tech Museum of Innovation from a small to a mid-size science center. The city offered a prime downtown location for the building and provided significant funding. The city also provided the management for the design and construction of the building. This process involved many players and was complex; however, the city did reduce the project management burden and protect The Tech from the liability for cost overruns. The building lease with the city also provided annual funding for building maintenance as well as up-front funding to assist in launch and operational expenses.

Another important factor was the engagement of corporate sponsors in funding the exhibits, often coupled with in-kind equipment donations. The Tech's board became an active fund-raising and contributing board. Corporate participation together with large-scale city support combined to make this precedent-setting community project in Silicon Valley successful.

Finally, the second half of the decade of the 1990s was an economic boom period that made a large, nonuniversity-based fund-raising effort possible. It is not an exaggeration to say that the public and private revenues for this project would not have been raised during the kind of economic conditions experienced in the early 1990s or those that prevailed during 2000–2003.

LESSONS LEARNED

Advertising expenditures should be distributed over a longer period, especially in an expensive media region. Public relations and free media at the opening should have greater emphasis while reserving advertising dollars for years two, three, and four when publicity will be scarce and the audience needs a reminder to return. Especially in an area with multilingual demographics, media buys should be concentrated in ethnic media to build an interest in that audience.

Exhibition implementation should be phased in over time, and the "no prototype, no exhibit" approach enforced. Lean heavier on quality, with proven interactive potential and technical reliability, than on quantity, and then open a new gallery each year after the museum's opening. Rather than a "big bang" only at opening, try for a mid size "bang" followed by multiple smaller "bangs" at annual intervals and, if necessary, add traveling exhibitions to fill space. The Tech has become predominately a regional attraction dependent on repeat visits by the local residents, who must be enticed by new opportunities.

Interpretation staff, developing and implementing programs, should be emphasized along with exhibition development. Given the need to fill the entire exhibition space on opening day, project management priorities favored "hardware," exhibits with a capital cost and fabrication schedule, over program development. As a result, resources were heavily skewed toward exhibits, leaving inadequate creative attention to programming, and insufficient interpretation staffing. Fewer exhibits with more demonstrators and interpreters would correct this imbalance.

The capacity for rapid, frequent change must be given greater priority in all exhibition and program planning. Frequent updating requires more innovation in design and fabrication to achieve equivalent value with less time and less expense. Smaller galleries would create more thematic possibilities and offer a sense of anticipation as changes come. The Tech's exhibition was fabricated of materials of the highest quality, however, the need for frequent change will place more emphasis on interpretation and less on exhibit finishes.

CHALLENGES AHEAD

Looking to the future, The Tech must embed itself in the communities of the South Bay that are rapidly changing in their demographics. Among other things, this requires more bilingual interpretation (English and Spanish), staffing, volunteers, advertising, and an explicit partnership strategy to build relationships with organizations able to engage large numbers of diverse and particularly underserved communities. In addition, The Tech must succeed as a regional visitor attraction, even as it establishes a global presence through its Tech Museum Awards: Technology Benefiting Humanity, its education programs, and its web presence. To this end, achieving capital funding through corporate and individual support to renew exhibits on a minimum five-year cycle is key.

Another challenge is to reestablish a strategic planning process. The Tech began strategic planning early on; however, during the opening of the building and the starting of operations, strategic planning was deferred in order to focus primarily on annual planning. The organization needs an ongoing strategic planning process to drive annual planning.

Finally, The Tech has developed a signature pedagogy, Design in Mind Learning, an approach to learning science and technology that is becoming the basis for its work to inspire students, engage families, and empower educators. This teaching model embodies The Tech's mission to inspire the innovator in all learners and draws on the rich tradition of the engineering and design process. The Tech's goal is to infuse Design in Mind Learning, a project-based learning pedagogy, into all it does, both in the museum and through its school and community partners. The Design in Mind Learning approach encourages engagement in basic concepts of technological innovation that novices can readily grasp.

These challenges exist in an environment where funding pressures require greater than normal resourcefulness and partnering. Maintaining the vitality of The Tech will be critical to attracting consistent repeat visitation. Visitors will need to feel that understanding the technologies affecting their lives is important and that their understanding will grow in fun and entertaining ways with regular visits to The Tech. The need to attract repeat visitors is shared with other regional science centers. To successfully respond to this need, science and technology centers must find ways to move public perception of a science-technology center visit from an interesting option to a regular and vital activity for all families and individuals in the region.

CHAPTER 44

How to Foster Innovation within Your Science-Technology Center: Observations from Under the Seat Cushion

Susan B. F. Wageman

The need for innovation in science centers, large and small, is universal. If we want our visitors and members to come back over and over again, they have to feel that there will be opportunities to see and do something new or different each time. As our institutions evolve and grow—and as the world changes around us—we need to change the products and programs we offer, reach new markets, work in new ways, and even design new organizational structures to most effectively achieve our missions.

The experiences that I intend to share in this chapter are primarily from The Tech Museum of Innovation, but they should have relevance for all types and sizes of organizations.

I joined The Tech's staff in 1994. At that time, there were 35 employees, a budget of $1.8 million, and 111,000 visitors to the temporary 20,000-square-foot museum each year. Today we have 115 staff, a budget of $11 million, and 440,000 annual visitors to our 112,000-square-foot museum that opened in 1998. As a participant and observer in this growth process, I became so fascinated with the dynamics of change in the organization that I decided to pursue a master's degree in organizational management. As I learned more about organizational systems, leadership, change, and innovation, I recognized patterns of practice that I had noticed at The Tech that seemed to promote innovation.

FOSTERING INNOVATION

My research showed that innovative organizations tended to be defined by decentralized decision making, lateral communication, and few hierarchical distinctions (Morand 1995). Information flow was open and pervasive (Morand 1995) and the organizational structure relatively flat (Russell 1986). Organizations that were structured to be reflective about their practice, trying to learn from their mistakes (and successes),

seemed to have an advantage because they were more capable of considering different points of view (Brown and Duguid 1991), enabling innovation to flourish. Rigid formal structures did not recognize nor implement innovative ideas (Dougherty and Hardy 1996). Although my research suggested these approaches led to innovation, they had to be adapted to each unique circumstance because the solution for one particular institution did not always fully meet the needs of another.

Informal Systems and Formal Organizational Structure

You might not think of the space under the seat cushion as part of the chair; however, your chair would not be the same without this space. Although not part of the structure, it is integral to the chair's shape. It is also where you often find those small important things you have lost. The informal systems of an organization— lunch gatherings, water cooler conversations, unofficial collaborations between people, and such—are like the space under the seat cushion. They are an integral part of every organization, but not always recognized as critical to the entire system. Formal systems include the organizational structure, rules, job descriptions, defined processes, meetings, and so on, but informal systems are created by staff as ways to socialize and get work done.

Sherman and Schultz provide a theoretical framework for understanding the interrelation of informal systems and formal structure (1998). They propose that excessive infrastructure binds an organization into repeating existing patterns instead of enabling new ones to emerge. To enable innovation, it is necessary to step out of existing perspectives, reevaluate models in light of new understandings, and then adjust behaviors to align with the new understandings. When organizations try to close their systems and constrain behavior, they "constrain the best source of creativity available to them—the distributed knowledge of their workforce" (Sherman and Schultz 1998, 8). A system that encourages interactions between and among staff and that allows for open and pervasive communication provides the best opportunities for innovation.

Russell (1986) studied the influence of age and size on innovation. Younger, smaller, less formal organizations that operate primarily through informal systems tend to be better at coming up with innovative ideas. However, younger organizations do not always have the infrastructure needed to implement innovations (Russell 1986). Older and larger organizations with more rigid formal structures may be better at implementing innovation, but they may lose touch with or try to suppress informal systems that allow innovative ideas to emerge. The ability of small organizations to initiate innovation is balanced by their challenge with implementation, and larger organizations often have the opposite problem.

Connecting Formal and Informal Practices

Building on these ideas, I found that highly innovative organizations seem to have well-developed formal practices that tap into informal activities. Connecting formal and informal practices both prepares the environment for innovation and enables adoption and implementation. These connecting practices include (1) nurturing communication, socialization, and creativity in the informal systems and (2) creating links to bring the best of what emerges in the informal systems into the formal structure for development and implementation. For example, 3M allows researchers to spend 15 percent of their time working on whatever they want (Coyne 2001) and Tech staff returning from visiting other science centers host informal brown bag lunches to share what they learned. These are the types of practices that I have identified as "connecting practices."

CONNECTING PRACTICES IN ACTION

In reflecting upon my own contributions to The Tech's success, I realized that a significant portion of my efforts were through informal systems and that—consciously or not—the organization's leaders employed many connecting practices. Following are examples of three types of connecting practices: basic, proactive, and systematic. This is by no means a definitive list and there are undoubtedly connecting practices that I have not identified. On the basis of my research to date, it seems that the use of connecting practices is more important than the specific set of practices used (Wageman 2002). I encourage you to try these out, even make up your own.

Basic Connecting Practices

UTILIZING SOCIAL NETWORKS— Nearly all the informal activities that take place in an organization occur within informal social networks, complex relationships among people within an organization and beyond. Individuals within this network facilitate the movement and sharing of information and ideas. If an organization understands how its informal networks work and who the key players are, they can use this knowledge to get the pulse of the organization, learn about new ideas, and target messages for dissemination (Gladwell 2000).

Utilizing social networks is one of the easiest connecting practices to implement because most people interact with others informally as they work, perhaps even more frequently than they interact through formal systems. I participate in the informal system at The Tech in a very conscious way. I seek out conversations with my colleagues before and after staff meetings, at luncheon get-togethers, or in other natural gathering spots—to learn what is happening outside my immediate area of focus and what staff is thinking and talking about. When there are issues or ideas I believe should be considered, I'll bring them up in conversations with staff or volunteers who I know will be interested and are likely to share with others. Special events that bring diverse staff and volunteers together are great opportunities for learning about more remote parts of the organization and for sharing one's own work and ideas.

DESIGN FOR INFORMAL COMMUNICATION— Traditional offices with doors reinforce strongly hierarchical organizational structures, leaving little room for informal interactions from which innovation can emerge. IDEO (a design company that specializes in innovation) designed their offices with spaces for informal gatherings and clustering around shared work (Kelley and Littman 2001). The Theatre de la Jeune Lune included a kitchen in their new facility to serve as an informal meeting place for their staff (Light 1998).

Creating spaces for informal gatherings is definitely an art—especially in the museum setting. There are several spaces at The Tech that are designed for informal communication. The lunch room with its coffee and vending machines and the workroom with mail slots, copy machines, worktable, and fax have accommodations that nearly everyone uses—and they are separated enough from everything else so that informal conversations do not intrude on others. Some cubicle spaces are joined to facilitate collaborative work, and others have small tables for gathering without having to reserve a meeting room. One vice president located the printer station in his office and encourages group members to use his conference table for meetings. This helps him stay connected while he works on his own priorities.

MANAGEMENT BY WALKING AROUND— David Packard coined this term. His philosophy was that managers should get out into the business and talk to people at all levels, so they will really know what is going on—and pick up on ideas that supervisors or middle managers may not recognize as valuable (Packard, Kirby, and Lewis 1995). In our first museum, many of the offices were laid out so that we had to regularly travel through the public areas of the museum. This helped all staff stay in touch with our visitors. Today, some staff members rarely leave the museum, the shop is off-site, and the rest of us work in cubicles in another building. Many museums require employees who do not regularly work with the public to take a turn working on the front lines so they become more connected to the organization's mission. Managers are encouraged to walk around and talk to staff both within and beyond their domain of responsibility. Even before I became a manager, I realized that this practice was a great opportunity for me to communicate my ideas to the leadership when they were actively listening. I am always prepared with questions or ideas to discuss when members of the leadership stop by on their rounds.

INNOVATIVE IDEA SEEKING— Peter Drucker maintains that conscious, purposeful searching is frequently responsible for innovation (2000). General Electric holds regular town meetings to solicit ideas (Stewart 1997). For this practice to be effective, it is essential that the possibility for adoption be obvious and that decisions be followed swiftly by honest feedback clearly justifying the action taken.

One forum that The Tech has created to seek out ideas and innovations is the Food for Thought luncheon. The president invites a diverse group of staff from different departments to an informal lunch (hosted by a generous board member) to discuss whatever they wish. Some of these conversations have resulted in immediate changes to practices that had not been recognized by the leadership as needs.

Proactive Connecting Practices

RESOURCES FOR TINKERING— Devoting resources to tinkering and exploring new ideas contributes signif-

icantly to innovation. IDEO's Tech Box is a library of parts, pieces, and other interesting things that anyone can borrow to inspire them to think about design (Kelley and Littman 2001). Pfizer scientists have free rein in following leads "wherever they go," since, as CEO and chairman William Steere and R&D senior vice president John Niblack say, "We cannot discover tomorrow's drugs if we freeze everyone in rigid procedure" (Kanter, Kao, and Wiersema 1997, 141).

The Tech's website began as a semi-personal project of Tech volunteers and staff. Managers provided time and some resources to nurture the work, but overall it was initially developed outside of the formal system before it was incorporated into the formal structure of The Tech.

STORYTELLING— Storytelling is a powerful tool for sharing knowledge and culture. Most of us use stories to explain what we do and why. For example, IDEO provides spaces for teams to exhibit their work and share stories with colleagues and others (Kelley and Littman 2001). Having informal displays available to handle and talk about inspires information sharing and new ideas. However, stories can promote innovation or can lock an organization into existing practices (Sherman and Schultz 1998), so it is important for organizations not to get attached to stories that promote the attitude "it's always been done this way."

The challenge is to strategically capture and tell the stories that truly guide and illustrate the mission and organizational culture. One of my favorite stories is from a day I was facilitating a computer take-apart activity in the museum. A young girl had been there for quite a while. Finally, her mother said, "We *must* go now. I'm sure we can find something in the garage for you to take apart." I tell this story to illustrate what we mean by "compelling." Tech exhibit developers use informal systems to inform their concept development process. They leave out objects and posters with questions. Anyone can participate and the conversations and feedback help developers find ideas they might not have thought of themselves. Stories also can illustrate the work process. For example, I share my stories of how I have been successful in connecting informal systems and the formal structure to help others understand the value of these connecting practices.

Systematic Connecting Practices

EVANGELIZING— Gareth Morgan describes what I call evangelizing as the work of strategic termites (Morgan 1993). By discussing new ideas and approaches informally, building understanding and support, and just doing things a new way (without waiting for permission), it is possible to create a groundswell of interest leading eventually to formal adoption. Although it may seem that this practice is setting the organization up for unpleasant surprises, by enabling strategic evangelizing, you can build support for changes in the informal systems that prepare the way for changes in the formal structure.

When I joined The Tech, we rarely used evaluation, except for big projects that provided funding for outside consultants. As grants coordinator, I recognized the need for better evaluation but found few receptive ears. When talking to people about their challenges or discussing our vision for the future, I shared what I was learning about evaluation in the context of their challenges. Over time, a growing consciousness about evaluation developed. (Of course, major stakeholders contributed to this growing sense of need as well.) Just before we opened our new museum in 1998, I was asked to develop and lead a program that would integrate evaluation throughout the institution. Although there is still more to do, managers now incorporate evaluation into their work.

EMBRACING EMERGENT ROLES— Most organizations define specific roles for each employee. A few organizations have established structures that enable new roles and relationships to develop as work evolves. Embracing such emergent roles means that organizational structure is modified to fit the patterns that emerge from the work. For example, staff volunteers do not have fixed departments for each new project on the basis of their knowledge and interests (Gladwell 2000). By regularly changing project team groupings, these organizations increase the range of knowledge-sharing interactions among people with diverse ideas, experience, and approaches.

The Learning Experiences Group at The Tech embraces emerging roles more frequently than any other part of the organization. As we focused our work and responded to economic and audience changes, we merged three separate outreach programs to reach key target audiences. As our work changed, staff roles shifted and responsibilities evolved. Over time, the dynamics of program delivery changed and different staff found themselves working together more frequently. Instead of setting an organizational structure and struggling to fit the work within it, we change formal organizational structure to fit how we are accomplishing the work.

INSPIRING THE INNOVATOR

I hope that these ideas and examples are useful as you attempt to integrate innovative practices into your science center. I have learned that in many ways The Tech is extraordinary in its ability to change and innovate as it matures and the environment changes. At the same time, The Tech struggles as it grows to build the formal structures necessary to survive, while trying to maintain the innovative culture of its youth. I believe that The Tech's greatest advantages in fostering innovation are that we all are committed to one mission "to inspire the innovator in everyone,"[1] constantly guided by our core values of learning, innovation, inclusiveness, integrity, and teamwork. Connecting practices are ways to foster broad communication at all levels in an organization and to ensure that there is space and openness towards creativity and new ideas. In the context of a small or growing science center, paying attention to connecting practices may be just what you need to keep your innovative edge.

NOTE

1. "Inspire the innovator in everyone" is memorable shorthand for the official mission, "to be an educational resource that engages people of all ages and backgrounds in exploring and experiencing technologies affecting their lives and to inspire young people to become innovators in the technologies of the future."

REFERENCES

Brown, J. S., and P. Duguid. 1991. *Organizational learning and communities-of-practice: Toward a unified view of working, learning and innovation*. The Institute of Management Sciences (now INFORMS). Available from www2.parc.com/ops/members/brown/papers/orglearning.html.

Coyne, W. E. 2001. How 3M innovates for long-term growth: Planning for serendipity—six positive steps and three "don'ts" contribute to building a tradition of innovation. *Research Technology Management* 44 (2): 21.

Dougherty, D., and C. Hardy. 1996. Sustained product innovation in large, mature organizations: Overcoming innovation-to-organization problems. *Academy of Management Journal* 39 (5): 1120–53.

Drucker, P. F. May 1, 2000. The discipline of innovation. *Harvard Business Review*.

Gladwell, M. 2000. *The tipping point: How little things can make a big difference*. Boston: Little Brown.

Kanter, R. M., J. J. Kao, and F. D. Wiersema. 1997. *Innovation: Breakthrough ideas at 3M, DuPont, GE, Pfizer, and Rubbermaid*. New York: HarperBusiness.

Kelley, T., and J. Littman. 2001. *The art of innovation: Lessons in creativity from IDEO, America's leading design firm.* New York: Currency/Doubleday.

Light, P. C. 1998. *Sustaining innovation: Creating nonprofit and government organizations that innovate naturally.* San Francisco: Jossey-Bass.

Morand, D. A. 1995. The role of behavioral formality and informality in the enactment of bureaucratic versus organic organizations. *Academy of Management Review* 20:831–73.

Morgan, G. 1993. *Imaginization: The art of creative management.* Newbury Park, Calif.: Sage Publications.

Packard, D., D. Kirby, and K. R. Lewis. 1995. *The HP way: How Bill Hewlett and I built our company.* New York: HarperBusiness.

Russell, R. D. 1986. The effect of environmental context and formal and informal organizational influence mechanisms on the process of innovation: Toward an integrated theory of innovation. PhD diss. Pittsburgh: University of Pittsburgh.

Sherman, H. J., and R. Schultz. 1998. *Open boundaries: Creating business innovation through complexity.* Reading, Mass.: Perseus Books.

Stewart, T. A. 1997. *Intellectual capital: The new wealth of organizations.* New York: Doubleday/Currency.

Wageman, Susan B. F. 2002. Under the seat cushion: Find the keys to innovation connecting the formal structure with informal organizational systems. Master's thesis. Santa Barbara, Calif.: Fielding Graduate Institute.

Part XI

OVERVIEW OF SCIENCE CENTERS

CHAPTER 45

Science Center History

ASSOCIATION OF SCIENCE-TECHNOLOGY CENTERS (ASTC)

SCIENCE MUSEUMS HAVE ROOTS IN THE 19TH CENTURY and before, but the science center field is largely a product of the 1960s and early 1970s, a time of social ferment in North America and Europe. Against a backdrop of student uprisings, antiwar protests, and voter registration campaigns, curriculum reform projects engaged scientists like physicist Frank Oppenheimer. The hands-on approach to education they were pioneering, and the populist spirit of the times, soon gave rise to visions of a new-style museum.

The Pacific Science Center, one of the first to use the term, opened in a Seattle World's Fair building in the mid-60s. The Smithsonian Institution invited visitors into a new Discovery Room where they could touch and handle formerly off-limits specimens. In 1969, Oppenheimer's Exploratorium opened in San Francisco, and the Ontario Science Centre opened in Toronto. By the early 1970s, COSI, the Center of Science and Industry in Columbus, Ohio, had run its first camp-in.

ASSOCIATION OF SCIENCE-TECHNOLOGY CENTERS (ASTC) FOUNDED IN 1973

It didn't take long for these new-style museums to band together for mutual support. In 1971, 16 museum directors gathered to discuss the possibility of starting a new association, one more specifically tailored to their needs than the older American Association of Museums. ASTC held its first Board of Directors meeting on March 29, 1973, and was officially incorporated that year.

ASTC had 23 founding members, including the Exploratorium, Ontario Science Centre, Pacific Science Center, and COSI. A number of older natural history and technology museums that were beginning to adopt the hands-on exhibit approaches of newer centers—including Philadelphia's Franklin Institute, Chicago's Museum of Science and Industry, Boston's Museum of Science, Pittsburgh's Buhl Planetarium,

and the Oregon Museum of Science and Industry in Portland—were also founding members.

Federal support was crucial in getting the fledgling science center organization up and running. The American Museum of Science and Energy in Oak Ridge, operated by the federal government, had hosted the first organizing meeting in 1971. In 1974, the National Science Foundation provided $75,000 in general operating support, and ASTC was able to open its Washington, D.C., office, publish its first newsletter, and launch its first traveling exhibition. Additional grants from NSF, in 1975 and 1977, provided crucial support for workshops, publications, and traveling exhibitions.

Throughout the late 1980s and 1990s, the science center idea took hold around the world. ASTC had 27 members in 1974, 170 members in 1984, 438 members in 1994, and has approximately 540 today, including more than 420 operating or developing science centers and museums. ASTC is the largest organization of interactive science centers in the world and has members in more than 40 countries. In addition to science centers, members also include planetariums and space theaters, natural history and children's museums, aquariums and zoos, and related organizations and professional associations.

ASTC'S MISSION

ASTC is an organization of science centers and museums that are dedicated to furthering the public understanding of science among increasingly diverse audiences. ASTC encourages excellence and innovation in informal science learning by serving and linking its members worldwide and advancing their common goals. Through a variety of programs and services, ASTC provides professional development for the science center field, promotes best practices, strengthens the position of science centers within the community at large, and fosters the creation of successful partnerships and collaborations.

- The ASTC Annual Conference is the world's leading meeting for science center professionals. More than 1,800 science center directors, scientists, educators, and exhibit creators attend more than 100 conference sessions, professional development seminars, interest group meetings, and educational field trips. Throughout the year ASTC offers online and regional workshops providing professional development on a variety of topics.
- ASTC assists science centers and youth museums in developing effective educational experiences and mentoring programs for adolescents who have the fewest opportunities. With local community partners, the centers guide youth in learning about science and developing important work and personal skills.
- ASTC serves as liaison between the science center community, the United States Congress, and federal agencies to ensure support for science learning. ASTC organizes and circulates hands-on science exhibitions that are available to museums throughout North America.
- ASTC maintains a website featuring content resources for the field, publishes a bimonthly news journal and other print publications, and tracks trends and compiles statistics to guide science center operation.

Although ASTC remains the largest and most global of the science center networks, others have emerged to serve the needs of science centers in Europe (ECSITE), Latin America (Red-POP), the Asia-Pacific region (ASPAC), and other areas. Every three years, beginning in 1996, these networks have organized a Science Centre World Congress, an opportunity for lively discussion and commitment to the mission common to all. Within this strong global network, ASTC's focus has been on serving science centers' needs for information, advocacy, and professional development.

ABOUT SCIENCE CENTERS

Science centers connect people with science. Science centers give science a presence in the community and offer people of all ages and backgrounds the opportunity to ask questions, discuss, and explore. Science center visitors encounter hands-on exhibits such as giant levers, wave tanks, and walk-in kaleidoscopes. They may go to a demonstration, watch a sky show or big-screen film, participate in a workshop, or even take part in a debate about a current issue such as bioethics. In

the process, they can experience the pleasure of lifelong learning, whether with family, friends, or on their own. At science centers, everyone is welcome.

Science centers provide firsthand experience and an opportunity to develop intuitions about the natural world. In science centers, people can feel infrared radiation, and experience angular momentum—so when they encounter these concepts in other settings, they'll be likelier to understand them. That's why schools rely on science centers for memorable field trips and auditorium programs, hands-on curriculum, science kits, and even training for teachers.

Science centers encourage curiosity. Exhibits that are beautiful or surprising—or even funny—can encourage visitors to approach new phenomena and ideas. In the words of Frank Oppenheimer, founder of San Francisco's Exploratorium, "No one ever failed a museum." For some, the interests awakened by science center experiences have turned into a passion for science, and the beginning of a lifetime devoted to teaching or research.

KEY FINDINGS FROM THE ASTC SOURCEBOOK OF SCIENCE CENTER STATISTICS 2004

Updated data based on a June 2005 report of an ASTC survey is available on ASTC's website under Science Center Highlights.[1]

- ASTC members are located in more than 40 countries. The 185 respondents to the 2004 *Sourcebook* survey included 154 from the United States, 7 from Canada, and 24 from 20 other countries.
- Total attendance reported by all respondents was 63,207,103. In the United States, total attendance reported by all institutions was 42,311,024. Extrapolating from reported figures to all ASTC members, we estimate that 83.3 million people were served by ASTC's 420 science center and museum members worldwide during their most recent fiscal year. This includes an estimated 60.3 million served by the 342 science center and museum members in the U.S.
- School groups make up a significant percentage of science center and museum attendance. Worldwide, respondents reported serving 16.5 million students in school groups. Much of the service to schools is off-site, through auditorium and other outreach programs. Among the 119 institutions that reported both school off-site attendance and total off-site attendance, 96.1 percent of total off-site attendance was made up of students in school groups.

- Science centers and museums also offer programs designed for after-school hours; the number of institutions in the United States reporting such programs increased from 47 percent at the time of the last survey (2002) to 57 percent in 2004.
- Membership programs are offered by 88.1 percent of survey respondents. 153 institutions reported a total of more than 773,777 individual, family, and senior memberships.
- Survey respondents reported 18,558 paid employees: 181 institutions reported 10,768 full-time employees, and 178 reported 7,790 part-time employees. On, average, personnel costs constitute 55.9 percent of operating expenses.
- Most science centers and museums (90.3 percent) have volunteer programs. 162 institutions reported 41,598 volunteers. The number of volunteer hours reported by 146 institutions worldwide totaled 2,435,075. Among 129 U.S. respondents, volunteer hours totaled 2,026,790—a contribution worth $41.9 million.
- Large-format theaters were reported by 33 percent of survey respondents. Also common are planetariums (35 percent have them) and outdoor exhibit areas, or science parks (32 percent).
- For every operating dollar that survey respondents take in, 27 cents comes from public funds; 24 cents from private donors; 3 cents from endowment income; and the remainder, 46 cents, from ticket sales, program fees, facility rentals, and other sources.
- Most respondents (87.6 percent) charge for general admission. The median general admission ticket price is $7 for adults and $5 for children.
- Nearly two-thirds of respondents have endowments. However, endowments are less common in science centers (57.5 percent have them) than in other kinds of museums participating in this survey (73.4 percent).
- Finally, operating expense variation is great among the science centers and museums that participated in this survey. At one end of the scale, is an institution reporting annual operating expenses of $7,460, and at the other end, one reporting expenses of $123,780,000.

Association of Science-Technology Centers Incorporated (ASTC)
1025 Vermont Avenue, NW, Suite 500
Washington, DC 20005-3516

ASTC's website, www.astc.org, provides links to science centers worldwide and resources for the field. It is the central source for information on members' news and on developments in the science center field.[2]

Association of Science-Technology Centers Incorporated (ASTC)
website: www.astc.org

SCIENCE CENTER NETWORKS

Global

Association of Science-Technology Centers Incorporated (ASTC)
1025 Vermont Avenue, NW, Suite 500
Washington, DC 20005-3516
United States of America
phone (202) 783-7200
fax (202) 783-7207
e-mail: info@astc.org
website: www.astc.org

ICOM International Committee of Science and Technology Museums (CIMUSET)
c/o Paul F. Donahue, President
Canada Science and Technology Museum Corporation
2421 Lancaster Road
P.O. Box 9724, Station "T"
Ottawa, ON K1G 5A3
Canada
phone (1)(613) 993-8365
fax (1)(613) 990-3635
e-mail: pdonahue@technomuses.ca
website: www.cimuset.net

Asia

Asia Pacific Network of Science and Technology Centres (ASPAC)
c/o Questacon—The National Science and Technology Centre, King Edward Terrace
Canberra ACT 2600
Australia
phone (61) (2) 6270 2811
fax (61) (2) 6273 4346
e-mail: bhoneyman@questacon.edu.au
website: www.aspacnet.org

Australasian Science and Technology Exhibitors Network (ASTEN)
c/o Scienceworks
2 Booker Street
Spotswood VIC 3015
Australia
phone (61)(3) 9392-4801

fax (61)(3) 9392-4848
e-mail: ghamilto@museum.vic.gov.au
website: www.astenetwork.net

National Council of Science Museums (NCSM)
Sector V, Block GN
Bidhan Nagar
Calcutta 700091
India
phone (91)(033) 2357-9347
fax (91)(033) 2357-6008
e-mail: ncsmin@giascl01.vsnl.net.in
website www.ncsm.org.in

Europe
European Collaborative for Science, Industry &
   Technology Exhibitions (ECSITE)
70 Coudenberg, 5th Floor
B-1160 Brussels
Belgium
phone (32)(2) 649-7383
fax (32)(2) 647-5098
e-mail: wstaveloz@ecsite.net
website: www.ecsite.net/new/index.asp

Ecsite-UK
Wellcome Wolfson Building
165 Queensgate
London SW7 5HE
United Kingdom
phone (44)(0) 20 7019 4955
fax (44)(0) 870 770 7102
website: www.ecsite-uk.net

Irish Science Centres Association Network (iSCAN)
c/o Annette McDonnell
Royal Dublin Society
Ballsbridge
Dublin 4
Ireland
e-mail: annette.mcdonnell@rds.ie
website: www.iscan.ie

Nordisk Science Center Forbund (NSCF)
NSCF Administration 2001-2003
Experimentarium
Tuborg Havnevej 7
Box 180
DK-2900 Hellerup
Denmark
phone (45) 3927-3333

fax (45) 3927-3395
e-mail: nscf@experimentarium.dk
website: www.nordicscience.org

Latin America/The Caribbean
Associação Brasileira de Centros e Museus de Ciência
   (ABCMC)
José Ribamar Ferreira, President
Espaço Museu da Vida
Fundação Oswaldo Cruz
Av. Brasil, 4365 - Manguinhos
21045-900 Rio de Janeiro RJ
Brazil
phone (55)(21) 3865-2121
fax (55)(21) 3865-2131
e-mail: riba@coc.fiocruz.br
website: www.abcmc.org.br

Red de Popularización de la Ciencia y la Tecnología
   para América Latina y el Caribe (Red-POP)
c/o UNIVERSUM
Museo De Las Ciencias
UNAM—Universidad Nacional Autónoma De México
Edif. Universum, 2 Piso
Zona Cultural De Ciudad Universitaria
CP 04510, México
Distrito Federal
phone (52) (5)6654136 / 6227265
fax (52) (5)6654652
e-mail: juliatag@redpop.org
website: www.redpop.org

North America
Asociación Mexicana de Museos y Centros de
   Ciencia y Tecnología (AMMCCyT)
Dirección General de Divulgación de la Ciencia,
   UNAM
Universum, edificio C, Tercer piso, Circuito Cultural
C.U. 04510, Coyoacán
México
Distrito Federal
phone (52)(5) 622-7277
e-mail: ecra@servidor.unam.mx
website: www.ammccyt.org.mx

Canadian Association of Science Centres (CASC)
100 Ramsey Lake Road
Sudbury, ON
Canada
P3E 5S9
phone (705) 522-3701 ext. 296

fax (705) 522-4954
e-mail: ross@sciencenorth.ca
website: http://canadiansciencecentres.ca

Africa
The Southern African Association of Science and
    Technology Centres (SAASTEC)
Alfred Tsipa, President
Unizul Science Centre
Richards Bay
South Africa
phone (035)797-3204
fax (035)797-3204

e-mail: atsipa@iafrica.com
website: www.saastec.co.za

NOTES

1. Reprinted with permission from the ASTC Sourcebook of Science Center Statistics 2004. Copyright © Association of Science-Technology Centers Incorporated.
2. Copyright © Association of Science-Technology Centers Incorporated, 2005. All rights reserved. (a) "Science Center History" by Wendy Pollock, available at www.astc.org/about/backgrounders.htm; (b) "About Science Centers," available at www.astc.org/sciencecenters/index.htm; (c) "Science Center Networks," available at www.astc.org/profdev/networks.htm.

CHAPTER 46

Fanning the Flames: The Exploratorium at the Birth of the Science Center Movement

Robert Semper

IN DECEMBER 1977, I CAME TO THE EXPLORATORIUM AS the project director of a new U.S. Department of Education's Fund for the Improvement of Postsecondary Education (FIPSE) project designed to foster the development of new science centers on university campuses around the country. I soon found out that I had joined an organization that had as its core mission not only the creation of a unique public learning environment with interactive exhibitry in the airplane hangar–like Palace of Fine Arts in San Francisco but also the more ambitious goal of fostering the development of like-minded informal science education institutions nationwide. Universities were just a new venue for the spread of the educational vision of the founding director, Frank Oppenheimer.

The Exploratorium's roots, like some of the other science centers that were formed in North America at the end of the 1960s, grew directly out of the science education reform efforts of the time. Prior to this, science museums were primarily object and collections based institutions. The science and industry museums of the twentieth century, descendents of the history of science and technical museums of the 1700s and 1800s, had begun to add models and interactive exhibits demonstrating the science principles behind their objects (the Museum of Science and Industry in Chicago was one such institution, inspired by the Deutsches Museum in Munich, as was the Evoluon in Eindhoven and the old Children's Gallery in the London Science Museum. These institutions tried to connect more directly with visitors in a variety of ways, including adding hands-on exhibitry). What was different about the institutions such as the Lawrence Hall of Science in Berkeley, the Ontario Science Centre in Toronto, and the Exploratorium in San Francisco is that they began with no physical collections at all but rather a set of interactive exhibits, often based on school curricula, which presented natural and scientific phenomena to the public in an experiential and manipulative manner.

Oppenheimer believed in the power of learner-centered science education and sought to implement this approach by providing individuals with direct access to the phenomena of nature through carefully designed exhibits and experiences. He initiated a focus on basic everyday science and human perception. He fostered the incorporation of the artistic process in the work of science education. And he believed in the opportunity that environments such as the Exploratorium provided individuals to experience this way of learning.

Part laboratory, part workshop, part playground, the Exploratorium's playful and nonpretentious environment encouraged exploration and inquiry. Its "always under development" look, with an exposed shop and handmade exhibitry, made it seem approachable to visitors even as the subject matter of the exhibits was on the cutting edge of scientific discovery. Most important, it encouraged the visitor to do things themselves.

Almost from the day it opened, people came by the Palace of Fine Arts to meet Oppenheimer and to visit the fledgling Exploratorium with an eye to developing something similar for their own community. Oppenheimer's impassioned presentations about an educational museum, one with collections of phenomena and ideas rather than objects, stimulated a lot of interest in the 1970s. His charismatic charm and proselytizing zeal convinced many that they could realize their dream of creating an engaging public educational institution in their hometown.

The museum saw itself as actively having a role in stimulating the creation of new institutions and influencing the work of existing ones. Far from being a side issue, the museum took on dissemination activities as a core element of its work. This interest in supporting the growth of the science center movement

Figure 46.1.   Exploratorium floor, San Francisco, Calif.

led the Exploratorium to develop a number of different programs and tools to help people create their own institutions. From its early days, the museum actively developed programs and solicited funds to support dissemination efforts. By 1982, this work constituted over 10 percent of the institution's budget. This conscious effort at dissemination, coupled with its user-friendly style and its research-and-development approach, created the opportunity to influence the development of science centers worldwide.

OPPORTUNITIES FOR DISCUSSION

In addition to general visitation, the museum early on developed organized programs to provide opportunities to study its workings in depth. The FIPSE project mentioned at the start of this article was based on Oppenheimer's own experience at the University of Colorado in the 1960s prior to his starting the Exploratorium. While teaching in the physics department, he developed a freshman physics lab with ten different individual open-ended laboratory settings. Rather than having every student do the same lab on a given week, he designed an experience where they could choose a set-up on the basis of their own interest. This new kind of freshman physics laboratory was one of the experiences that he used to set up the Exploratorium.

As he thought of how to help new science centers get started, he realized that universities had the science faculties and infrastructure that could support the development of exhibits and the creation of an exhibition space for both students and the public. In

its three years of existence, the FIPSE project worked with twenty-one different museum projects and hundreds of visiting professionals, providing workshops at the Exploratorium on exhibit development and museum design. Exploratorium staff also visited the participating museum locations to talk with staff and supporters. Of the twenty-one projects that were part of the program, many are still in existence today, including the Ann Arbor Hands-On Museum, the Discovery Place in Charlotte, the Science Museum of Virginia, the Reuben H. Fleet Science Center, the Children's Museum of Manhattan, Lakeview Museum in Peoria, and Omniplex in Oklahoma City, to name a few.

The early institutional development of work by the Exploratorium was continued for the museum community through the offering of exhibit development workshops for exhibit builders and educators from small and large museum. Sponsored as one of three museum improvement projects in the 1980s by the Kellogg Foundation and directed by Exploratorium staff member Sally Duensing, these workshops involved hundreds of museum staff members from many different kinds of museums from the United States and around the world. The workshops included opportunities to meet with Exploratorium exhibit developers and staff, to discuss particular museum development issues with one another, and to develop exhibit ideas in the Exploratorium exhibit development shop.

TOOLS FOR DISSEMINATION

People often said in those early years that they wished that they could take back home the experience that they had when visiting the Exploratorium to show to others what they were talking about. The visit experience was simply too difficult for people to put into words. So in 1974, Oppenheimer wrote a proposal to the National Science Foundation (NSF) and the National Endowment for the Arts (NEA) to support the development of a film designed to give people an experience of visiting the museum. Produced and directed by a talented young filmmaker Jon Boorstin, the final twenty-three-minute film "The Exploratorium" was composed of scenes from the exhibit floor of visitors playing with exhibits filmed using a hidden camera. The film had a specially composed score but no narration and was designed to offer the sense of the visit experience in a way that no amount of text could provide. This film was circulated by the Exploratorium to countless developing projects where it was shown to potential staff members, funders, board members, and community officials of these fledgling enterprises. The Exploratorium film with its equal educational and artistic goals was one of the first projects ever jointly funded by NSF and the NEA, and it received a 1974 Academy Award nomination for short subject documentary—a first for NSF.

The Exploratorium film was followed by the creation of an equally important dissemination tool, the Exploratorium Cookbook, a 1975 publication designed to help science centers construct educational interactive exhibitry. The Cookbook consists of a set of recipes for exhibit development that included schematic plans for creating exhibits based on existing Exploratorium developments. The first book, containing eighty-two exhibits, was written by Ray Bruman, under an NSF grant, and was quickly followed by volumes II and III written by Exploratorium staff member Ron Hipschman. Along with the intended audience of museum developers, the book was developed with an educational audience in mind as well. The original book was developed in loose-leaf form so that teachers could buy a single recipe for a class project.

The Cookbooks have been used by many developing science centers all over the world. Over 10,000 copies have been sold to date, and the Exploratorium still sells over 100 copies of each volume per year, twenty-five years after their initial production. At Oppenheimer's death, a companion booklet called *Working Prototypes* was created to discuss the exhibit development methodology of the Exploratorium. While a few people used the Cookbooks to build specific exhibits, the much more common use was to stimulate the thinking of exhibit development and museum design at various settings. The straightforward nature of the design principles helped many smaller museums adopt a do-it-yourself exhibit development philosophy and realize that they could develop an exciting environment for learning.

Another important dissemination vehicle was filmmaker Jon Else's 1982 *NOVA* PBS documentary on the Exploratorium titled "The Palace of Delights." Spawned through discussions at the 1981 Exploratorium Science Media Conference, a meeting that brought together scientists, journalists, and television and radio producers to discuss the issues of informal public science education, the hour-long documentary presented a working view of the design, philosophy, and operation of the Exploratorium. Besides its repeated broadcast on PBS, the film was also loaned out by the Exploratorium to many different museums to help start-up efforts.

These tools have been used by people all over the country and the world to develop and expand their own vision of creating a place of discovery. It is interesting to note that each of these places has developed their own style and sensibilities and none of them turned out to be the same as the Exploratorium.

CURRENT DEVELOPMENTS AT THE EXPLORATORIUM

As other institutions have flourished and the science center field has developed, the Exploratorium has also continued its developmental trajectory. It has learned much from what other colleagues and institutions have been doing. Coming full circle from its roots as an outgrowth of the school science curriculum development efforts of the 1960s, the museum developed a significant set of teacher-professional development programs through the 1970s and 1980s, which taught how to use the resources and ideas of the museum to support science teaching in the schools. This work has led more recently to an international effort to support the work of museums with schools through the NSF-funded Center for Informal Learning and Schools, a collaboration of the Exploratorium, University of California Santa Cruz, and Kings College London. An important part of the center is the Informal Learning Certificate program, which provides an opportunity for staff development for museum educators.

The development of the Internet has allowed the Exploratorium to create an entirely new learning opportunity for a worldwide audience. In the early 1990s, it discovered that the advent of the World Wide Web meant that it could extend its efforts in providing learner-centered education to a remote, nonvisiting audience. We found that our experience with physical exhibit development naturally translated into the world of this new electronic domain, and this medium allowed us to develop science center experiences for a new audience of the general public, teachers, and students.

In 1991, French physicist and science educator Goéry Delacôte became the new executive director of the Exploratorium. Delacôte was himself in residence at the Exploratorium for a few months as he developed his thinking for the new national science museum at la Villette in Paris, the Cité des Sciences et de l'Industrie. During his tenure at the Exploratorium, he has encouraged the institution and the field as a whole to extend its impact by adding research capacity and adopting a network strategy. His idea is to connect the strengths of individual institutions in different physical locations by linking them in a knowledge, support, electronic, and physical network to become a significant force for educational reform. It is his view that currently, our individuality, a deeply imbedded and often valuable feature of our institutions, limits our ability to be a serious player in K-12 education. We present a fragmented, disorganized, and uneven partner to the world of formal schooling. By coming together in a network of sharing, standards, and communication, our field can become a major player in the educational world. This will require the creation of a new culture of learning for our field. (Note: Delacôte left the Exploratorium and became chief executive at At-Bristol in the United Kingdom in September 2005.)

To implement this vision, the Exploratorium with three original partner museums developed ExNET, the Exploratorium Network for Exhibit-based Teaching. With a rotating set of leased exhibits, staff development opportunities, and an ongoing electronic network, the ExNET partnership now involves eight museums in the U.S. along with many additional international representatives. Recent work has seen the network partners develop their own networks in their region to carry on similar work. It is through partnerships and collaborations like these that science centers will be able to extend their influence.

In thinking back on the history of the Exploratorium and the development of small science centers, I think it is useful to remember that the Exploratorium was actually once a small science center itself. Like many other places, it started small, in a borrowed building, with volunteer help, a set of loaned exhibitions, a few hand-built exhibits, a tiny start-up grant, and few public expectations. What it did have was an inventive spirit, a flexible style, a frugal operation, a strong philosophy, and a huge ambition. In short, it had that start-up, entrepreneurial energy that is so important for sustainability. Throughout its development, the key feature for its success was its tenacious adherence to its mission, despite many seemingly attractive opportunities that would have moved it off the mark. It is this development and maintenance of a strong personality that shows through to visitors and supporters alike that sustains any institution, large and small alike. (Note: Dennis Bartels, PhD, became the executive director of the Exploratorium as of May 1, 2006.)

CHAPTER 47

SESAME Program: The Impact of a PhD Program on Science Museums

Vicki Breazeale and Watson "Mac" Laetsch

ORIGIN AND HISTORY OF THE SESAME GROUP

This chapter describes the origin and history of the Search for Excellence in Science and Mathematics Education (SESAME) Group at the University of California, Berkeley (UCB) from the perspective of one of its founders and a student who participated in the program. We discuss how the Group uniquely prepared science museum and postsecondary education professionals to become leaders in their fields, how this program continues to have an impact on those fields, and its relevance as a model for other graduate programs for science museum professionals.

The SESAME Group was created in a time of renewed national emphasis on science education and because the regents of the University of California wished to establish a memorial to honor E. O. Lawrence upon his death in 1958. The launch of Sputnik by the Soviet Union on October 4, 1957, is cited as the main event that instigated nationwide reform in science education, and E. O. Lawrence was Berkeley's first Nobel laureate, the first from a public university, and a vital contributor to the development of atomic energy.

The memorial selected for Lawrence was the Lawrence Hall of Science (LHS): "an educational center with displays to be viewed by young and old and facilities for teacher training." The regents appointed physics professor Harvey White as the first director of LHS in November 1959. White was an obvious candidate for this position, since he was the star of a popular science television series, *Continental Classroom*, viewed by millions and broadcast by over 150 television stations worldwide. He spent several years developing science exhibits for LHS in a temporary campus building before its official opening in 1968. The Exploratorium was founded a year later in San Francisco by Frank Oppenheimer, who had conducted research with his brother, Robert, at Lawrence's Radiation Laboratory.

An interdisciplinary committee of Berkeley faculty submitted a proposal in 1967 for the Graduate Group in Science and Mathematics Education to the Graduate Council. Alan Portis, professor of physics, led efforts to create the Group. He and other members of the committee became key faculty in the Group, including physics professor Robert Karplus, mathematics professor Leon Henkin, botany professor Watson Laetsch, and physics professor Frederick Reif.[1] The proposal was approved in sufficient time for the Group to admit its first graduate students in fall 1968. The Group was housed in the Physics Department in Le Conte Hall until the mid-1980s.

The SESAME faculty collectively agreed to require a master's degree in an area of science represented by a faculty member of the Group as well as high academic standards for admission to the Group. The Group's original curriculum required students to close gaps in their particular science discipline and to acquire a comprehensive knowledge of educational theories and research methodologies prior to conducting original research into an aspect of teaching and learning in a specific science.

The goal of the Group was to train scholars with the background to communicate with scientists as well as educational researchers and practitioners. At the time of its creation, there were no programs like SESAME at any college or university in the nation. Even now, there are only a few programs that resemble SESAME, such as the science and mathematics education programs at University of Washington in Seattle and the University of Georgia in Athens. Two relatively new master's/doctoral/professional certification programs, funded by the National Science Foundation (NSF) and focused on informal science education research, include the Center for Informal Science and the Schools (CILS), a collaboration of the Exploratorium; the University of California at Santa Cruz; Kings College,

University of London; and the Center for Inquiry in Science Teaching and Learning (CISTL), Washington University at St. Louis. However, in both cases the focus is on relationships between K-12 education and informal settings such as museums and science centers, rather than the unique contributions of such learning, including learning from the Internet, community-based organizations, libraries and books, magazines, and film and television, all possible within the SESAME model.

The program had tremendous impact at UCB. Professor Richard White, a member of the Department of Electrical Engineering and Computer Sciences (EECS) and the SESAME chair from 1980, became interested in SESAME in the early 1970s and learned about self-paced instruction through the Group. He subsequently developed self-paced courses in his department. Development of NSF-funded self-paced courses in mathematics, physics, computer science, and biology exemplifies the special collaboration between SESAME faculty, LHS, and campus academic departments, who made contributions to the university curriculum. It also demonstrated a unique interaction between a science center and university education.

Another example of the impact of LHS and SESAME on instruction on the Berkeley campus is provided by the early use of computers. Lawrence Hall of Science offered the first general public access to computers in the Bay Area; the LHS time-share computer was the only one on campus. This computer enabled students to experience live computer use instead of batch processing, and this capability was made available to the 800 students in the introductory course for biology majors (Biology 1) by means of terminals supplied by LHS. SESAME student Ruth von Blum developed instructional modules for Biology 1 and analyzed their effectiveness. This was the first use of computer-assisted instruction (CAI) on the campus for undergraduates outside of the Computer Science Department. Biology 1 is now available online. In turn, experience gained in this course influenced computer activities for school children and the public at LHS.

The program also influenced science education nationally and internationally, particularly in the area of innovative curriculum development. NSF began funding university-sponsored curriculum development projects for the first time in the late 1950s and the 1960s to improve both elementary and secondary school science. Karplus, one of the founders of the SESAME program, initiated one of these NSF-funded curriculum development projects in 1962. The success of the Science Curriculum Improvement Study (SCIS), both nationally and internationally, prompted Karplus to propose the creation of the Center for Research in Science Education to the Berkeley Campus administration in 1965. The Center would house SCIS and foster analogous activities by other faculty. In July 1967, the chancellor's office decided to move SCIS to the soon-to-be-opened Lawrence Hall of Science.

Laetsch became intellectually interested in informal science education while he was director of the U.C. Botanical Garden (1969–1973). He led the development of the Outdoor Biology Instructional Strategies (OBIS) curriculum at the garden and the LHS, the first NSF grant to support informal science education efforts. John Falk was hired as a curriculum developer for the OBIS project, and his thesis research provided the basis for a number of OBIS activities. OBIS is a useful resource to many small nature centers and science museums with outdoor biology programs and continues to be distributed by LHS and used internationally.

The move of SCIS to LHS, its involvement with OBIS, and the tenures of Portis (1970–1972) and Laetsch (1972–1980) as directors of LHS began a long and productive cross-fertilization between the SESAME Group and LHS, which continues today. Demonstrating the continued strong connections between faculty members in the Graduate School of Education and SESAME is the work of Lawrence Lowery, who became an active member of the Group in the early 1970s. He developed the Full Options Science System (FOSS), a nationally and internationally acclaimed hands-on science curriculum for grades K-8, at LHS. Growth of FOSS continues at LHS with Lowery as the principal investigator. These interactions encouraged both science and education faculty to become involved with SESAME, and a number of them advised SESAME degree candidates.

The SESAME program was a national and international leader in other ways as well, forging a number of strong links nationally and internationally. Laetsch was one of the Association of Science-Technology Centers (ASTC) founders in 1973 and he served as its first president. The close interaction between ASTC and the Indian Council of Science Museums began when Laetsch was a member of the Indo-U.S. Subcommittee on Education and Culture. LHS staff, SESAME graduate, Falk, Laetsch, and a number of staff from ASTC institutions consulted with Indian science museums in the 1980s.

Table 47.1. SESAME Thesis Categories

Category	Number
Informal Education	7
General Cognition	18
Mathematics	13
Computers/Technology	18
Physics	7
Biology	7
Chemistry	2
Nutrition	2
Totals	73

The partnership between the SESAME Group and the Lawrence Hall of Science also created the first academic center for informal science education in the early 1970s. A number of directors, senior staff, and consultants to today's science museums worked at LHS and were closely associated with SESAME. They include Alan Friedman, director of the New York Hall of Science; George Moynihan, past director of the Pacific Science Center (PSC); Dennis Schatz, associate director of education at PSC; Ted Kahn, president and CEO of DesignWorlds for Learning, Inc.; and Cary Sneider, vice president for programs at the Museum of Science in Boston.

Throughout its existence, the SESAME Group has remained small, intimate, and interdisciplinary. A total of seventy-four PhDs were awarded in the Group from 1972 through to 2003, with an average graduation rate of 2.4 students per year. A summary of thesis categories is provided in Table 1.

Information is available for the current professions of forty-seven SESAME graduates (64 percent of the total from above). The majority of these graduates (53 percent) are on faculties of colleges and universities where they are engaged in curriculum and program development, educational research, and teaching. Of this group, most (80 percent) are faculty in departments of their original science discipline (mathematics, physics, chemistry, biology, and computer science), while the remainder (20 percent) are faculty in education departments. A significant number of these graduates (21 percent) have started their own for-profit and nonprofit organizations focused on a variety of educational issues, needs, levels, and modes. A similar number of graduates (19 percent) have spent most of their postgraduate careers as science museum professionals. The remainder (7 percent) worked in the educational technology industry as chief administrators and directors responsible for program training and development.

Many of the students who conducted research in informal education, as well as several in biology and one in nutrition, were students of Laetsch during his tenure as director of the Lawrence Hall and as a vice chancellor. These graduates who worked in informal education either for or during their graduate research are, in chronological order, John Falk, Jeffry Gottfried, Judy Diamond, Sherman Rosenfeld, Vicki Breazeale, and Samuel Taylor. Falk conducted his research on lawns at the LHS and around his home. Gottfried and Diamond conducted their research at LHS with Diamond using research tools from the scientific study of animal behavior (ethology) to conduct her research. This methodology was subsequently used by Rosenfeld at the San Francisco Zoo and Taylor at California Academy of Sciences in San Francisco's Golden Gate Park where he gained considerable insight into visitor behavior and learning in a natural history museum. Breazeale worked at LHS while conducting research on the public's concepts about nutrition. This research led to the development of public nutrition exhibits, one of which traveled with other exhibits to shopping malls throughout the country in the early 1980s; these were the first traveling science exhibits developed specifically for major shopping malls.

The SESAME graduates who became science museum professionals have made significant contributions to both the development of science museums of all sizes and the elucidation of informal learning, increasingly referred to as free-choice learning. Falk is the most cited author and coauthor of the most cited publication in science museum literature. Diamond's (now a professor and associate director of the University of Nebraska State Museum) and Rosenfeld's (now at the Weizmann Institute of Science in Israel where he works on projects in both formal and informal science education) doctoral theses are among the top publications cited in museum literature. Until recently, Taylor was editor of *Curator* (1991–2004) and director of Samuel Taylor Museum Consulting in Morristown, N.J. He has served as director of exhibitions at the American Museum of Natural History in New York and curator of the Education Department at the California Academy of Sciences in San Francisco. The work of these leaders has been useful to professionals starting and growing science centers.

Several SESAME students besides the original group advised by Laetsch are active in science museums. Sue Allen is director of visitor research and evaluation at the Exploratorium. Marcelle Siegel works as an educational specialist with the Science Education for Public Understanding Program (SEPUP) at the Lawrence Hall. Sherri Hsi is director of research at the Exploratorium's Center for Learning and Teaching. And the program continues to support students interested in learning in these contexts. Two current SESAME students are conducting research in informal education: Scott Randol and Timothy Zimmerman. Randol is interested in assessing and fostering inquiry in science centers, and Zimmerman is investigating the use of technology to integrate classroom learning and information gained in field trips to the Monterey Bay Aquarium.

HOW SESAME UNIQUELY PREPARED SCIENCE MUSEUM AND SCIENCE EDUCATION PROFESSIONALS

Several attributes of the SESAME Group are key factors in its continued vitality and influence. The founders of the Group were tenured science faculty prior to the creation of SESAME. Serendipity was certainly a factor in their initial encounters, but their shared intellectual interest in science education fostered their association. The SESAME Group became the perceived intellectually acceptable forum for continuation and growth of this interdisciplinary collaboration of scientists, educational researchers, and science educators.

There is no algorithm for the unique interdisciplinary collaboration of scholars that created the SESAME Group. Many, perhaps most, educational innovations are difficult to sustain beyond their creation, but SESAME has been an exception. Barbara White, current chair of the SESAME Group, says, "SESAME is a well-developed and highly respected PhD program that is well-known internationally. We believe we have one of the richest, most diverse and multidisciplinary contexts available in the world."

SESAME students participate in a distinctive learning process. Science is a major part of their intellectual framework upon entrance into the Group, since they have completed a minimum of six progressive years obtaining the required bachelor and master's degrees. The SESAME curriculum takes students through a comparable instructional process of rigorous study in a compressed amount of time, beginning with concentrated immersion into educational philosophies, theories, and practices. The study of science from general to specific followed by study of education from general to specific is highly effective in cultivating creative questions and insights. The value of a science background is both the knowledge and a well-developed sense of intellectual inquiry, which fosters the practice of constant questioning. These intellectual values and practices are particularly important for the health of the science center community.

SESAME is an interdisciplinary graduate group rather than a department. This allows the Group an element of fluidity and spontaneity that is rarely possible in a discipline-oriented department. Graduate groups however are more transitory than departments. The long existence of SESAME is an affirmation of its value and effectively demonstrates how a small, intellectually based program can have a significant impact on a field or fields. Given these characteristics, SESAME represents a useful model for the creation of similar programs focused on the graduate training of science museum professionals.

NOTE

1. Reif conceived of the name and acronym SESAME— Search for Excellence in Science and Mathematics Education.

CHAPTER 48

Science and Discovery Centers: The European Perspective

MELANIE QUIN

HISTORICAL BACKGROUND

The modern European science center has an interesting family tree. One line of heritage can be drawn from the grand museums of science and industry established at the start of this century; here are examples:

- The London Science Museum, whose origins can be traced to the Great Exhibition of 1851 and which opened in its present building in 1928
- The Deutsches Museum, Munich, founded in 1903, opened to the public in 1925
- The Palais de la Découverte, Paris, opened in 1937 for the International Exhibition of Art and Technology
- The Museum of Science and Industry, Chicago, 1933

The other line derives from the 'hands-on' philosophy developed in North America through the decade of science-education reform that followed Sputnik's launch in 1957.

Since the 1960s, the educational philosophy and methods of the Exploratorium, the Ontario Science Centre, and half a dozen other pioneer North American science centers have, in turn, provided inspiration for institutions around the world. New science centers have been developed, and established science museums have borrowed the techniques of the science centers to give themselves a face-lift.

In October 1987, the Nuffield Foundation established a three-year in-house project (the Interactive Science and Technology Project) to help the development of interactive exhibitions—by providing a focus for information exchange both in the United Kingdom and internationally—and to promote the development of hands-on ideas and methods.

The Nuffield project served as a resource for the science centers and, building a strong network of contacts stretching from the British Broadcasting Corp. (BBC) and the British Association for the Advance-

ment of Science (The BA) to science centers worldwide, itself served as a launch pad, in 1989, for ECSITE—the European Collaborative for Science, Industry and Technology Exhibitions (www.ecsite.net). The foundation of ecsite also reflected the growing interest of traditional museums in a new medium of science communication—the flavor of interactivity is increasingly pervading institutions formerly best known for their historic collections and elegant glass cases. The ecsite membership (some 300 voting and associate members) includes independent science centers, museums committed to the interactive approach, educational and scientific research institutions, and the exhibition and software designers who supply the sector.

ECSITE IN A NUTSHELL

Ecsite's main functions are the following:

- Publishes a newsletter
- Runs a website
- Convenes meetings and conferences
- Provides an infrastructure for training and staff exchanges
- Provides guidelines on developing travelling exhibitions
- Maintains a database of experts
- Fund-raises and runs projects using European Community (EU) money
- Generates private sponsorship for specific projects, for example, Chemistry for Life, a travelling exhibition sponsored by large chemical companies

Ecsite facts and figures are as follows:

- Formed in 1989, with the support of the Nuffield Foundation
- Permanent post of executive director created in 1995

- Permanent executive office established in Brussels in 1997
- Operational budget of €120,000 per annum (projects excluded)
- Membership income represents 20 percent of income
- 300 members worldwide; fifty countries represented
- 100 institutions based in the European Union; receiving 30 million visitors per year

RECENT U.K. DEVELOPMENTS

Science centers were introduced to the U.K. in the early 1980s. They built on earlier exhibitions with working models in Europe, and on the hands-on philosophy developed in North America. The first U.K. centers to exemplify this exploratory approach—the Exploratory in Bristol, Techniquest in Cardiff, Technology Testbed in Liverpool, and the Launch Pad gallery at the Science Museum in South Kensington—were soon followed by science centers that opened both in museums and as stand-alone facilities. The hands-on method has proved extremely popular with the public and the approach has been adopted in a wide range of fields.

From the mid-1990s, the sector expanded on an unprecedented scale through Millennium Commission funding, its excellence reflected in the level of investment. The Millennium Commission's investment of £250 million was matched by equivalent funding from other public and private sources. The Wellcome Trust was the single largest private supporter with approximately £45 million invested in eight centers.

Fourteen major new regional science centers have been created—opened in 2000–2001—and are well distributed throughout the U.K. They radically enlarge the U.K.'s science center resource.

The creation of millennium centers and the initiative of the Wellcome Trust to encourage collaboration between the Wellcome-funded centers led to the call for a formal national network, to act as an advocate for science centers—established and new—throughout the U.K.

Ecsite-uk, a formally constituted U.K. chapter of ecsite, was established in Autumn 2001. It is supported by the Government's Office for Science and Technology (OST) and the Wellcome Trust. Ecsite's executive office is hosted by the BA in central London.

The aims of ecsite-uk and the Network of Science Centers and Museums are to achieve recognition for science centers, to integrate them in the framework of government funding for programs, and to attract additional funding for revenue and capital activities.

Ecsite-uk represents a wide range of centers, from new to well established, small to large, national to local—over forty are independent science centers; a similar number are discovery centers in museums, botanic gardens, aquariums, and planetariums. Together, they attract over thirteen million visitors annually.

FROM UNDERSTANDING TO ENGAGEMENT

Once we spoke of public understanding of science. Today we speak of public engagement with science. The tools to be used are dialogue and debate—though our North American colleagues use the words *argumentation* and *discourse*.

The European Union (EU) Action Plan (European Commission 2001), and the U.K. House of Lords Science and Technology Select Committee report (*Science and Society* 2000), both identified a crisis in confidence in science that can only be solved through real dialogue with the public. This is an opportunity and also a challenge. It raises questions of practical development and of personal skills and attitudes:

- How to add to our exhibitions spaces for discussion and debate? How to devise dialogue exhibits?
- How to ensure our visitors are prepared for brains-on activity? How to turn scientists and communicators from talk mode to listen mode?

Concerning development, some outstanding examples were highlighted during the ecsite 2002 conference: live webcasts with CERN particle physicists through Live@Exploratorium; Futures opinion exhibits in the Science Museum's Wellcome Wing; opportunities to discuss issues of biodiversity with taxonomists at the Natural History Museum's Darwin Center.

But what of skills and attitudes? How many visitors arrive with the thinking and debating skills to progress from exchanging misinformed opinion to engaging in meaningful dialogue? The roles of the scientist and communicator are crucial—they must have not only the humility and genuine interest to listen but also the ability to animate dialogue with the skill of a chat-show host.

The sea change from promoting understanding to stimulating and supporting engagement has profound implications for exhibition and program development and science center staff training.

SMALL U.K. SCIENCE CENTERS: CASE STUDIES

	Total Indoors Exhibition Area Square meters	Total Visitors per Annum	Visitors in School Groups	Full-time-equivalent Staff	Operating Cost per Visitor
Curioxity	100	12,500	32%	$1^{1}/_{2}$	£3.20
The Look Out	360	98,000	9%	15	£5.90
The Observatory	900	43,000	25%	7	£6.50
Satrosphere	900	55,000	18%	16	£9.00
Sensation	1,000	62,000	18%	14	£8.06

CURIOXITY, OXFORD

Curioxity is a tiny center, opened in 1990 and operated by the Oxford Trust, which also runs an annual Oxford Science Festival and extensive regional outreach of shows and workshops.

It is a simple one-to-two-hour visit in Curioxity, with a themed demonstration that is most valued by teachers. Oxfordshire has a large number of small rural schools with limited resources for science, and many Oxfordshire primary schools now have a visit included as an integral part of their schemes of work to enhance and extend what they can deliver in their classrooms. Teachers often use a visit to the center to introduce a new topic or to revise and summarize at the end of a classroom topic.

Curioxity is a very small science center (100 square meters or 1,076 square feet). The thirty-six exhibits (please note that on this side of the Atlantic, we say "exhibit" for one unit and "exhibition" for a complete show) follow three themes:

- Light and Sound
- Forces and Motion
- Maths and Problem Solving

The center is specifically designed for primary age children and is seen by many teachers in Oxfordshire as an extra classroom. Described by the *Times* Educational Supplement as "ideal for a teacher with a class group: the hands-on science center without fear," its smallness is its greatest strength.

The key features of a visit to Curioxity for schools are the following:

- One school group has exclusive use of the center at any one time.
- The smallness of the center makes it non-threatening for teachers and pupils particularly in their first years of school—all pupils are within view of the teachers at all time.
- The small number of exhibits means that after the initial excited rush around, pupils spend quality time on each exhibit.
- As a standard part of a visit, a qualified teacher guide leads a demonstration workshop on one of the themes—the layout of the room means that they can use all the relevant exhibits in the demo without having to move the pupils around.
- Equipped with more knowledge about each of the exhibits and the science behind them, pupils then have time to revisit the exhibits.

THE LOOK OUT DISCOVERY CENTRE, BRACKNELL FOREST

Operated as a hands-on center since 1996 by Bracknell Forest Borough Council, most of the Look Out's hands-on exhibits are designed and built by Techniquest in Cardiff, and scripts and training for their shows come from Techniquest, too. The center also offers 2,600 acres of Crown Estate woodland with walks, trails, and orienteering adjoining the center.

The Bugs and Beasties Mini Beast Safari, held annually during the summer term, is typical of their programming, attracting some 5,500 children in sixty days. The forty-minute voyage of discovery into the fascinating world of mini beasts and their environment combines both indoor teaching and outdoor exploration with the aim of fostering a greater understanding and enjoyment of nature and increased sensitivity towards the natural world. The aim is to show the children how to experience and explore nature and to consider how all parts of the woodland ecosystem are linked, so that change in one part affects other parts. It's hands-on, fun, and educational, and has both indoor teaching and outdoor exploring. An Explainer is assigned to each group for the length of the session, ensuring that children receive individual attention. Children learn to be sensitive to nature as well as experiencing an insect's perspective of the world.

THE OBSERVATORY SCIENCE CENTRE, HERSTMONCEUX
See Stephen Pizzey, chapter 10, The Observatory Science Centre, Herstmonceux, and Inspire Discovery Centre, Norwich.

SATROSPHERE, ABERDEEN
Satrosphere today occupies Aberdeen's old Tramsheds. It opened in shop-front premises over ten years ago and moved to the Tramsheds in 2001. There are some 100 exhibits on (changing) themes that include light, energy, human body, communications, and earth and space.

Satrosphere has well-established links with other regional providers of science, technology, engineering, and math activities. To extend and enrich the service to schools, Satrosphere coordinates joint visits to the science center and other local Science Technology, Engineering, and Mathematics (STEM) providers, for example, Satrosphere and the Conoco Natural History Centre, Aberdeen, together offer four topics: Night and Day; the Body; Flight and Feathers; Senses. At Satrosphere, the children (aged five through to seven) attend a thirty-minute show and spend time exploring the hands-on exhibits on the exhibition floor. At the Conoco Centre, the children are involved in various hands-on activities, using the natural history museum, botanic gardens (outdoors and greenhouses), and classroom. The Day and Night topic covers the following:

- At Satrosphere—space; changing day length; sounds of night and day
- At Conoco—night and day animals

These joint visits help schools: a two-center visit reduces transport costs and provides a higher quality experience by combining the strengths of the two centers involved.

SENSATION, DUNDEE
As distinct from the case study centers above, Sensation is a millennium center. Rather than a slow build up, growing as you go from a few tens of thousands of pounds-sterling initial investment, Sensation represents £5 million pounds capital investment over a project-development period of around three years. The building is undeniably a flagship of modern architecture. The same high production values apply to design of the interior and the exhibits. Content—in terms of good science—parallels that of other small U.K. centers (and other millennium centers in the medium and large categories). Sensation's exhibits are based on the senses and the biology of humans, animals, and plants. It also includes an IT resource center, multimedia auditorium, and Live Lab.

Like the dozen other millennium science centers in the United Kingdom, Sensation opened in a blaze of local publicity and enjoyed a honeymoon period of high-level visitation. When the project-development director left, the board was lucky to secure very rapidly a new chief executive with solid experience in the leisure-attraction sector. He set to meet the following challenges:

- Living within a reduced admissions income in the period following the high-profile introduction to the market
- Working harder to create sponsorship opportunities with the center no longer the newest attraction in town
- Establishing a predictable figure for building management and operating costs
- Resolving potentially costly disputes with contractors for buildings and contents

Now in year three, costs have been brought in line with income and future efforts lying in increasing visitors. Sensation is increasingly introducing headlining temporary exhibitions (two to three months) to tell its target audiences that when they next visit they will see and experience something new. This does not, however, remove the need to upgrade the core exhibition.

Sustainable in the short term, the major issue going forward will be capital renewal of exhibits, but today Sensation is one of the few millennium centers (of any size) on track to deliver on its educational mission without recourse to more than a few percent annual operating subsidy from government.

SUSTAINABLE OPERATION

In October 2002, Ecsite-UK conducted a survey. Twenty-eight U.K. centers provided financial and other quantitative information, together with information about the educational programs and exhibitions they provide for visitors to engage with contemporary science and with the issues associated with scientific development.

Examination of operating cost and of staffing levels relative to the scale of operation reveals differences between small and medium-large centers and highlights the science centers' rich educational programming—or lack thereof.

It is indicative of the diversity of the sector that science centers report a gap between income and operating expenditure of 3–86 percent. This must be bridged each year from nontrading sources. For the U.K. sector as a whole, the average is 30 percent as compared with a U.S. sector average of 43 percent.

Those medium-large centers that are near the sector staffing average are millennium centers—lottery funding financed the capital investment, but makes no contribution to running costs. Today, beyond the opening-year visitor high, they have stripped back to make sure they break even—they are operating on a survival-staffing zero-innovation budget. The result is a cutback to very basic education programs and little depth in public programs.

Techniquest, Cardiff, is more highly staffed than average, and those staff (thirty-five FTEs [full time equivalents] of eighty total) are directly related to education and public programming. Techniquest enjoys a worldwide reputation for quality and innovation in the science center sector. Its breadth and depth of education and public programming is supported by funding from the Welsh Assembly Government (www.techniquest.org). Aside from the national museums, Techniquest is the science center best able to deliver on the mission that unites the sector simply because it is in receipt of a level of public funding that recognises the contribution it makes. Techniquest typically earns 65 percent of its operating income and receives 35 percent from government (the Welsh assembly).

Science and discovery centers in the U.K. are bringing in 70 percent of their operating expenditure from trading income (including admission charges), as compared with 57 percent in the United States and 31 percent Europewide.

It should be stressed that this operating expenditure includes exhibit maintenance and rolling development of programming activities but does not include capital renewal. The U.K. sector—the major national museums included—cannot rely on public support for capital renewal and must always deploy a dedicated development team to raise the funds for gallery renewal and any expansion or refurbishment of specialist facilities. This is not a financially sound situation but it is the current U.K. reality.

THE WAY FORWARD

Science and discovery centers are entrepreneurial places, earning a significant proportion of their operating costs. However, their fundamental purposes—in Europe as in North America—are not deliverable on a purely commercial model. They must be commercially efficient but must also secure other kinds of support— not least from government—if they are to make the kind of contribution that was envisaged when considerable capital funding was invested in their creation.

Ecsite-uk has a very clear role in this respect: to work with government officials to secure funding for the sector in return for efficient and effective performance.

Science and discovery centers have a vital role to play as centers of informal learning in support of the formal education sector, and in delivering the *Science and Society* agenda. They already contribute substantially to the delivery of government objectives in several key areas. They have the buildings, facilities, and core staff to contribute further. But they will only be able to continue— and to do more—if they are funded appropriately.

For a science center to pay its own way, it must rely on establishing a subsidiary business that has a significantly greater annual turnover than the center itself. Only in a few special circumstances is this possible. In all other cases, there must be a call on public funds—a call that is justified by the extent to which public agendas are being met by the few centers which do indeed have the support needed to deliver government objectives.

REFERENCES

British Council. Public understanding of science. Briefing Sheet 6. www.britishcouncil.org/science-briefing-sheet-06-public-understanding-july01.doc.

Cordis. 2002. Benchmarking the promotion of RTD culture and public understanding of science.

Dierking, L., and J. H. Falk. 1994. Family behavior and learning in informal science settings: A review of the research. *Science Education* 78 (1): 57–72.

Ecsite, the European Network of Science Centers and Museums. www.ecsite.net.

Ecsite-uk, the Network of Science Centers and Museums. www.ecsite-uk.net.

Ecsite-uk, Data Collection 2001. www.ecsite-uk.net/about, click on <reports>.

Ecsite-uk Member Centers and Museums. www.ecsite-uk.net/centers.

European Commission. 2001. Europeans, science and technology. *Eurobarometer 55.2*. Luxemburg: Office for Official Publications of the European Communities. europa.eu.int/comm/research/press/2001/pr0612en.html.

OST/Wellcome Trust. 2000. Science and the public: A review of science communication and public attitudes to science in Britain. London: Wellcome Trust.

Science and Society. House of Lords Science and Society Select Committee Report, Feb. 2000. www.parliament.the-stationery-office.co.uk/pa/ld199900/ldselect/ldsctech/38/3801.htm.

Association of Science-Technology Centers. *Sourcebook of science center statistics 2001*. Washington, D.C.: ASTC.

WEBSITES Ecsite, the European Network of Science Centers and Museums: www.ecsite.net

Ecsite-uk, the Network of Science Centers and Museums: www.ecsite-uk.net

BIG, the British Interactive Group: www.big.uk.com

The Nordic Science Center Association: www.nordicscience.org/indexeng.htm

Part XII

SCIENCE CENTERS AND THE FUTURE

Raceway into the future. Visitors design and build their own race cars, then try them out on W5's giant Raceway. Photographed by Oscar Williams.

CHAPTER 49

Developing a New Business Model For Science Centers[1]

JOHN H. FALK

BOOM TIMES FOR MUSEUMS, THE ARTICLES IN THE *New York Times, Washington Post,* and *Museum News* proclaim.[2] The museum community, including particularly the science center community, experienced unprecedented growth during the last quarter of the twentieth century—growth in number, size, and popularity. Even into the twenty-first century, more new science centers are built and many existing science centers are involved in aggressive expansion. Although there have been a few valleys and plateaus, the overall trend over these past years has been steadily up. So why are so many within the science center community nervous?

Many are anxious because they feel that the boom has peaked. In the early years of the new century, most science centers reported that their attendance numbers were flat or actually declining.[3] Although new science centers continue to open, each with a flush of new attendance, the situation at individual institutions is not so rosy. Established institutions are struggling just to keep their attendance numbers constant. This fall-off in visitor numbers is making many within the science center community extremely concerned, and for good reason. Attendance is at the heart of most science centers' business model. Attendance numbers drive budgets; either directly through the revenues generated by gate and gift shop, or indirectly as a key part of the grant and contribution formula. Like most organizations in our society, for-profit and nonprofit alike, science centers operate on a growth model. Given that the yardstick of success is visitor numbers, annual growth in attendance figures indicates success, declines represent failure.

I believe that the prevailing business model of science centers is increasingly out of alignment with a rapidly changing world. The longer we cling to this old model, the more frequent will be closures and crises. Of course, there is no single business model that will fit each and every science center, but in this chapter, I will attempt to provide the outlines of an alternative business model framework; an alternative way to do business in these changing times. In this brief chapter, I will describe what a business model is, briefly summarize the foundations of most science centers' current business models, and suggest some of the shortcomings of these current models in light of the rapid changes in our society; shortcomings that are exacerbated when an institution is small. Finally, I will conclude by suggesting what I believe should be the foundation of a new science center business model. Space does not permit me to highlight in detail institutions that are utilizing this new model, but over the past several years, a number of institutions are beginning to try out various aspects of what I propose here and by the time you read this chapter, I suspect there will be dozens of institutions using variations on these ideas. However, I feel confident in saying that the best models are yet to be invented.

WHAT IS A BUSINESS MODEL?

Business model is a term that emerged rather recently in the business vernacular and is sometimes disparaged because it has often been used in a fast and loose manner. But when defined and used properly, it is a term that can provide valuable insights. Whereas business strategy is primarily about the overall positioning of a business within the business ecosystem, the term *business model* also includes key structural and operational characteristics of a business. In other words, when used properly, business model is a broader description of a business than just its strategy. Nonprofits like science centers have business models just as certainly as do for-profits; it's just that they are not always aware of it. And just as in the for-profit world, poor nonprofit business models spell trouble for the organization.

In brief, a business model is a description of the operations of a business including the purpose of the business, components of the business, the functions and values of the business, and the revenues and expenses that the business generates. A business model is the mechanism by which a business intends to manage its costs and generate its outcomes—in the case of for-profits, revenues, and in the case of nonprofits, public good. It is a summary of how an organization plans to serve the needs of its customers. It involves both strategy and implementation.

THE CURRENT, INDUSTRIAL AGE BUSINESS MODEL

Although every science center develops its own unique business model on the basis of the specific realities of its community, its mission, and its assets, nearly all science centers start from a traditional museum business model; a model with its roots firmly planted in the Industrial Age. Like any business, science centers create products, goods, and services, in return for support. Unlike a for-profit, science center success can not simply be measured as profits—revenues generated minus revenues spent. As nonprofit organizations, it is expected that the science center's measure of success will be a benefit to society. In return for these benefits, individuals, other organizations, or government will provide fiscal support. Thus in order to understand the basic outlines of a science center's business model, one needs to know what are the outcomes or benefits the organization hopes to achieve; what are the products produced by the organization in pursuit of those outcomes; how and from whom are revenues generated to produce those products; and what are the measures of success that the organization and its funders use to determine whether those benefits have been delivered.

Figure 49.1 shows how these elements were traditionally organized by science centers into a business model; this is a typical Industrial Age Business Model. This model is top-down and very linear. The most important asset is the building; the more iconic the better. The main goal is to fill the building with interactive science exhibitions and visitors. Ideas for how to accomplish this task flow from the head of the organization (director and core staff) down to the consumers (general public). There are precious few feedback loops from the consumers back to the director and core staff. Evidence of success is that people use the products or services. The quantity of units delivered or purchased (for example, exhibitions mounted, number of visitors) are sufficient evidence of success. What is important is that this model suggests that the organization is only tangentially, if at all, dependent upon those outside the organization—collaborators, competitors, economic cycles, and so on—the science center intersects with the outside world through outside contractors (science experts, evaluators), fund-raising, and marketing; otherwise it functions essentially as an island separate from the larger world. This is a classic "build it and they will come" approach to doing business.

This is an Industrial Age, mass consumption model of science education—provide as many people as possible with a generic science education service. The goal of the Industrial Age model is to create, as efficiently and cost-effectively as possible, a limited number of products or services that can be sold to as many people as possible with demand maintained by skillful branding and marketing. Most, if not all, science centers still operate as if they are trying to satisfy a mass audience, "all people," with one-size-fits-all products and generic exhibitions and programming.

In order to better understand where I'm coming from, allow me to provide a brief history of the twentieth-century Industrial Age model. Although the basic ideas of the Industrial Age were created during the seventeenth century, it was only in the twentieth century that all facets of the industrial economy clicked into place—the mass consumption of goods and services, supplied by the efficiencies of mass production and sustained by mass marketing. Few of us today fully appreciate that at the dawn of the twentieth century, most goods and all services were still custom created, by individuals for individuals. The distribution and sale of these goods and services were almost exclusively local, with only a relatively small percentage, and then only for the most isolated or most affluent customers, sold by catalogue and shipped.[4] This all changed in the twentieth century. By century's end, virtually all goods and services were mass-produced and nationally, if not internationally distributed and sold. Whereas at the beginning of the twentieth century, less than one percent of the cost of a product was allocated to marketing—advertising, promotions, and packaging—today this aspect of a product represents the single largest cost, as much as 60 percent for some products.[5] Today, only a tiny percentage of the products we buy are locally produced, and even the services we seek are more likely than not provided by large corporations through national chains.

The Industrial Age model became so pervasive, and successful, that eventually all sectors of twentieth-century society adopted its methods. Its imprint can be seen in not just the for-profit sector, but in the way

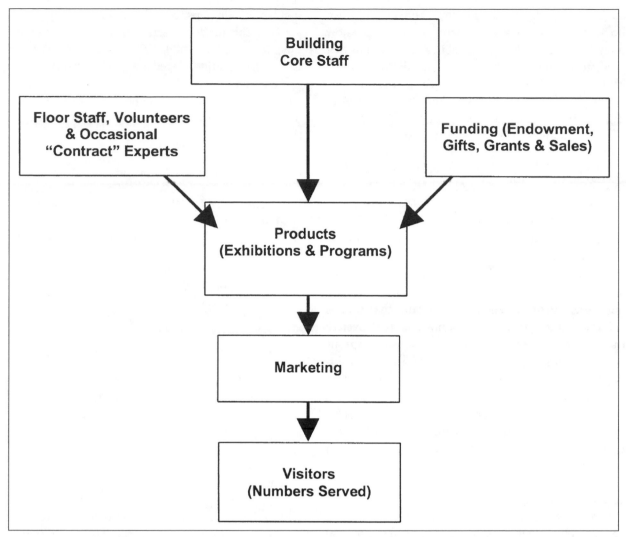

Figure 49.1. Industrial Age Business Model (for Science Centers).

nonprofit businesses such as health care, government, and education are produced and run. The best example of how to apply the twentieth-century industrial model to a community service is the public school system. It is no accident that schools are often accused of using a factory model; they do. The idea was to develop a system that provided the same standardized education to everyone, regardless of background, interest, or intelligence. This problem was solved using the same methods as the auto assembly line—reduce the task to its constituent pieces and engineer how each worker will attack those constituent pieces. Educators accomplish this by assuming that all first grade children have roughly the same basic needs, defined as the skills (benchmarks, standards) necessary to graduate and enter the second grade. In turn, all second grade children have the same basic needs, which is to have the necessary skills to graduate and enter the third grade.

And so on, until the child pops out at the end of the assembly line (once eighth grade, then twelfth grade, and now sixteenth grade) educated and graduated.

At the beginning of the twenty-first century many of the twentieth century's most successful organizations, including mass production pioneers such as McDonald's, Disney and the public schools were in deep trouble. The world today had become too complex to succeed by just being the most prolific provider of a service or product; too many others can produce the same service or product; and there will always be someone who claims they can do it cheaper, and more important, better. The issue for McDonald's is that you can now buy better quality food, with greater selection, equal convenience, and often at a comparable price elsewhere. The issue for Disney is similar. The public can now access, comparable quality entertainment, with comparable selection, right in their own backyard; why go all the way to L.A. or Orlando? The schools, by contrast, no longer can get away with merely talking about how many students graduate; the public now demands that graduating students actually be educated! What good is a high school diploma if you lack basic literacy, numeracy, and problem-solving skills and knowledge? And if the public schools can't or won't provide that level of quality, someone else will—private schools, charter schools, corporate schools, or homeschools.

Whether we realize it or not, most science centers, too, are still using this same Industrial Age business model that strives to provide the widest and largest-possible public with one, brief, basically identical encounter with the ideas of science and technology—just move 'em in and move 'em out! Everyone hopes the experience is fun and educational; but institutional leaders are rarely, if ever, held accountable for achieving these outcomes. The model dictates that success is measured in quantity served, not quality delivered; the greater the attendance, the greater the accomplishment. The assumption is high visitor attendance indicates that something good must be happening. This assumption also assumes that you're the only game in town; which, if it was ever true, is unlikely to be true now. This is because the fundamental nature of society is changing, changing in ways that run counter to an Industrial Age business model.

TOWARD A NEW, KNOWLEDGE AGE BUSINESS MODEL

The world is rapidly transitioning from the Industrial Age to the "Knowledge Age." In the Knowledge Age, the public wants quality not quantity, personalization not standardization, long-standing relationships with its institutions, not being just another nameless, faceless cog. In the Knowledge Age, the public is increasingly demanding high-quality, learning-rich experiences, experiences that help them move along their lifelong learning journey. Science centers are capable of being these kinds of institutions, but currently most fall short. Science centers need to strive for personalized learning rather than mass education.

So, how can we create a new Knowledge Age science center business model, particularly one suited to the needs and realities of small institutions? The short answer is, "with great caution and humility." We should all avoid thinking that there's a silver bullet out there, an easy solution. Although I don't know exactly all the details of this new model, I think I have a sense of what its broad outlines should look like. And as I suggested earlier, there are a number of institutions

that have made steps toward accommodating bits and pieces of this model (several are included in this volume: for example, Science North, Sci-Port Discovery Center, Exploration Place, and COSI Toledo).

BASIC FRAMEWORK To be successful in the Knowledge Age, every business will need to create a business model designed to address and answer four key questions:

1. Why do you exist? Who are you serving? In the case of museums, who is your public and what specific needs do they have that you are uniquely positioned to satisfy?
2. What assets do you bring to the table? What are the internal assets your institution brings to its business, such as the human resources of staff, board, and supporters; also the assets of collections, building, and brand?
3. How will you forge and maintain partnerships and collaborations with like-minded organizations in the community in order to leverage your impact?
4. How will you support and sustain your business? What is your business strategy? What is the unique combination of products and services you can provide in order to satisfy specific public needs and generate sufficient revenues to keep your doors open?

The answers to these four questions cannot be made in a vacuum; the answers must always be situated within the context of the larger world or ecosystem. Each organization must understand the realities of the economic, social, and political context in which it operates, both the ever-evolving marketplace of the business world but equally changing values, needs, and desires of the larger society in which the organization exists. Finally, a business model is a dynamic, not a static, thing; it must always be evolving and changing because the world in which a business operates is constantly changing. At the heart of any successful business model is a strategy for judging the marketplace, evaluating current successes and failures, and predicting the future so the institution can keep one step ahead of the competition (see figure 49.2).

DEFINING AND MEASURING SUCCESS IN THE KNOWLEDGE AGE First and foremost, the new business model must begin by redefining what constitutes success. Individually and collectively, science centers will need to invest in maximizing the quality of the learning experiences they provide. However, this is not like the objectified quality of the assembly line—each part being exactly identical within predefined tolerances. No, the quality

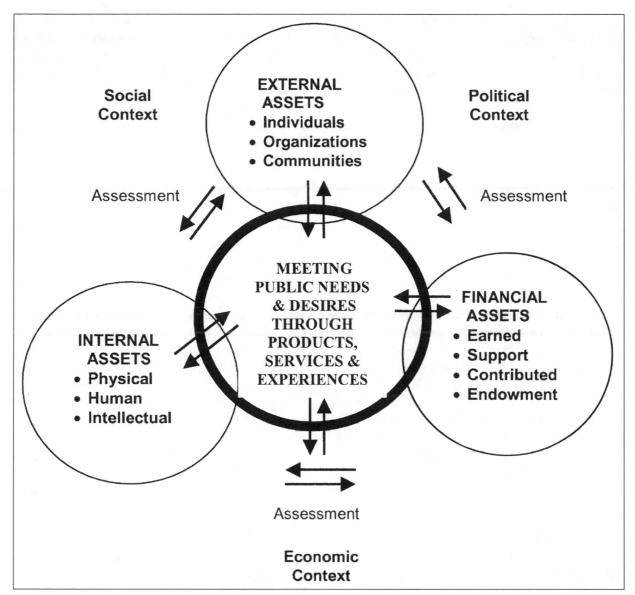

Figure 49.2.   Knowledge Age Business Model.

we need to aspire to is more like the subjective, individualized quality of the master craftsman; it's back to the future. In the Knowledge Age, everything begins and ends with an individualized, personal quality. Each and every science center in the land should be asking themselves, what can I do today to improve the quality of the learning experience we provide each and every visitor?

Historically, the wish of every new, small science center was to be just like the big guys—to be like the Exploratorium, Museum of Science, or Pacific Science Center—just like them, only smaller; in the Industrial Age, bigger was better. In the Knowledge Age, bigger is no longer better.[6] Today, both larger and smaller institutions need to determine what size and activities best

support the quality outcomes they aspire to. From this perspective, being smaller and more nimble is likely to be an asset rather than a liability. Smaller institutions have the ability to be integral to their communities, a place where everyone can know everyone else and treat people accordingly. This is more readily accomplished in a small, semiclosed community such as Ann Arbor, Mich., Ithaca, N.Y., or Sudbury, Ont., than it is in a large, totally open system such as New York, Los Angeles, or Chicago. With the exception of a handful of very large institutions, it is likely that science centers will need to become places where visitors feel compelled to visit not once or twice a year, but constantly, month after month; dare I even suggest, week after week. Science centers need to strive to be places where staff provide personalized support and actually know the visitors as individuals, so that they can say things like, "Remember last week, you were stuck on this problem, but today you seem to have come up with a solution. Tell me about it." This is the future. It is an approach to business that places the individual (or the family, or the small social group) at the center of what success looks like. It's like the financial company's ad campaign that says, "We measure success one customer at a time."

Speaking of measuring success, how will this new century science center measure and record its accomplishments so that it can demonstrate its success to funders, trustees, and the public? Clearly, the new measure of success of a twenty-first century science center should be lives changed, not bodies served. Documenting individual change and growth is not as easy as counting clicks on the turnstile, but it is possible. There are an increasing number of new methodologies being developed that allow science centers to measure even subtle changes in people's knowledge, attitudes, interests, and even behaviors.[7] What we need to eschew, once again, are the simple, shallow tools we've used in the past—the exit questionnaire or satisfaction survey. In fact, an increasing number of businesses have realized that satisfaction is actually a poor indicator of customer loyalty or financial benefit.[8] In the future, our annual reports will need to boast that in the past year, the science center allowed these thousands of visitors to solve these personal problems, other thousands to explore their individual questions and curiosities, and still more thousands to pursue their personal, lifelong quest for understanding in these hundreds of areas of science and technology; and then we'll back up these statements with real data and real statistics.

FREE-CHOICE LEARNING How do you create transformative learning experiences for the many; how can you know what will move people along their personal learning journey? I believe the key that unlocks this challenge is *free-choice learning*—free-choice learning is not another word for "informal learning"; it is not just out-of-school learning. Free-choice learning is about providing choices, it is about supporting learning based on the needs of the learner rather than the needs of the educator, it is about empowering the learner rather than serving the learner, it is an active rather than a passive process. And more than anything, free-choice learning is about negotiating ownership and control of learning, it is about *sharing* both the decision making and responsibility for what to learn, where to learn, and with whom to learn. In the twentieth century, we

believed that knowledge was the province of the few, and the masses were provided only what was deemed they needed. Educational institutions were paternalistic, all-knowing, and in charge. In the twenty-first century, the need for learning will be so pervasive that all institutions, at some level, will need to be educational institutions. Educational institutions will need to increasingly become structured to be facilitators of learning rather than purveyors of learning. Places such as science centers will need to figure out how to help individuals gain access to information, provide them with the ability to know how to sort out the wheat from the chaff, and, perhaps most important, provide skilled and knowledgeable support in helping them accomplish these tasks. The one thing we can predict about the future of knowledge acquisition in general, and science and technology knowledge acquisition in particular, is that none of us will have the ability or time to be sufficiently expert to make sense of all the available information in each and every issue and topic we are likely to find important or interesting. Hence, all of us will become increasingly dependent upon those who can filter and synthesize information for us; dependent upon those who will serve as our "personal information trainers."[9] In other words, the successful educational institutions in the twenty-first century will be those who can provide individuals with the ability and support necessary to be effective lifelong, free-choice learners; it will be about supporting individuals in their need to know while at the same time guiding those individuals in knowing what is important to know and learn. In the Knowledge Age, the greatest and most satisfying experiences an individual can have are ones that support and facilitate their intellectual interests and curiosities; and not just for a moment, but across a lifetime. That's a ride that no amusement park can ever duplicate.

INCOME FROM VISITORS So how is this going to work? How are we going to create a financially viable enterprise on the basis of investing in more and more quality experiences? Won't this limit the number of people we can serve, which reduces our financial base? Space does not permit me to explore the myriad strategies that might be implemented to begin moving toward a more individual-centered approach, but I will discuss one aspect of science center financing at the heart of the current model, which I believe can shed some light on how this transformation might actually work.

Currently, science centers financially penalize themselves for having visitors come frequently. Currently, most science centers encourage frequent visits by selling memberships. In return for a slightly higher, one-time price (usually something on the order of two to three times what it would cost to enter once), members are allowed unlimited entry. The visitor who uses their membership wisely can come dozens, if not hundreds, of time, and except for the first couple of visits, all subsequent visits are free; the institution loses money every time these members show up. Meanwhile, the majority of earned revenue projections are based upon maintaining a high level of first time–one time visitors—since they pay full freight. What's wrong with this picture? The best customers provide limited revenues while those with limited commitment provide the bulk of revenues. Marketing 101 says it's easier to keep a customer than to get a new one, so why do we constantly build our business models around the need to continually attract new or infrequent visitors? A Knowledge Age, quality business model would base pricing on services received—the more you receive, the more it costs (although there are rewards for frequent use).

We need to shift to a strategy that rewards frequent use without penalizing the institution. Frequent flyers don't get cheaper tickets than one-time flyers, but they do, in theory, get better service. Instead, of encouraging our best customers to spend less, we should be encouraging our best customers to spend more. What would make them spend more? The answer is simple, the best customers get the best service—read, more personalized service.

Currently, the primary benefit of being a science center member is the ability to get into institution repeatedly, for free. There is virtually no service benefit to being a member, except in some places not having to wait in the same line for tickets as others. Sure you get invited to a few openings and get a newsletter (most of which is just promotion for the institution), but once you get inside the science center, you're treated like everyone else. What a deal! Then we wonder why it's so difficult to retain these individuals as members? We should not be surprised; we have confused buying a membership with loyalty for the organization; actually, all the member is doing is shopping for a bargain. There is nothing in the "contract" that feels like a quid pro quo of loyalty between institution and member. That's because consciously (or unconsciously) we know that our members provide us with relatively little income, thus we give them relatively little in return. However, if we shifted our funding schemes so that those who came frequently provided the greatest income, we could afford to provide these frequent visitors with the best

service. Of course, we'd have to be able to recognize these very important people when they walk around the building. We could provide them with a distinctive ID tag, have staff who recognize them, or perhaps both. If this model was enacted, I'd like to believe that it would be a self-reinforcing system—the better, more individualized the attention the repeat visitor gets, the better the experience that visitor has; the better the experience, the more frequently that visitor will come back; the more frequently they come, the more revenue generated. That's how the local gym works; and the best-paying customers get personal trainers. Or is it that those willing to hire a personal trainer generate the most revenues?

ELITISM Finally, we need to address the question of elitism. If the strategy advocated above is enacted, won't this limit the number of visitors and won't it drive the institution toward serving only the few rather than the many? The answer to all of these questions is likely to be yes; however, there's nothing inherently elitist about this approach as long as the institution is truly open to all citizens. Entry and participation should be determined not by age, income, race or ethnicity, or economic situation, but by the individual's interests and desires relative to learning about science. There should be ways, particularly as a nonprofit, to subsidize those who cannot afford premium experiences while charging those who can. There are many health systems that successfully scale health care costs to financial capability. There are other health care systems that attempt to provide all users equal quality care, regardless of finances. These latter systems require significant public subsidy, and the jury is still out on whether they really can provide all things to all people. It would seem a mixed approach: one that subsidizes those who can't and charges those who can but encourages all to participate is likely the best answer.

CONCLUSION

The world is changing and science centers need to change with it; they need to accommodate rather than buck the major trends outlined above. The single most important change science centers need to make in the twenty-first century is to move from a fixation on numbers of individuals served to an emphasis on delivering personally transforming learning experiences. Ideally, science centers should strive to meet as many people's needs as possible, but in the long run, quality will be more important than quantity. Institutions need to ensure that each individual they work with has a personally compelling and individually relevant science and technology learning experience; not just in rhetoric but in reality. This is a nontrivial change. It will require a shift of major proportions in the ethos of all parts of the institution—staff, management, and trustees. It is a shift that has already begun to happen outside the community. More and more funders are saying that output measures of performance (that is, numbers served) are insufficient proof of success. In order to secure funding, organizations must describe how this effort will benefit the learners in a community and conduct evaluation that demonstrates accomplishment of these outcomes (that is, evidence of changes in learners). I believe that science centers that fully make this transition can anticipate a future of unprecedented growth and prosperity; those that do not will find the going increasingly challenging. In the Knowledge Age, the race will be won by those organizations who understand their communities, commit to serving individuals rather than the masses, and who are capable of quick and intelligent responses to change. Smaller institutions have a unique advantage in this new business landscape. So capitalize on your small size and bring on the future!

NOTES

1. This chapter is based upon a forthcoming book by John H. Falk and Beverly K. Sheppard. *Thriving in the knowledge age: New business models for museums and other cultural institutions.* Lanham, Md.: AltaMira Press, 2006.

2. Trescott, J. 1998. Exhibiting a new enthusiasm: Across U.S., museum construction, attendance are on the rise. *Washington Post*, sec. 1, June 21.

   Lusaka, J., and J. Strand. 1998. The boom—And what to do about it. *Museum News*, 77 (6): 54–60.

   Lowry, G. D. 1999. The state of the art museum, ever changing. *New York Times*, Arts and Leisure, January 10.

3. Wendy Pollock. 2003. Pers. comm. ASTC.

4. Zuboff, S., and D. Maxmin. 2002. *The support economy.* New York: Viking.

5. Ibid.

6. Drucker, P. F. 2001. *The essential Drucker.* New York: HarperBusiness.

7. Ellenbogen, K., J. Luke, and L. D. Dierking. 2004. Family learning research: An emerging disciplinary matrix? *Science Education* 88:S48–S58.

   Falk, J. H. 2004. The director's cut: Towards an improved understanding of learning from museums. *Science Education* 88:S83–S96.

   Rennie, L. J., and D. J. Johnston. 2004. The nature of learning and its implications for research on learning from museums. *Science Education* 88:S4–S16.

   Adams, M., J. H. Falk, and L. D. Dierking. 2003. Things change: Museums, learning, and research. In *Researching visual arts education in museums and galleries: An international reader*, eds. M. Xanthoudaki, L. Tickle, and V. Sekules. Amsterdam: Kluwer Academic Publishers.

8. McEwen, W. J., and J. H. Fleming. 2003. Customer satisfaction doesn't count. New York: Gallup Organization.

9. Falk, J. H., and L. D. Dierking. 2002. *Lessons without Limit: How free-choice learning is transforming education.* Walnut Creek, Calif.: AltaMira Press.

CHAPTER 50

Reality, Variety, and Ingenuity: Futures for Science Centers

PETER A. ANDERSON

WHEN INVITED TO WRITE ON THE FUTURE OF SCIENCE centers, it is wisest to decline gracefully and disappear quickly. No one knows what the future will be. But having been privileged to read all of the texts for this book, I am reminded of the competence, dedication, intelligence, and energy of so many fine people in this field. It gives every reason for great optimism. I am convinced that science centers will continue successfully if we do the following:

- Stay focused on our core realities
- Maintain the individual characters of our centers and the richness and variety of our programs
- Continue to probe and invent new modes of operation
- Remember where we fit in the lives of our many stakeholders

What Are We?

The term *science centers* covers a great range of institutions. Among science centers themselves, each is different and that is not only good—it is vital to the strength of the field. The rather loose working definition used here is,

science centers are institutions that attract members of the public to visit them, and to engage informally with real things and real processes mostly related to science and technology. They inspire interest in science and technology and improve people's understanding and acceptance of these fields. Visiting them is an interesting and pleasant experience.

Of course, they do lots of other things too, such as outreach, relating art to science, showing IMAX movies, and teaching and helping teachers—the list is almost endless. But the core is the educational mandate and people's own contacts with, and explorations of, real things and real processes. And what applies to science centers also applies by and large to science museums,

and substantially to children's and history museums and aquariums, and in many ways to art galleries and zoos.

WHAT IS OUR KEY FINANCIAL REALITY?

The educational mandate essentially ensures that we cannot live on revenues from our visitors alone. We need a subsidy to maintain a program attractive enough to sustain our attendance. We are therefore vulnerable to the decline of visitation and to the loss of other support. Few science centers need a subsidy of less than 30 percent of necessary expenditures, and most need more.

Why Are We Now Worried about Long-Term Survival?

Science centers have been going for thirty-five years at the time of writing. There are hundreds of them, and the number is growing. We have our ups and downs, and just now we are in a pretty general down. We are very vulnerable to economic swings, especially in the United States, where more support comes from the private sector than in most other countries. And worldwide, it seems that people visit less during recessions. This must be in part because of the cost of traveling and admission—science museums with free admission don't suffer as much. And, we all stare forward, through a glass darkly, looking for long-term trends in what the public likes and does (see John H. Falk, chapter 49, Developing a New Business Model for Science Centers) and what supporting agencies will look upon favorably.

Are science centers past their peaks? Will our attendance wane, or continue to wane? Will our funding diminish in the long term? What changes should we be making, to survive? Operating a science center is not easy, anywhere in the world. They are complicated organizations even when small. Balancing the books each year is a trial for most. Even those centers around the world that receive generous funding from governments or banking institutions cannot rest complacently. They must always be concerned that their funders continue to see them as worthy of such support when compared with many other competing needs or politically attractive causes.

Yet, the overwhelming majority of centers continue to hold their own. There are ups and downs, but an overall upward trend as new centers appear and older centers expand. There is no reason why this should change drastically, if the centers both stick to core values and retain the variety, versatility, and vitality that are so evident in the chapters of this book.

CORE AND ENDURING REALITIES

As science centers continue our competition for approval and visits, it will be vital to work always with some core and pretty unchanging facts and realities of what we are, and what we value.

A Place to Visit

We are places, outside of home and school, where people visit. We are attractive places to go to (or drop into, if we are very small) with family and friends. We are not home or school; we are a trip; an outing; somewhere special to spend a few interesting hours in comfort (and safety, where that is a concern). This property of science centers is a foundation stone. All our nondestination programs, such as web presence, broadcasts, outreach, and in-house schools, are secondary to our nature as visitor destinations. We cannot replace this without changing the science center into a media center or a school or a school service—and other institutions already do those things better.

An Informal Educational Institution

Our long-term goal is educational, or learning, if you prefer that view. We want our visitors to end up understanding science and technology better and generally thinking more about lots of things. Whatever we want them to do, think, or feel during their visits, greater knowledge and understanding is the desired long-term product. This educational goal provides a litmus test for most of the things we do—other than purely money-raising activities. We could not abandon education, and still be a science center.

We specialize in informal, learner-driven learning, without set curricula or exams. In this, we share space in our visitors' lives with all the other informal learning channels available, including input by other people, books, television, and just general experience of the world around us. Viewed this way, we are specialized extensions of that general experience—and we do best when experiences in our centers relate to and extend the rest of our visitors' lives. That, again, is a quality that will wear well.

About Science and Technology—Mainly

We are science and technology centers. This is for a number of reasons, not least of which is that all nations need scientific and technological skills for their economic well-being. We know well that scientific creativity and skill are not separable from other skills and creativity, and we include art, music, and other facets of life in our programs. Our core, though, is science and technology, and we view the world through that lens. This is a continuing asset.

Showing Real Things and Real Processes

That core of engagement with real things and real processes has been key to past success and will be just as vital in the future. People want to be with, to see, and to touch these real things and operate the real processes. Every time we unnecessarily use a simulation or contrived experience, or separate the visitor from a direct experience, we lose some of our basic quality—and some of our value to our visitors. This does not preclude the use of environmental "theming" and designed atmospheres, where that enhances the real experiences. And novel environments, such as a submarine or carefully simulated coal mine, stimulate thought and observation. But, the furniture of these designed environments must overwhelmingly be real. If we abandon this "realness," we lose much of our appeal—especially in a world filled ever more with simulations and virtuality.

Experiences contrived for people, such as movies or spectator sports or virtual reality, gain much of their validity from the onlookers' real personal experiences. It is impossible to assess the skill of an athlete unless you have had some athletic experiences as bases for extrapolation. Nor can you comprehend the feelings in a love story if you haven't been there to some degree yourself. People know this—at some level, anyway—and truly value the real. (See Claude Faubert, chapter 20, The Use of Objects in a Small Science Center). Reality is an enduring quality. It's a good rack to hang your hat on, in terms of engaging people.

We Work Largely in the Affective Domain

Though our end goal may be mainly cognitive understanding, we work principally by stimulating interest and imagination. We motivate our visitors to learn more, and give them the confidence that they can do so. This is our special role as informal education centers. We cannot teach much structured science and technology in two or three hours a year. If we don't send our visitors away fascinated, or at least intrigued, we haven't done much for their knowledge of science and technology.

That is not to demean cognitive gains. A wonderful cognitive insight—the Aha! or Eureka! phenomenon—is marvelously inspiring: "That's beautiful!" and "*I* understood it! Probably, I can understand more." We need to choose our subjects and experiences to yield the most of these insights, rather than to be scientifically complete. If we don't, we will quickly lose our value as sources of interest and inspiration—and lose many of our visitors, too.

For science center people, *inspiration* and *fascination* should be germinal words. And this again, is something that will endure.

STRENGTH IN VARIETY

Variety Inside Science Centers

We must keep great variety in our centers, both in the offerings of each center, and among our several centers. There are two main reasons for this. First is the refreshed feeling in visitors as they move from one kind of environment or activity to another. Second is that the visitors themselves are very different, with an enormous variety of interests and capabilities. However broad the appeal of any one experience, none can cover the whole range of visitors. It is vital to work with many modes of engagement and a wide range of subject matter.

Exhibitions—Phenomena and Packaging

There is a tension in our business between pure phenomena exhibits and those that are more dressed-up—visually clustered, with area treatments, controlled ambiance or stage-set—the "scenography" of European designers. In fact, each appeals to different visitors, so there is every reason to have some of each. This writer prefers some level of environmental treatment, on the grounds that it can increase the interest level for any visitor. It broadens the appeal to visitors with different thinking and learning styles and lesser intrinsic interest in science and technology. This in turn can bring in more paying visitors. And many visitors with little interest in science bring in children who may become scientists.

Variety in Content

Science and technology are very broad subjects, with enormous ranges of content and many possible approaches or entry points. And the ideas and experiments of science are created, as much as works of art are created. In terms of creativity, art is a very valid adjunct to science, as the Exploratorium and Ontario Science Centre recognized in the early 1970s. Technology has countless extensions into our lives and lifestyles, and these afford many different ways of catching the interests of our many different visitors. Blue screen weather stations are good examples.

Learning and Loose Parts

Children's playgrounds were influenced some years ago by Nicholson's Theory of Loose Parts. This was noted at the time by Taizo Miake,[1] who introduced it into the thinking of the Ontario Science Centre. Miake says that, "In any environment, both the degree of inventiveness and creativity, and the possibility of discovery, are directly proportional to the number and kind of variables in it." In other words, the more loose parts and degrees of freedom, the greater the learning potential. It is a very useful notion. When we "harden" exhibits to keep them strong and maintainable, we usually restrict the degrees of freedom. As we move to the future, we must try always to maintain the open-endedness, multiple outcome potentials of our exhibits. This is easiest when there are many floor staff members available.

FLOOR STAFF

Of all our tools, live presentation is the most powerful for holding attention and conveying concept or complex scenarios. People are interested in people, and give them more attention than they give to other types of exhibition. Live shows are also good at handling crowded times by packing people closely together.

In fact, exhibit floor staff, as readers well know, have an enormous impact on the quality of the visitor experience. This applies not only as demonstrators and explainers (hosts, guides, and so on) but also as pleasant and welcoming people who can help visitors in many ways. The same staff members also conduct or assist with camps, sleep-ins, public events, theater, and countless other programs. They permit much more open-endedness in the exhibits. Yet, when financial pressures mount, these lower-paid staff members are usually the first to be cut back. This is bad practice.

At the Tech Museum of Innovation, a very good summative study examined inter alia what visitors said affected most their enjoyment of their visits. The top four were the following:

1. The number of staff they met
2. The courtesy of those staff members
3. The number of broken exhibits
4. Their access to exhibits (probably related to crowding)

Staff can generate change and variety, and can give more individually tailored experiences to visitors (see John H. Falk, chapter 49, Developing a New Business Model for Science Centers). This writer believes that more funds should go to live programming and floor staff, and less to an ongoing struggle to change exhibits constantly. For example, Techniquest in Cardiff has had great success in dealing with peak seasons by enhanced programming, not temporary exhibitions.

Showing Current Science

To maintain relevance and currency, many science centers present current scientific achievements and research. This is a worthy goal, but it takes considerable staff time and is another reason to spend more on program staff at the cost of reducing temporary exhibits and even easing up on exhibition renewal. (See Beryl Rosenthal, chapter 29, Promoting Public Understanding of Research through Family Science Programs at MIT.)

Education Programs and Schools

A science center's relationships with student groups, teachers, and the school system does much to define it. Through the schools, science centers get the best access to their most important clients—children. Science centers in most countries can help teachers—especially elementary teachers—to teach hands-on, rather than books-on. School systems can be difficult to deal with, as they are often very large and subject to many pressures. The relationship needs constant work, but this can be richly rewarded (see Colleen Blair, chapter 26, Building Capacity to Work with Schools).

Variety in Programs

In addition to school programs, there is a great range of other activities that either bring in revenue directly or raise the profile of the science center, attracting support indirectly. Many of the best of these programs are collaboratives and attract financial and other support from collaborators or agencies who value collaborations. Outreach, camps, web presence, artists in residence, community events, and competitions—all these and many more are used by science centers. The range represents great vigor and versatility, and gives strength to the many centers.

Collections

Collections and the preservation of our material heritage are the defining features of traditional museums, and many science centers are museums in part. Collections provide a great richness of real things, and many of these are interesting to a large part of the public. Collections can enrich the exhibitions and offer a wealth of programming opportunities, as seen in the Science Museum of Minnesota, for example, or the Discovery Museum in Newcastle in Britain. The curatorial staff expands the knowledge base of the institution. The proper management of a collection can, of course, be costly, and usually must be supported by tax-based funding. (See Claude Faubert, chapter 20, The Use of Objects in a Small Science Center.)

It is the firm belief of this writer that science centers should make much more use of artifacts in their exhibitions, whether or not these artifacts are part of a collection. A charming grandfather clock with its practical function and period design would expand the meaning of pendulum exhibits and extend their appeal to more people. A telescope or theodolite or stereoscope would do the same for a lens table. Many of our phenomena exhibits are visually boring and conceptually narrow. Collections or purchased artifacts can help to make the rich garden of things and ideas appeal to the broadest range of people.

Variety between Science Centers

One often hears the complaint that science centers are too much the same, but that is not the experience of this writer. Much of the strength of our field lies in the differences between one center and another. Of course, many exhibit and program elements are duplicated—and why not, if they are good? But the whole package that constitutes a science center generally differs significantly from its sister institutions, and these differences keep our profession on its toes.

INGENUITY

It is impossible to read the pages of this book without being impressed by the energy and ingenuity of the people in this field. In many cases, it may be relearning and reinventing what others have done before—at least, in part. But that is alright, as also is the ability to recognize a good idea, and to copy it, preferably with improvements. Science center staffs contain high proportions of inventive people who are alert to the world around them and who react creatively to changes in their environments. So long as this continues, science centers will be able to survive the bad times and flourish in the good ones.

Devising Programs

Science centers frequently devise programs that are new to them, and often novel to the field. See for instances, Creating Developmentally Appropriate Early Childhood Spaces in Science Centers by Kim Whaley (chapter 22), or The Why and How of Doing Outreach Programming: Fulfilling My Fantasy by Dennis Schatz (chapter 27). This is a great strength. It is essential to success, though not sufficient in itself.

MARKETING

This is a key area of science center endeavor, and one in which our performance is uneven. Yet, there is an encouraging record of ingenuity and achievement here, too (see Kim L. Cavendish, chapter 15, Marketing Basics: Applications for Small Science Centers). Advertising is expensive, and to be used only when specifically advantageous—and of course, for a large screen theater. Science centers often undervalue public relations—press reports of what is going on in the center. Some exceptional ingenuity has been shown by public relations officers in persuading somewhat jaded reporters to read press releases or to attend press receptions. One of the more difficult tasks of a good PR person is to sensitize a science center's staff to good PR opportunities—to realize that the scientist or engineer coming to help with an exhibit or program can be photographed in action, and be the basis of a press release. Or, to see that many other seemingly small events can make useful press notices. They don't have to be earthshaking to be good.

Some of the more effective marketing programs have been joint promotions. A science center's public and educational character makes a valuable companion to many commercial ventures' advertising programs, and can be used to very good effect.

Collaborations

Collaborations are in vogue. Science centers are involved with schools, universities, and industry to an unprecedented degree. The National Science Foundation and other foundations now look for evidence of collaborations, with good reason. Different groups bring varied perspectives, and the resulting programs tend to be richer. Working with other groups or organizations with compatible goals can be very energizing, and cross-fertilization of ideas is a foundation of creativity. Science center people have embraced this thrust, and it has enhanced their own ingenuity. It is a strength for the future. (See Sarah Wolf, chapter 40, Collaborations: From Sharing a Museum Site to Winning NSF Grants.)

An interesting and unusual example of collaboration is the International Centre for Life in Newcastle, in Britain. The science center there is part of an organization and building complex that it shares with a research arm of the University of Newcastle and a biotech business incubator. The complex is in the center of town and includes bars and a nightclub. There are many mutually supportive intellectual, commercial, and social activities there. The complex as a whole seems to be more stable than its parts, like an ecosystem.

Closing the Financial Gap

Last but not least: the key demand on ingenuity is the need, year after year, to close the financial gap between what a science center can raise from its visitors and the total operating cost, including capital renewal, that it must spend in order to attract those visitors. The first need is excellence in the exhibitions and other programs. After that, it is a matter of keeping up the interest, trust, and enthusiasm of the many subsidiary funders from members through individual, corporate, and other sponsors; foundations; and governments. And just maintaining the interest is not enough. The sands shift, and former supporters drop away. New sources of support must constantly be sought. It will never cease, but it has gone on successfully for thirty-five years now, and there is no doubt in this writer's mind that our colleagues are up to the challenges.

Where Do We Sit in Our Visitors' Lives?

There is no better tool for staying relevant and effective than to ask frequently, "Where do we sit in the lives of our visitors?" The answers are not simple, as visitors differ so greatly from one another. Yet, as Falk and Dierking point out in The Museum Experience, everyone comes with a personal agenda, and most come in groups with social agendas. They work through these agendas in our centers, and when we are aware of this and use it, we can be successful. Science center people know this pretty well, and if we keep up this awareness we will do much for our futures.

Two Cases in Point

The Exploratorium and the Ontario Science Centre are the oldest science centers in North America, opening within weeks of each other in 1969. Their current developments illustrate the several theses in this chapter. They show real things and real processes. They have great variety in their endeavours. They are ingeniously reinventing themselves as we prepare this book.

The Exploratorium, under the leadership of Goéry Delacôte, has developed to a point where the well-known exhibition in the Palace of Fine Arts is not the major part of its overall program. Its research and development aspect now dominates. It develops collaborative educational programs; it develops web-based engagement, and it develops collaborative exhibition arrangements. It offers other centers (twenty of them worldwide) exhibits on a rotating basis—so that most of its exhibits are outside its premises. The Exploratorium is also engaged with the University of California at Santa Cruz and Kings College, London, in a long-range initiative funded by the National Science Foundation. This graduate and research program in informal learning has an ultimate goal of helping science centers generally to develop alliances with their local schools, leading to better science learning practices.

The Ontario Science Centre (OSC), led by Lesley Lewis, has embarked on a major restructuring of its approach to its visitors and to the community at large. Its Agents of Change program is based on the belief that science centers have important roles to play in the cultivation of creative citizens and should be prepared to create new solutions to the challenges the world faces. The program includes an intense visitor engagement mediated by staff and a series of long-range partnerships with industry and other educational institutions. They seek to create and nurture a culture of innovation in which to grow the next generation of leaders. They have started a move from being a place to visit, to building relationships beyond the site and beyond the visit. By 2006, over 30 percent of the OSC's exhibition space will be devoted to this initiative. Diverse activities and programs are designed to inspire creativity and "encourage problem solving and collaboration skills, teach the importance of risk taking (and forgiveness), and promote science literacy."

CONCLUSION

Worldwide, science centers have a bright future. In North America, they are maturing and have to work hard to maintain their niches as competing operations grow. If they continue to be as competitive as they have been, they should succeed in this. European and Australian centers are following along this evolution, on average some years behind in maturing, though not lagging in quality. In Africa, several small, grassroots centers are growing despite great difficulties, driven by dedicated people. There are several science centers in the Middle East and others on the drawing board. India has well-developed science centers and science parks and is building more. There is an active science center and children's museum network in Latin America. China is recognizing its acute need for informal science learning in spectacular fashion, with science centers in Hong Kong, Beijing, and a giant one in Shanghai. Others are being developed or planned in several cities, including Guangzhou and Ningbo. Some of these will be larger than any existing science center, as the nation works to strengthen the science training in its 1.3 billion population, and to support its 35 million schoolteachers.

The past thirty-five years has been just the beginning. The outlook is brilliant!

NOTE

1. Taizo Miake was director of programs at the Ontario Science Centre.

REFERENCES

Nicholson, S. 1971. How not to cheat children: The theory of loose parts. *Landscape Architecture* 62.

Falk, J. H., and L. Dierking. 1992. *The museum experience*. Washington, D.C.: Whaleback Books.

The future and growth of small science centers are in her/your hands.

Resources for Exhibit Fabrication

Kathleen R. Krafft, Paul Orselli, and David Taylor

LOCAL SOURCES

1. Visit your local stores, and set up accounts; you may get contractors' rates. Check out plumbing and electrical and hardware and lumber and paint supply stores. Sometimes places such as plumbing supply stores will let you behind the counters to look in their bins. Most stores are very supportive of local nonprofit organizations and enjoy the challenges of helping you when you are doing weird things in building exhibits.
2. Find out when it is quiet to get extra suggestions— not first thing in the morning when contractors are getting the parts they need for the day.
3. Never categorize or stereotype your stores—in exhibit fabrication you may well find what you need at strange, unexpected places. So visit, and see what is in stock at auto supply places (12-volt fans for your hand-powered generator, for instance), floor covering stores, fabric stores, office supply places, etc.
4. Addresses and telephone numbers often change! Use websites to confirm contact information.

THE BIG THREE NATIONAL SOURCES

If you don't have these catalogs, get them or check them out online! If you order online or call them, your order is usually delivered the next day for minimal shipping charges. These suppliers also have local branches throughout the country. Check the phone book or the website to locate your nearest outlet.

- Grainger: www.grainger.com, motors, blowers, plumbing, and electrical
- McMaster-Carr: www.mcmaster.com, 3,500 pages of hardware, plumbing (including clear PVC pipe and fittings), electrical, materials (metal, plastics, etc.)
- MSC: www.mscdirect.com

ASSISTIVE DEVICES

- Enabling Devices: www.enablingdevices.com
  385 Warburton Avenue
  Hastings-on-Hudson, NY 10706
  (800) 832-8697
- FlagHouse: www.flaghouse.com
  601 FlagHouse Drive
  Hasbrouck Heights, NJ 07604
  (800) 793-7900
- Maxi-Aids: www.maxiaids.com
  42 Executive Boulevard
  Farmingdale, NY 11735
  (800) 522-6294
- Sammons Preston: www.sammonspreston.com
  4 Sammons Court
  Bolingbrook, IL 60440
  (630) 226-1300 Fax: 630-226-1389

CHEMICALS, SAFETY, AND LABORATORY EQUIPMENT

- Cole-Parmer: www.coleparmer.com
  625 East Bunker Court
  Vernon Hills, IL 60061
  (800) 323-4340
- Fisher: www.fisherscientific.com
  Liberty Lane
  Hampton, NH 03842
  (603) 926-5911 Fax: (603) 929-2379
- Flinn Scientific: www.flinnsci.com; Flinn catalog is a must for information on chemical safety, storage, and disposal, and for laboratory safety.
  PO Box 219
  Batavia, IL 60510
  (800) 452-1261 Fax: (866) 452-1436
- Lab Safety Supply: www.labsafety.com; Materials handling, safety, 55-gallon drums, storage units, etc.
- Sargent-Welch: www.sargentwelch.com
  PO Box 5229
  Buffalo Grove, IL 60089
  (800) 727-4368 Fax: (800) 676-2540
- Sigma-Aldrich: www.sigmaaldrich.com; Unusual chemicals.

DIGITAL IMAGES
- Corbis: www.corbis.com
- Getty Images: www.gettyimages.com

EDUCATIONAL/CLASSROOM MATERIALS
(Visit your local schools—they have lots of catalogs.)

- Childcraft: www.childcrafteducation.com
  PO Box 3239
  Lancaster, PA 17604
  (800) 631-5652
- Creative Publications: www.wrightgroup.com/index.php
- Discount School Supply: http://www.discountschoolsupply.com
- Educational Innovations: www.teachersource.com
  362 Main Avenue
  Norwalk, CT 06851
  (203) 229-0730 Fax: (203) 229-0740
- Edmund Industrial Optics: www.edmundoptics.com;
  Lenses, optical parts
- Edmund Scientific: www.edsci.com; Magnets, polarizing sheet, all kinds of science stuff
- ETA/Cuisenaire: www.etacuisenaire.com; Math manipulatives, posters
  500 Greenview Court
  Vernon Hills, IL 60061
  (800) 875-9643
- Health Edco: www.healthedco.com; Health: models, posters, etc.
- Lakeshore: www.lakeshorelearning.com; Early childhood materials,
- TC Timber: www.haba.de/T_C_Timber.903.0.html

ELECTRONICS
- Allied: www.alliedelec.com
  7410 Pebble Drive
  Fort Worth, TX 76118
- Anatek: www.anatekcorp.com; Video and TV related electronics.
  PO Box 1200
  100 Merrimack Road
  Amherst, NH 03031
  (603) 673-4342 Fax: (603) 314-0350
- Digi-key: www.digikey.com
- Happ Controls: www.happcontrols.com;
  Pushbuttons, pinball accessories, etc.
  106 Garlisch Drive
  Elk Grove, IL 60007
  (888) BUY-HAPP
- Hosfelt Electronics: www.hosfelt.com
  2700 Sunset Boulevard
  Steubenville, OH 43952
  (888) 264-6464 Fax: (800) 524-5414

- Jameco: www.jameco.com
  1355 Shoreway Road
  Belmont, CA 94002
  (800) 831-4242
- Markertek: www.markertek.com; Cameras, cables, tools, audio equipment; Great source for video production equipment and unusual stuff
  812 Kings Highway
  PO Box 397
  Saugerties, New York 12477
  (800) 522-2025 Fax: (845) 246-1757
- Mouser: www.mouser.com
  1000 North Main Street
  Mansfield, TX 76063
  (800) 346-6873
- Newark: www.newark.com
  4801 N Ravenswood
  Chicago, IL 60640-4496
  (773) 784-5100 Fax: (888) 551-4801
- Radio Shack: www.radioshack.com
- Ramsey Electronics: www.ramseyelectronics.com
  Good source of electronics kits that can be turned into exhibits.
  590 Fishers Station Drive
  Victor, NY 14564
  (800) 446-2295 Fax: (585) 924-4886
- Solid State Advanced Controls: www.ssac.com; Sometimes the only source for hard-to-find electronic timers and other modules that do switching, current measuring, etc., generally for 120VAC circuits.
- Supercircuits: 207.207.29.130/store/home.asp;
  Video and security equipment
  One Supercircuits Plaza
  Liberty Hill, TX 78642

FAKE FOODS
- Barnard, Ltd: www.themedecor.com; Decorations, displays, and props.
  375 W Erie
  Chicago, IL 60610
  (888) 5-THEMES Fax (312) 475-0215
- Fake-Foods.com: www.fake-foods.com
  204 North El Camino Real, #432
  Encinitas, CA 92024
- Hubert: www.hubert.com; Display supplies
- Incredible Inedibles: www.incredibleinedibles.net
- Iwasaki Images: www.iwasaki-images.com
  630 Maple Avenue
  Torrance, CA 90503
  (800) 323-9921 Fax: (310) 618-0876

FIBERGLASS AND MOLDMAKING
- Aircraft Spruce & Specialty: www.aircraft-spruce.com;
  Fiberglass supplies, Kevlar, aviation instruments, the entire world of aviation fasteners.
  (877) 4-SPRUCE

- Fiberglas Coatings: www.fgci.com; A great source for fiberglassing supplies, casting resins, and knowledge
  3201 28th Street N
  St. Petersburg, FL 33713
  (727) 327-8117 Fax: (727) 327-6691
- Fibre Glast: www.fibreglast.com
  95 Mosier Pkwy
  Brookville, OH 45309
  (800) 330-6368 Fax: (937) 833-6555
- Polytek: www.polytek.com; Rubber moldmaking supplies, casting materials

FRAMING AND MOUNTING MATERIALS
- Light Impressions: www.lightimpressionsdirect .com
  PO Box 787
  Brea, CA 92822
  (800) 828-6216

FURNITURE
- Hafele: www.hafeleonline.com; Huge assortment of hardware for furniture making
- Highland Park: http: www.highlandparkfurniture. com/ip.asp
  505 Laurel Avenue, Suite 102
  Highland Park, IL 60035
  (866) 473-8764
- Mockett: www.mockett.com/default.asp?ID=2; Hardware, pulls, wire grommets, etc.
- Worthington Direct: www.worthingtondirect.com/
  6301 Gaston Avenue, Suite 670
  Dallas, TX 75214
  (800) 599-6636 Fax: (800) 943-6687

GEARS, CLUTCHES, SHAFTS, ETC.
- Atlanta Belting: www.atlbelt.com; Conveyor belt— smooth, textured
- Boston Gear: www.bostongear.com
  14 Hayward Street
  Quincy, MA 02171
  (888) 999-9860 Fax: (800) 752-4327
- Emerson/Morse/Browning: www.emerson-ept.com
- WM Berg: www.wmberg.com
  499 Ocean Avenue
  East Rockaway, NY 11518
  (800) 232-BERG Fax: (800) 455-BERG

GRAVITY WELLS
- Divnik International: www.divnick.com/wishing_ well.htm
  321 Alexandersville Road

Miamisburg, Ohio 45342
(937) 384-0003
- Whitaker Center: www.whitakercenter.org/wonders/ gravity.asp
  (717) 214-ARTS

GREEN BUILDING AND MATERIALS
- Build it Green: http://builditgreen.org/
- Center for Neighborhood Technology: http:// building.cnt.org/resources
- Green Exhibits: http://www.greenexhibits.org

HARDWARE, WOOD, AND TOOLS
- 2SQ: www.2sq.com; Huge selection of bits, tools, sup- plies
- Ballew Tool and Saw: www.ballewtools.com; Sharpens saw blades, sells blades and bits
  325 S. Kimbrough
  Springfield, MO 65806
  (800) 288-7483
- Carbide.com: www.carbide.com; Router bits, etc.
- Cherry Tree: www.cherrytree-online.com; Wood balls, parts
  408 S Jefferson Street
  Belmont, OH 43718
  (800) 848-4363
- Enco Tools: www.use-enco.com; Tools, general selec- tion, and large tools
- Fastenal: www.fastenal.com; Industrial and construc- tion supplies
  2001 Theurer Boulevard
  Winona, MN 55987
  (507) 454-5374 Fax: (507) 453-8049
- Grizzly: www.grizzly.com; Large and small tools, bits, supplies, wood samples, etc.
- Hafele: www.hafeleonline.com; Huge assortment of hardware for furniture making
- Harbor Freight: www.harborfreight.com; Inexpensive tools, variable quality on some brands
- JC Whitney: www.jcwhitney.com; Automotive supplies
- Klingspor: www.klingspor.com; Woodworking: sandpaper in bulk (belts, drums, disks, sheets)
- Lee Valley: www.leevalley.com; Woodworking tools, also cheap source for small neodymium magnets
  PO Box 1780
  Ogdensburg, NY 13669
  (800) 871-8158
- Lehman's: www.lehmans.com; Old-time tools, blacksmithing supplies
  One Lehman Circle
  PO Box 321
  Kidron, OH 44636
  (888) 438-5346

- Northern Tools: www.NorthernTool.com
  2800 Southcross Drive W
  Burnsville, MN 55306
  (800) 221-0516 Fax: (952) 882-6927
- Roberts Plywood: www.getwood.com; Curved plywood, large wooden tubes
  150 Rodeo Drive
  Brentwood, NY 11717
  (516) 586-7700
- Saunders Brothers: www.saundersbros.com; Dowels (small to large), wood turnings
  170 Forest Street
  PO Box 1016
  Westbrook, ME 0409
  (800) 343-0675
- Small Parts Inc: www.smallparts.com; Bearings, hardware, small tools
  13980 NW 58th Court
  PO Box 4650
  Miami Lakes, FL 33014
  (800) 220-4242
- Southco: www.southco.com; Latches, cabinet hardware
- Tool Parts Direct: www.toolpartsdirect.com; Parts for tools—with diagrams for identifying the part
  6620 F Street
  Omaha, NE 68117
  (866) 597-3850
- West Marine: www.westmarine.com; Marine supplies
- Woodcraft: www.woodcraft.com; Tools and supplies
  (800) 535-4482
  Woodworker's Supply: woodworker.com/

LIGHTS
- Bulbman: www.bulbman.com
- Interlight: www.interlight.biz/
  7939 New Jersey Avenue
  Hammond, IN 46323
  (800) 743-0005
- Topbulb: www.topbulb.com
  5204 Indianapolis Boulevard
  East Chicago, IN 46312
  (866) TOP-BULB

MAGNETS
- Adams Magnetic: www.adamsmagnetic.com
- Dowling-Miner: www.dowlingmagnets.com
- Force Field: www.wondermagnet.com
  2606 West Vine Drive
  Fort Collins, CO 80521
  (877) 944-6247
- Kling Magnetics: www.kling.com; Magnetic paint.
  343 Rt. 295

PO Box 348
Chatham, NY 12037
(518) 392-4000 Fax: (518) 392-8191

METALS
- See McMasterCarr for small quantities
- McNichols: www.mcnichols.com; Perforated sheet metal, steel grating
  5505 West Gray Street
  Tampa, FL 33609-1007
  (813) 282-3828 x 2100 Fax: (813) 287-1066
- Murphy-Nolan: www.murphynolan.com
- Nolan Supply: www.nolansupply.com
  111-115 Leo Avenue
  Syracuse, NY 13206
  (877) 529-6846

MISCELLANEOUS
- Archie McPhee/Accoutrements: www.mcphee.com; Wacky products! Einstein action figures
  2428 NW Market Street
  Seattle, WA 98107
  (206) 297-0240
- Displays 2 Go; Small sign holders, stands, displays.
  55 Broad Common Road
  Bristol, RI 02809
  (800) 572-2194
- Ecospheres: www.eco-sphere.com; Self-contained ecosystem spheres
  4421 N Romero Road
  Tucson, AZ 85705
  (800) 729-9870
- Forbex: www.forbex.com; Fake grass
- Gerbert Limited: www.gerbertltd.com; Recycled flooring materials
  715 Fountain Avenue
  PO Box 4944
  Lancaster, PA 17604-4944
  (800) 828-9461
- Hobby-Lobby International: www.hobby-lobby.com
  5614 Franklin Pike Circle
  Brentwood, TN 37027
  (615) 373-1444 Fax: (615) 377-6948
- Hobby People: www.hobbypeople.net Small motors, controllers for models
- Jestertek (formerly Vivid Group): www.jestertek.com; Virtual reality games and supplies
- JML Direct Optics: www.jmloptical.com Parabolic mirrors
  76 Fernwood Avenue
  Rochester, NY 14621
  (585) 342-8900 Fax: (585) 342-6125

- Large Plastic Gears with Magnetic Bases
  John Bowditch c/o Ann Arbor Hands-On Museum
  220 E Ann Street
  Ann Arbor, MI 48104
  (734) 995.5439
  jbowditch@aahom.org
- Light Stick Art: www.subliminaryartworks.com/
  Bill Bell
  139 Davis Avenue
  Brookline MA 02445
  (617) 277-4719
  billbell@subliminaryartworks.com
- Oriental Trading Company: www.orientaltrading.com; Cheap multiples. Craft and party items
- Pirelli Flooring: www.artigo.com; Interesting flooring products
- Rheoscopic Fluid: www.kalliroscope.com
- Rhode Island Novelty: www.rinovelty.com
  19 Industrial Lane
  Johnston, RI 02919
  (800) 528-5599
- Sand & Solutions: waupacasand.com/rubber_mulch_products.htm; Rubber mulch
  (For clean sandboxes and playgrounds.)
  (715) 258-8566
- Stella Color: www.stellacolor.com; Images on carpet; interesting mural wallpaper
- Toysmith: www.toysmith.com
- Ultrasonic Mistmakers: www.mainlandmart.com/foggers.html
  2535 Durfee Avenue
  El Monte, CA 91732
  (626) 258-2928
- Yemm & Hart: www.yemmhart.com; Recycled building materials
  1417 Madison
  Marquand, MO 63655
  (573) 783-5434 Fax (573) 783-7544

PHOSPHORESCENT SHEET MATERIALS
- ABET Laminati: www.abet-laminati.it; Lumiphos laminate material
- Flinn Scientific: www.flinnsci.com; Small sheets and paint
  PO Box 219
  Batavia, IL 60510
  (800) 452-1261 Fax: (866) 452-1436
- Hanovia: www.hanovia-uv.com
  825 Lehigh Avenue
  Union, NJ 07083
  (800) 229-3666
- Shannon Luminous Materials: www.blacklite.com
  304 A North Townsend
  Santa Ana, CA 92703
  (800) 543-4485

PLASTICS
- See McMasterCarr for small quantities.
- AIN Plastics: www.tincna.com; Sheet goods, smaller quantities (will cut), rod, tubing, etc.
  22355 West 11 Mile Road
  Southfield, MI 48034-4735
  (248) 233-5600 Fax: (248) 233-5699
- Curbell Plastic: www.curbell.com; Sheet goods, smaller quantities (will cut), rod
  7 Cobham Drive
  Orchard Park, NY 14127
  (716) 667-3377
- Outwater Plastics: www.outwater.com; Weird architectural stuff, tee molding in all sizes and shapes and colors, etc.
  4 Passaic Street
  Wood-Ridge, NJ 07075
  (888) OUTWATER (688-9283) FAX: (800) 888-3315
- United States Plastic: www.usplastic.com; Lots of plumbing parts, tubing
  1390 Neubrecht Road
  Lima, OH 45801-3196
  (800) 809-4217 Fax (800) 854-5498

SAFETY RESOURCES AND MATERIALS
- MSDS on line: http://www.msdssearch.com
- Arts, Crafts & Theater Safety: http://www.artscraftstheatersafety.org

SCIENCE MATERIALS SUPPLIERS
(also Check Educational/School Supplies)

- American 3B Scientific: www.a3bs.com
  2189 Flintstone Drive, Unit O
  Tucker, GA 30084
  (770) 492-9111
- Arbor Scientific: http://www.arborsci.com
  PO Box 2750
  Ann Arbor, MI 48106
  800) 367-6695 Fax: (734) 477-9373
- Carolina Biological: www.carolina.com; Microscope slides, fruit flies and other critters, lots more
- Edmund Scientific: www.edsci.com; Magnets, polarizing sheet, all kinds of science stuff
  Kelvin Scientific: www.kelvin.com/
  280 Adams Boulevard
  Farmingdale, NY 11735
  (800) 535-8469
- NASCO: www.nascofa.com; A site for multiple supply catalogs; fake food (nutrition teaching aids), magnifiers, health education, etc.
- PASCO: www.pasco.com; Excellent physics supplies
  10101 Foothills Boulevard

Roseville, CA 95747
(800) 772-8700
- Pitsco: www.pitsco.com; Large variety of kits, meters, etc. for all kinds of activities
915 E Jefferson
PO Box 1708
Pittsburg, KS 66762
(800) 835-0686
- Science Kit & Boreal Laboratories: www.sciencekit.com; Large variety of kits and supplies
- SEIDAM: www.seidam.com
Kelvin Building
University of Glasgow
University Avenue
Glasgow, SCOTLAND
G12 8QQ
+44 (0) 141 330 2047
- Ward's Natural Science: www.wardsci.com
PO Box 92912
Rochester, NY 14692
(800) 962-2660

SCROLLING IMAGE SIGNS AND LIGHTBOXES
- Bowman Displays: www.bowmandisplays.com
648 Progress Avenue
Munster, IN 46321
(800) 922-9250
- Dick Blick: www.dickblick.com
PO Box 1267
Galesburg, IL 61402
(800) 828-4548
- Warwick Products Company: www.warwickproducts .com/displays.htm; Store fixtures, displays

SURPLUS SUPPLIERS
- American Science and Surplus: www.sciplus.com; Weird collection of small parts
PO Box 1030
Skokie, IL 60076
(847) 647-0011
- Herbach and Rademan: www.herbach.com; Cheap motors, blowers, power supplies etc.
353 Crider Avenue
Moorestown, NJ 08057
(800) 848-8001

THEATRICAL SUPPLIES
- Dazian Fabrics: www.dazian.com; Theatrical and outdoor fabrics

- Pro Sound & Stage Lighting: http://www.pssl.com; Audio, video, party lights
1070 Valley View Street
Cypress, CA. 90630
(800) 268-5520
- Rosco: www.rosco.com; Specialized lighting fixtures and gels (colored Mylar sheets), hardware
- Rose Brand: www.rosebrand.com; Theatrical supplies
- Sam Ash: www.samash.com/home; Musical instruments, sound equipment
(800) 4-SAMASH
- Seattle Fabrics: www.seattlefabrics.com; Theatrical and outdoor fabrics
Seattle, WA. 98103
(206) 525-0670 Fax (206) 525-0779

WRITTEN RESOURCES
AAM Bookstore: www.aam-us.org/bookstore
Are We There Yet? Conversations about Best Practices in Science Exhibition Development. Kathleen McLean and Catherine McEver, editors, The Exploratorium, 2004
ASTC Bookstore: www.astc.org/pubs
Cell Lab Cookbook: A Guide for Building Biology Experiment Exhibits. Susan Fleming, Science Museum of Minnesota, 2003
Cheapbooks 1, 2, 3 Paul Orselli, editor, ASTC, 1995, 1999, 2004
Exhibit Builder Magazine: www.exhibitbuilder.com
Exhibit Labels: An Interpretive Approach. Beverly Serrell, AltaMira Press, 1996
Experiment Bench: A Workbook for Building Experimental Physics Exhibits. Colleen M. Sauber, editor, Science Museum of Minnesota, 1994
Exploratorium Cookbooks I, II, and III. Raymond Bruman and Ron Hipschman, The Exploratorium, 1976, 1980, 1987
Explore and Discover: The Ann Arbor Hands-On Museum Exhibits Guide. H. Richard Crane, Ann Arbor Hands-On Museum, 1992
How to Make a Rotten Exhibition Jan Hjorth, *Curator,* Vol. 20, no 3, 1977. Though nearly twenty-five years old, this article, justifiably deserves to be regarded as a classic.
Planning for People in Museum Exhibitions. Kathleen McLean, ASTC, 1993
Try It! Improving Exhibits through Formative Evaluation. Samuel Taylor, editor, New York Hall of Science, 1992
User Friendly: Hands-On Exhibits That Work. Jeff Kennedy, ASTC, 1990

ASTC and Related Organizations

ASTC and other science center networks offer professional development opportunities specific to the science center field. The following organizations also provide programs and publications in areas relevant to various aspects of science center work. All addresses are in the United States, unless otherwise noted.

American Association of Botanical Gardens and
  Arboreta
351 Longwood Road
Kennett Square, PA 19348
phone: 610/925-2500
fax: 610/925-2700
website: www.aabga.org

American Association of Museums (AAM)
1575 Eye Street, NW, Suite 400
Washington, DC 20005
phone: 202/289-1818
fax: 202/289-6578
e-mail: info@aam-us.org
website: www.aam-us.org

In addition to AAM's annual meeting and professional development seminars, the regional museum associations and AAM professional interest and standing professional committees also offer meetings, workshops, and publications. Of particular interest to science centers are the National Association of Museum Exhibitions and the Committee on Audience Research and Evaluation.

ALI-ABA Committee on Continuing Professional
  Education
4025 Chestnut Street
Philadelphia, PA 19104
phone: 800/253-6397
fax: 215/243-1664
website: www.ali-aba.org

American Zoo and Aquarium Association
8403 Colesville Road, Suite 710
Silver Spring, MD 20910
phone: 301/562-0777
fax: 301/562-0888
website: www.aza.org

Association for Volunteer Administration
P.O. Box 32092
Richmond, VA 23294-2092
phone: 804/346-2266
fax: 804/346-3318
e-mail: info@avaintl.org
website: www.avaintl.org

Association of Children's Museums
1300 L Street, NW, Suite 975
Washington, DC 20005
phone: 202/898-1080
fax: 202/898-1086
e-mail: acm@childrensmuseums.org
website: www.childrensmuseums.org

Association of Fundraising Professionals
1101 King Street, Suite 700
Alexandria, VA 22314
phone: 703/684-0401
fax: 703/684-0540
website: www.nsfre.org

The Banff Centre School of Management
Box 1020
Banff, Alberta T0L 0CO
Canada
phone: 403/762-6121
website: http://www.banffmanagement.com

BoardSource (formerly the National Center for
   Nonprofit Boards)
1828 L Street, NW, Suite 900
Washington, DC 20036-5114
phone: 202/452-6262 or 800/883-6262
fax: 202/452-6299
website: www.boardsource.org

British Interactive Group (BIG)
c/o Techniquest
Cardiff, CF10 5BW
United Kingdom
phone: (44)(0 29) 2047 5475
fax: (44)(0 29) 2048 2517
e-mail: secretary@big.uk.com
website: www.big.uk.com

Canadian Museums Association (CMA)
280 Metcalfe Street, Suite 400
Ottawa, Ontario K2P 1R7
Canada
phone: 613/ 567-0099
fax: 613/233-5438
e-mail: Can-cma@immedia.ca
website: www.museums.ca

Center for Education and Museum Studies
Smithsonian Institution
MRC 427
Washington, DC 20560
phone: 202/357-3101
fax: 202/357-3346
website: http://museumstudies.si.edu

Center for Informal Learning and Schools (CILS)
3601 Lyon Street
San Francisco, CA 94123
website: www.exploratorium.edu/cils/index.html

Cultural Resource Management Program
Continuing Studies, University of Victoria
P.O. Box 3030, STN CSC
Victoria BC V8W 3N6
Canada
phone: 250/721-6119
fax: 250/721-8774
e-mail: lmort-putland@uvcs.uvic.ca
website: www.uvcs.uvic.ca/crmp

Deutsches Museum Management Courses
Hauptabteilung Programme
Deutsches Museum
D-80538 Munchen
Germany
phone: (0049) 89/2 17 92 94
fax: (0049) 89/2 17 92 73
e-mail: h.a.programme@extern.lrz-muenchen.de
website: www.deutsches-museum.de/bildung/
fortbild/e_kk.htm

Duke University Certificate Program in Nonprofit
   Management
Duke Continuing Education and Summer Session
Box 90708
Durham, NC 27708-0708
919/684-6259
e-mail: learn@acpub.duke.edu
website: www.learnmore.duke.edu

Getty Leadership Institute
1200 Getty Center Drive, Suite 300
Los Angeles, CA 90049-1681
phone: 310/440-6300
fax: 310/440-7765
e-mail: gli@getty.edu
website: www.getty.edu/about/leader

Giant Screen Theater Association
Piper Jaffrey Plaza
444 Cedar Street, Suite 810
St. Paul, MN 55101
phone: 651/292-9884
fax: 651/292-9901
e-mail: gsta@uswest.net
website: www.giantscreentheater.com

International Museum Theatre Alliance, Inc.
   (IMTAL)
c/o Central Park Zoo
830 Fifth Avenue
New York, NY 10021
phone: 212/439-6542
e-mail: jellers@wcs.org
website: www.mos.org/IMTAL

International Museum Theatre Alliance, Inc. (IMTAL),
   Europe
IMTAL/Europe
c/o Museum of the Moving Image
Southbank, Waterloo
London SE1 8XT, England
phone: (017) 815-1336
e-mail: andrew.ashmore@bfi.org.uk
website: www.imtal-europe.org/index.asp

International Planetarium Society (IPS)
c/o Museum of the Rockies
Montana State University
Bozeman, MT 59717
phone: 406/994-6874
fax: 406/994-2682
website: http://www.ips-planetarium.org/

Museum Computer Network
232-329 March Road, Box 11
Ottawa, Ontario K2K 2E1
phone: 888/211-1477
e-mail: info@mcn.edu
website: www.mcn.edu

Museum Education Roundtable (MER)
621 Pennsylvania Avenue, SE
Washington, DC 20003
phone: 202/547-8378
fax: 202/547-8344
e-mail: info@mer-online.org
website: www.mer-online.org

Museum Management Program
University of Colorado
250 Bristlecone Way
Boulder, CO 80304
phone: 303/443-2946
fax: 303/443-8486

Museum Store Association
4100 E Mississippi Avenue, Suite 800
Denver, CO 80246-3055
phone: 303/504-9223
fax: 303/504-9585
e-mail: info@museumdistrict.com
website: www.museumdistrict.com

Museum Trustee Association
2025 M Street NW, Suite 800
Washington, DC 20036-3309
phone: 202/367-1180
fax: 202/367-2180
website: www.mta-hq.org

Natural Science Collections Alliance
1725 K Street, NW, Suite 601
Washington, DC 20006-1401
phone: 202/835-9050
fax: 202/835-7334
e-mail: general@nscalliance.org
website: www.nscalliance.org

Techniquest/University of Glamorgan MsC in
    Communicating Science
Stuart Street
Cardiff CF10 5BW
e-mail: suec@techniquest.org
website: www.techniquest.org/msc.htm

Theater in Museums Workshop
Science Museum of Minnesota
120 West Kellogg Boulevard
St. Paul, MN 55102
phone: 651/221-2587 or 651/221-4560
website: www.smm.org/educationprograms/
sciencelivetheater/TheatreinMuseums.php

Visitor Studies Association
8175-A Sheridan Blvd., Suite 362
Arvada, CO 80003-1928
phone: 303/467-2200
fax: 303/467-0064
e-mail: info@visitorstudies.org
website: www.visitorstudies.org

Directory of Science Centers, Institutions, and Individuals Represented

SCIENCE CENTERS, INSTITUTIONS, AND MUSEUMS

American Association of Museums
1575 Eye Street NW, Suite 400
Washington, DC 20005
www.aam-us.org

Association of Science-Technology Centers
Incorporated
1025 Vermont Avenue NW, Suite 500
Washington, DC 20005-6310
www.astc.org

Ann Arbor Hands-On Museum
220 E. Ann Street
Ann Arbor, MI 48104
www.aahom.org

Canada Science and Technology Museum
1867 St. Laurent Boulevard
Ottawa, ON K1G 5A3
Canada
cts@technomuses.ca

COSI
333 West Broad Street
Columbus, OH 43215
www.cosi.org

COSI-Toledo
1 Discovery Way
Toledo, OH 43604
www.cositoledo.org

Curious Kids' Museum
415 Lake Boulevard
St. Joseph, MI 49085
www.curiouskidsmuseum.org

Discovery Center Museum
711 N. Main Street
Rockford, IL 61103
www.discoverycentermuseum.org

Discovery Science Center
2500 North Main Street
Santa Ana, CA 92705
www.discoverycube.org

Ecsite-uk
Wellcome Wolfson Building
165 Queen's Gate
London SW7 5HD
United Kingdom
www.ecsite-uk.net

Experimentarium
Tuborg Havnevej 7
2900 Hellerup
Denmark
www.experimentarium.dk

Exploration Place
300 N. McLean Boulevard
Wichita, KS 67203
www.exploration.org

Exploratorium
3601 Lyon Street
San Francisco, CA 94123
www.exploratorium.edu

Fort Worth Museum of Science and History
1501 Montgomery Street
Ft. Worth, TX 76107
www.fwmsh.org

Hands On! Inc.
689 Central Avenue, #200
St. Petersburg, FL 33701
www.hofl.org

Headwaters Science Center
413 Beltrami Avenue
Bemidji, MN 56601
www.hscbemidji.org

Inspire Discovery Centre
Coslany Street
Norwich NR3 3DT
United Kingdom
www.inspirediscoverycentre.com

MIT Museum
265 Massachusetts Avenue
Cambridge, MA 02139
web.mit.edu/museum

Museum of Discovery and Science and
Blockbuster IMAX Theater
401 S.W. Second Street
Fort Lauderdale, FL 33312
www.mods.org

Museum of Life and Science
433 Murray Avenue
Durham, NC 27704
www.ncmls.org

Museum of Science
Science Park
Boston, MA 02114-1099
www.mos.org/

New York Hall of Science
47-01 111th Street
Queens, NY 11368
www.nyscience.org

The Observatory Science Centre
Herstmonceux
Hailsham
East Sussex BN27 1RN
United Kingdom
www.the-observatory.org

Pacific Science Center
200 Second Avenue
Seattle, WA 98109
www.pacsci.org

The Phoenix Zoo
455 N. Galvin Parkway
Phoenix, AZ 85008
www.phoenixzoo.org

Science Discovery Center of Oneonta
Physical Science Building
State University College
Oneonta, NY 13820
www.oneonta.edu/academics/scdisc

Science North
100 Ramsey Lake Road
Sudbury, ON P3E 5S9
Canada
www.sciencenorth.ca

Sciencenter
601 First Street
Ithaca, NY 14850
www.sciencenter.org

Science Projects
20 St. James Street
Hammersmith
London W6 9RW
United Kingdom
www.science-projects.org

Science Spectrum
2579 South Loop 289, #250
Lubbock, TX 79423-1400
www.sciencespectrum.org

Sci-Port Discovery Center
820 Clyde Fant Parkway
Shreveport, LA 71101
www.sciport.org

SciTech Hands-On Museum
18 W. Benton Street
Aurora, IL 60506
www.scitech.museum

SciWorks
The Science Center and Environmental Park
of Forsyth County
400 W. Hanes Mill Road
Winston-Salem, NC 27105
www.sciworks.org

The Tech Museum of Innovation
201 S. Market Street
San Jose, CA 95113
www.thetech.org

W5 whowhatwherewhenwhy
Odyssey, 2 Queen's Quay
Belfast BT3 9QQ
United Kingdom
www.w5.co.uk

The Works
5701 Normandale Road, 3rd Floor
Edina Community Center
Edina, MN 55424
www.theworks.org

INDIVIDUALS
Peter A. Anderson
Museum Consultant
Lilac Cottage
High Street
Laxfield
Suffolk IP13 8DZ
United Kingdom
paanderson@compuserve.com

Elsa B. Bailey
Education Consultant
1050 Noriega Street
San Francisco, CA 94122
ebbailey@earthlink.net

Lynn D. Dierking
Associate Director
Institute for Learning Innovation
166 West Street
Annapolis, MD 21401
www.ilinet.org

John H. Falk
President
Institute for Learning Innovation
166 West Street
Annapolis, MD 21401
www.ilinet.org

George E. Hein
Professor Emeritus
Lesley University
Cambridge, MA 02138
www.lesley.edu/faculty/ghein/george.html

President
TERC
Cambridge, MA 02140
www.terc.edu

Paul Orselli Workshop (POW!)
1684 Victoria Street
Baldwin, NY 11510
www.orselli.net

Robert "Mac" West
Informal Learning Experiences
P.O. Box 42328
Washington, DC 20015
www.informallearning.com

Cynthia C. Yao
Hands-On Museum Consultant
219 East Huron Street
Ann Arbor, MI 48104
cyao@comcast.net

identity. *See* branding; mission; naming

ILE (Informal Learning Experiences), 135–41

IMAX, 50, 75, 87, 88, 235

income, 54, 72, 81. *See also* budgets

incorporation, nonprofit, 4, 45, 56; tax exemption, 501(c)(3), 15, 78, 82

informal education, 265, 278; theory and practice 146–50, 154–55, 278. *See also* hands-on (interactive) philosophy

innovation, 219–21, 239–43

Inspire Discovery Centre, 58–59

Institute for New Science Centers. *See* Association of Science-Technology Centers

Institute of Museum and Library Services (IMLS), 10, 13, 47. *See also* grants

interactive. *See* hands-on (interactive) philosophy

International Year of the Child, xiii, 5

internship, 5

Junior League, 49, 86, 222–23, 234

KIDSPACE, 125–31. *See also* COSI Columbus

Knowledge Age, 272–76

Kresge Challenge, 9, 11, 13, 23. *See also* grants

Launch Pad, 222–26, 261. *See also* Discovery Center Museum

Lawrence Hall of Science, 252, 256

leases, 7, 81

lessons learned. *See* best practices

location, 6, 9, 130

mall, 5, 35–38, 39–43. *See also* storefronts, as startups

management: good practices, 13–14

managers, 15, 79–80

MAP (Museum Assessment Program), 47, 222. *See also* grants

marketing, 63– 64, 65, 79, 80, 84–85, 86–91, 229, 281; advertising, 89; branding, 90, 229, 230; product development, 187; promotion, 41; public relations, 84, 91

matrix, 200–201

memberships, 75, 77, 273–75

mentoring, importance of, 50

Millennium Projects, 97, 102, 261, 263

mission, 4, 22, 28, 56, 74, 81, 87, 98–99, 146, 153, 229, 238, 247, 278; statements 4, 28, 29, 81, 87, 98, 205, 242

MIT Museum, 167–71; FAST, 168–70

models, 10, 56, 259

Museum of Discovery and Science, 44, 86, 90

Museum of Life and Science, 217–18

Museum of Science (Boston), 121–24

Museum School, 162–66; curriculum, 162–63; impact, 165–66; staff, 164. *See also* early childhood education

naming, 4, 42–43, 74, 123, 229, 231–32; gift, as Taco Bell, 42–43; identity, 74, 97, 151–52, 224

nanotechnology, 170. *See also* exhibitions; technology

National Register of Historic Places, 3, 6, 7, 13

National Science Foundation (NSF), xi, 9, 13, 50, 108, 247, 256; impact of, 13, 50, 137. *See also* grants

networks, 152

objects, use of, 114–18, 121, 136, 137, 162, 278

Observatory Science Centre, 56, 57. *See also* Science Projects

Ontario Science Centre, 283

openings, 7, 12, 41–42, 76, 216

operations: general, 23, 41, 54, 75; sustaining, 75, 215–18, 225, 264

Oppenheimer, Frank, 29, 247, 252–53, 256

organization chart, 77

Orselli, Paul, 59

outdoor science parks, 58, 85, 132–34, 222–24

outreach, 10, 14, 108, 156–61, 176, 231; COSI on Wheels, 231; kits rental, 159; Science on Wheels, 156–58; virtual, 159. *See also* education programs

Pacific Science Center (PSC), xi, 33, 39, 50, 156–61; Science Carnival Consortium (SCC), xi–xii, 33, 39, 50, 60, 160

partnerships. *See* collaborations

performance tracking, 78–80

PERG (Program Evaluation and Research Group), 199

physicists, 11, 15, 17, 32

Phoenix Zoo, 147

Pizzey, Stephen, 63

plans, 10–11, 23, 24, 38; education, 145–50; business, 23; financial, 11, 23–24, 82; strategic, 11, 38, 47, 82, 175, 229, 238

preschool, 14, 121, 124, 125–31, 162–66, 229

prices, setting, 87–88

problems, 7, 9, 11, 34, 37

programs. *See* education programs; visitors: experiences

promotion, 41. *See also* marketing

prototyping. *See* exhibits

publicity, 3, 9, 26

public schools, 17, 153–55, 156–61, 271

public service, 209–10. *See also* ethics, code of

public relations, 84, 91. *See also* marketing

public trust, 233

public understanding of research, 167–71

public understanding of science, 17, 81, 24, 234. *See also* mission

recession, 7, 39

renovation, 6–12, 219–21

rentals, 72, 159

reports, 78–80; annual, 80; IRS, Form 990, 78

research, 86, 125–27, 149–50, 194–96

resources, 110–11

retreats, 10, 22, 25. *See also* boards

safety, 131, 132

schools, 6, 65, 77, 151–55; groups, 10, 280

institutions, and 1,700 corporate members. Individual members span the range of occupations in museums, including directors, curators, registrars, educators, exhibit designers, public relations officers, development officers, security managers, trustees, and volunteers.

Every type of museum is represented by the more than 3,100 institutional members, including art, history, science, military and maritime, and youth museums, as well as aquariums, zoos, botanical gardens, arboretums, historic sites, and science and technology centers.

Working Model: A Mechanism for the Effective Board, chapter 36, is reprinted with permission from *Museum News*, January/February 2003. Copyright 2003, the American Association of Museums. All rights reserved.

Code of Ethics for Museums, chapter 37, is reprinted with permission. Copyright 2000, the American Association of Museums. All rights reserved.

Association of Science-Technology Centers' (ASTC's) mission is to further the public understanding of science among increasingly diverse audiences. ASTC encourages excellence and innovation in informal science learning by serving and linking its members worldwide and advancing their common goals through a variety of programs and services. ASTC provides professional development for the science center field, promotes best practices, strengthens the position of science centers within the community at large, and fosters the creation of successful partnerships and collaborations.

ASTC has grown to more than 540 members, including more than 415 operating or developing science centers and museums. ASTC is the largest organization of interactive science centers in the world and has members in 39 countries. More than 158 million people visit ASTC member institutions annually, including more than 47 million schoolchildren. Bonnie VanDorn has been the executive director of ASTC for more than twenty years.

Elsa B. Bailey, PhD, has worked in the field of informal and formal education for over thirty years. Her professional work has included directing teacher education at the Miami Museum of Science and research and evaluation at Lesley University, Cambridge, Mass.; currently, she is an independent consultant in evaluation and museum-school collaboration. Bailey is on the National Science Teachers Association board as the director of the Informal Science Division, 2006-2009.

Mary Baske, along with friends who felt the need for an educational hands-on museum for children, founded the Curious Kids' Museum in 1989. She was the volunteer director of the museum for ten years and continues to be involved at the board level with fund-raising and long-range planning.

Bronwyn Bevan is the associate director of the Exploratorium Center for Learning and Teaching, where she oversees teacher professional development, youth and after-school programs, and higher education collaborations. Her background includes classroom-based research on partnerships between teachers and artists as well as cultural institutions and schools.

Colleen Blair, vice president for guest services, has been with the Fort Worth Museum of Science and History for twenty-seven years. She is passionate about finding innovative ways to build relationships with the formal education system to support learning by children and adults.

Vicki Breazeale, PhD, received a BS in nutrition and biochemistry, an MS in plant physiology, and a PhD (SESAME Group) from University of California at

Kirsten M. Ellenbogen, PhD, began her work in museums as a demonstrator at the Detroit Science Center eighteen years ago, later working in exhibition development and educational programming. Her research now focuses on the role of museums in family life and on the responsibilities of museums in systemic education reform.

Berkeley. She directed the Science Institute at New College of California from 1984 to 2002 and is currently vice president of Land Art, Inc., a company that creates public parks and restores habitats.

Kim L. Cavendish, president of the Museum of Discovery and Science, Fort Lauderdale, Fla., has led science museums since 1981; she also serves as CEO of Orlando Science Center and Virginia Air & Space Center. Cavendish is on the boards of Association of Science-Technology Centers and Giant Screen Theater Association and was two-term president of the Florida Association of Museums.

Adela "Laddie" Skipton Elwell, PhD, a New Yorker by birth, received a BS from University of Massachusetts and an MS and a PhD from Iowa State University. For thirty years, Elwell has helped provide resources for northern Minnesota science educators and students and has been executive director of Headwaters Science Center for eleven years.

Al DeSena, PhD, has been founding director of Exploration Place in Wichita, Kans., and of the Carnegie Science Center in Pittsburgh, Pa. He is currently a program director with the National Science Foundation. His bachelor's degree is in chemistry from Fordham University, N.Y., and his doctorate is in education from the University of Pittsburgh, where he also worked on the staff of the Learning Research and Development Center.

John H. Falk, PhD, is president and founder of the Institute for Learning Innovation, an Annapolis-based nonprofit learning research organization in Maryland. He is known internationally for his research on free-choice learning, particularly his efforts to understand why people visit places such as museums and libraries, what they do there, and how these experiences impact people's lives. He has authored over ninety scholarly articles and chapters as well as written or edited more than a dozen books, including *Thriving in the Knowledge Age: New Business Models for Museums and Other Cultural Institutions* with Beverly K. Sheppard (2006), *Learning from Museums* (2000), *Free-Choice Science Education* (2001), and *Lessons without Limit: How Free-Choice Learning Is Transforming Education* (2002).

Falk is an honoree in American Association of Museums's (AAM's) Centennial Honor Roll, which embodies the spirit of AAM's Centennial for leadership to the field and service to the public (2006).

Claude Faubert, a museum studies graduate, worked for the Ontario Science Centre as head of Astronomy and Earth Sciences. He became a science communication consultant and producer of science programs for TVOntario. In 1992, Faubert joined the Canada Science and Technology Museum as a project manager then became director of education; he has been director general since 2001.

Kit Goolsby, vice president of education at the Fort Worth Museum of Science and History, has twenty-five years' experience in the Museum School as teacher and administrator. With a degree in early childhood education, Goolsby serves as a museum liaison for children and their families, educational partnerships, and community-based organizations.

Hands On! Inc. helps clients conceive and develop a vision with creativity in process and product. Hands On! focuses on bringing knowledge and experience to people of all ages—to turn ideas into experiences through comprehensive master planning, design, and fabrication. Its strength is in developing extraordinary visitor experiences to support the sustainability of an organization.

Hands On! Team. Drawing by Lynn Wood.

Peter B. Giles was president and CEO of The Tech Museum of Innovation (1987–2005). Before coming to The Tech, Giles served as the first president of the Silicon Valley Manufacturing Group. He served on the boards of the Association of Science-Technology Centers and the Giant Screen Theater Association. Currently, he is executive director of Mauna Kea Astronomy Education Center in Hilo, Hawaii.

George E. Hein, professor emeritus, Lesley University, and president of TERC, is active in visitor studies and museum education as a researcher and teacher. He is the author, with Mary Alexander, of *Museums: Places of Learning* (1998) and of *Learning in the Museum* (1998). His primary current interest is the significance of John Dewey's work for museums.

Cassandra L. Henry is president of the Science Spectrum Museum and OMNI theater in Lubbock, Tex. She began her museum career as a volunteer, serving as one of the principal developers of the museum. Previous to the museum field, she taught high school English and French and served as first lady of Lubbock.

Karen Johnson was founder, president, and CEO of Discovery Science Center in Santa Ana, Calif., before retiring in 2003. She had primary responsibility for the $24.5 million capital campaign completed in 2002. Johnson received the following awards: Outstanding Non Profit Executive (1992), Woman of Excellence in Science (1998), and Visionary Award (1999).

Asger Høeg is the first and only executive director of Experimentarium since 1988. Høeg is one of the pioneers in the European science center activities. He has an MS in engineering and a BA in economics. He has worked for the Ministry of Transportation and The Royal Post and Telecompany. He conceived and carried out the first Danish Science Festival in 1998. Since 2004, he has been president of ECSITE, the European network of science centers and museums.

Maureen E. Kennedy has been a member of the board of directors of The Tech Museum of Innovation since its founding in 1983. She also served as treasurer and chief financial officer of The Tech for 20 years. She was chair of the Junior League of Palo Alto's Committee that formed The Tech.

Marilyn Hoyt joined the New York Hall of Science in 1985 to staff fund-raising for its 1987 public opening. Hoyt returned in 1997 to staff an expansion campaign, doubling the Hall. As of 2005, the Hall had completed a $91 million capital campaign and doubled its operating budget. She is currently president and chief operating officer of the New York Hall of Science.

Lucy Kirshner received an undergraduate degree in experimental psychology from Mount Holyoke College in 1970 and an MAT from Claremont graduate school in 1973. Her best education has come through informal experiences, often outdoors in the woods, where she's been allowed to explore and touch and follow her own curiosity. Her most valuable lessons come from being a parent and her work in the Ann Arbor Hands-On Museum and the Museum of Science in Boston.

Kathleen R. Krafft, PhD, is director of exhibits at Sciencenter. A low-temperature physicist by training, Krafft volunteered to build one outdoor exhibit during the Sciencenter's Community Build. This evolved into a full-time career. She is responsible for development and maintenance of interactive exhibits in the science museum and the outdoor science playground and has developed four sponsored traveling exhibitions with numerous volunteers.

Darlene Librero is the director of Exploratorium's High School Explainer Program, where she has designed and overseen the program's evolution over the past two decades. Also, she represents the museum in its relationship with new and expanding museums and in its creation and support of floor facilitator and youth programs.

Thomas Krakauer, PhD, has a science center career that includes twenty-nine years in the role of CEO, the last eighteen of which were at the Museum of Life and Science in Durham, N.C. He served on the board of the Association of Science-Technology Centers (ASTC) from 1984 until 2000 and received the ASTC Fellow Award in 2001.

Laura Martin, PhD, is executive vice president for experiences at the Phoenix Zoo, and has been a teacher and program administrator in formal and informal settings. Martin was previously vice president of education and research at the Arizona Science Center and research director at the Center of Teaching and Learning at the Exploratorium.

Watson "Mac" Laetsch, PhD, is vice chancellor and professor emeritus of plant biology at the University of California at Berkeley. He was formerly director of the university Botanical Garden and the Lawrence Hall of Science. He was a founder and a president of Association of Science-Technology Centers (ASTC) and worked with science centers in India for many years. Laetsch received the ASTC Fellow Award in 1991.

Ronen Mir, PhD, is the executive director of SciTech Hands-On Museum in Aurora, Ill., and a guest scientist at the Fermi National Accelerator Laboratory. Mir focuses on science education, developing science centers in the United States, South America, and Israel. He holds a PhD in physics and a museology certificate.

Alan Nursall joined Science North in 1984. He is currently the science director, responsible for exhibits, education programs, and operations for Science North and its earth sciences centre, Dynamic Earth. He also hosts a weekly science segment on Discovery Channel Canada's national science news show, Daily Planet.

Andrée Peek was president and CEO of Sci-Port Discovery Center in Shreveport, La., from 1997 to 2005. She built, opened, and operated the 67,000-square-foot science center, serving over 200,000 visitors annually. Previously, she was senior vice president of Discovery Place in North Carolina for seventeen years. Peek is a CPA and has an MBA from University of North Carolina at Charlotte.

Chuck O'Connor is a renowned museum planner and educator with widely recognized standards for excellence and over forty years of experience in the science center field. O'Connor served as owner representative for the architectural programming, design, and construction of new facilities for COSI Toledo and COSI Columbus.

Stephen Pizzey graduated in physics from Sheffield University in the United Kingdom and is founder and director of Science Projects Ltd., an educational charity established in 1986. Pizzey currently holds a fellowship from the National Endowment for Science, Technology and the Arts and is a Fellow of the Institute of Physics.

Paul Orselli has developed exhibits and education programs for museums around the world. He is also the editor and originator of the three best-selling *Cheapbooks* published by Association of Science-Technology Centers. Paul lives with his family on Long Island where he runs his exhibit and consulting company called POW! (www.orselli.net).

Melanie Quin, PhD, is executive director of ECSITE-UK, the National Science and Discovery Network (www.ecsite-uk.net). She has been in the science center field since 1987 as founding executive director of ECSITE (Europe), 1989–1991; exhibition developer at NewMetropolis (Amsterdam), 1992–1995; and science communication director at Techniquest (Cardiff), 1995–2001. Quin has edited the ECSITE quarterly newsletter continuously since 1989.

Albert J. Read, PhD, is professor emeritus of physics and director of Science Discovery Center of Oneonta State University of New York, College at Oneonta. Read taught introductory college physics for thirty-six years, making much of his own laboratory apparatus. For over fifteen years after retiring, he set up and operated a small science center at the college. He has gradually turned over much of the center's operation to another physics professor.

Beverly S. Sanford, PhD, has been executive director of SciWorks Science Center and Environmental Park, Winston-Salem, N.C., since January 1996. Prior to that time, she was vice president of programs at Discovery Place, Charlotte, N.C. She has served two terms on the board of the Association of Science-Technology Centers.

Beryl Rosenthal, PhD, is the director of exhibitions and public programs at the Massachusetts Institute of Technology Museum. She has a PhD in anthropology from SUNY Buffalo and a BFA from Parsons School of Design and the New School for Social Research. She has taught in a number of universities and served as an educator in nonprofits, state agencies, and museums for over twenty-five years.

Rebecca Schatz, a former engineering executive, started The Works in 1987, inspired by her love of gizmos, her love of children, and the remarkable work of the charismatic founders of the Exploratorium and the Ann Arbor Hands-On Museum. She treasures the talented, generous colleagues, children, donors, and volunteers who have made The Works succeed.

Christen E. Runge came to Informal Learning Experiences (ILE) after completing her master's degree in museum studies. Runge brought extensive database management experience to her responsibilities as former manager of ILE's Traveling Exhibitions Database, which she ran for six years.

Harold Skramstad is president emeritus of Henry Ford Museum and Greenfield Village, Dearborn, Mich., and

Susan Skramstad is former vice chancellor of institutional advancement at the University of Michigan-Dearborn. Both are consultants to nonprofits on issues of change, planning, fund-raising, and board and staff development. They are the authors of *A Handbook for Museum Trustees* (2003). Working Model: A Mechanism for the Effective Board, chapter 36, is reprinted with permission from *Museum News*, January/February 2003. Copyright 2003, the American Association of Museums. All rights reserved.

Harold Skramstad is an honoree in American Association of Museums's (AAM's) Centennial Honor Roll, which embodies the spirit of AAM's Centennial for leadership to the field and service to the public (2006).

Robert Semper, PhD, a physicist and science educator, joined the Exploratorium in 1977 and currently is executive associate director with responsibility for leading the institution's work in developing programs of teaching and learning using exhibits, media, and Internet resources. Semper is the principal investigator on numerous science education, media, and learning research projects.

David Taylor (1953–2005) died of liver failure at the early age of 52. He spent twenty-five years at Pacific Science Center (PSC) where he was director of science and exhibits for most of his career. He led the exhibit development effort that made the PSC a must-see institution for individuals starting science centers around the world. Taylor's creativity, love of science, and joy of learning are found in every "aha" moment that visitors experience while exploring PSC. He previously worked in planetariums and in public radio. He left PSC to pursue his PhD at the University of Washington, where he was especially interested in how to encourage girls to pursue an interest in science. All of us in the science center field were honored to work with him and were influenced by his work. Taylor's popular Ahha website will continue through PSC. *(Written by Dennis Schatz, vice president for education, PSC, Seattle, Wash.)*

Charlie Trautmann, PhD, received his PhD in engineering from Cornell and is executive director of Sciencenter, a hands-on museum in Ithaca, N.Y. At the museum, he oversees operations, program planning, finance, and institutional advancement. Trautmann has written widely on business planning, management, and phased development of museums.

Susan B. F. Wageman is director of community learning at The Tech Museum of Innovation. Wageman earned an MA in organizational management from Fielding Graduate Institute and a BA in history from Boston University. She has worked in the museum field for twenty-two years and has experience in evaluation, programs, exhibits, development, public relations, visitor services, collections management, and interpretation.

Charlie Walter is chief operating officer of the Fort Worth Museum of Science and History. He has served the museum field in many leadership positions including chair of the Association of Science-Technology Centers Annual Conference Program Planning Committee and board member of the

Association of Children's Museums; he has also been a member of the American Association of Museum's Accreditation Visiting Committee.

sulting firm, and through developing ILE's Traveling Exhibitions Database, he has become an international resource on this specialized business.

 **Robert "Mac" West,** PhD, is a museum veteran, depositing his doctoral thesis fossils at the Field Museum, researching at the American Museum of Natural History while he taught on Long Island, serving as curator of geology at the Milwaukee Public Museum, and directing The Carnegie Museum of Natural History and Cranbrook Institute of Science. Since 1992, he has operated Informal Learning Experiences (ILE), a broadly based museum con-

 **Kim Whaley,** PhD, is currently vice president of education and guest operations at COSI Columbus. She was previously associate professor of human development and family science at Ohio State University. Whaley helped with the design of little kidspace® and set the direction for programming for young children at COSI. She received her PhD in curriculum and instruction from Penn State University.